Carbon Mineralization
in Coastal Wetlands

Estuarine and Coastal Sciences Series: Volume 2

Carbon Mineralization in Coastal Wetlands

From Litter Decomposition to Greenhouse Gas Dynamics

Edited by

Xiaoguang Ouyang

Southern Marine Science and Engineering Guangdong Laboratory (Guangzhou), Guangzhou, China

Guangdong Provincial Key Laboratory of Water Quality Improvement and Ecological Restoration for Watersheds, School of Ecology, Environment and Resources, Guangdong University of Technology, Guangzhou, China

Simon F.S. Li Marine Science Laboratory, School of Life Sciences, The Chinese University of Hong Kong, Hong Kong Special Administrative Region, China

Shing Yip Lee

Institute of Environment, Energy and Sustainability, Simon F.S. Li Marine Science Laboratory, School of Life Sciences, The Chinese University of Hong Kong, Hong Kong Special Administrative Region, China

Derrick Y.F. Lai

Department of Geography and Resource Management, The Chinese University of Hong Kong, Hong Kong Special Administrative Region, China

Cyril Marchand

University of New Caledonia, ISEA, Noumea, New Caledonia

Series Editors

Steve Mitchell *and* **Michael Elliott** *on behalf of the Estuarine & Coastal Sciences Association (ECSA)*

ELSEVIER

Elsevier
Radarweg 29, PO Box 211, 1000 AE Amsterdam, Netherlands
The Boulevard, Langford Lane, Kidlington, Oxford OX5 1GB, United Kingdom
50 Hampshire Street, 5th Floor, Cambridge, MA 02139, United States

Notices
Knowledge and best practice in this field are constantly changing. As new research and experience broaden
our understanding, changes in research methods, professional practices, or medical treatment may
become necessary.

Practitioners and researchers must always rely on their own experience and knowledge in evaluating
and using any information, methods, compounds, or experiments described herein. In using such
information or methods they should be mindful of their own safety and the safety of others, including
parties for whom they have a professional responsibility.

To the fullest extent of the law, neither the Publisher nor the authors, contributors, or editors, assume
any liability for any injury and/or damage to persons or property as a matter of products liability,
negligence or otherwise, or from any use or operation of any methods, products, instructions,
or ideas contained in the material herein.

ISBN: 978-0-12-819220-7

For information on all Elsevier publications
visit our website at https://www.elsevier.com/books-and-journals

Publisher: Candice Janco
Acquisitions Editor: Louisa Munro
Editorial Project Manager: Andrea R Dulberger
Production Project Manager: R.Vijay Bharath
Cover Designer: Mark Rogers

Typeset by STRAIVE, India

Working together
to grow libraries in
developing countries

www.elsevier.com • www.bookaid.org

Contents

Contributors

Bin Chen
Third Institute of Oceanography, Ministry of Natural Resources, Xiamen;
Observation and Research Station of Coastal Wetland Ecosystem in Beibu Gulf,
Ministry of Natural Resources, Beihai, China

Guangcheng Chen
Third Institute of Oceanography, Ministry of Natural Resources, Xiamen;
Observation and Research Station of Coastal Wetland Ecosystem in Beibu Gulf,
Ministry of Natural Resources, Beihai, China

Frank David
Muséum National d'Histoire Naturelle, Station Marine de Concarneau,
Concarneau, France

Adrien Jacotot
ISTO, Université d'Orléans, CNRS, BRGM, Orléans, France

Erik Kristensen
Department of Biology, University of Southern Denmark, Odense, Denmark

Derrick Y.F. Lai
Department of Geography and Resource Management, The Chinese University
of Hong Kong, Hong Kong Special Administrative Region, China

Cheuk Yan Lee
Simon F. S. Li Marine Science Laboratory, School of Life Sciences, The Chinese
University of Hong Kong, Hong Kong Special Administrative Region, China

Shing Yip Lee
Institute of Environment, Energy and Sustainability, Simon F.S. Li Marine Science
Laboratory, School of Life Sciences, The Chinese University of Hong Kong, Hong
Kong Special Administrative Region, China

Audrey Leopold
Institut Agronomique néo-Calédonien, SolVeg, Noumea, New Caledonia

Jiangong Liu
Department of Geography and Resource Management, The Chinese University of
Hong Kong, Hong Kong Special Administrative Region, China

Cyril Marchand
University of New Caledonia, ISEA, Noumea, New Caledonia

Xiaoguang Ouyang
Southern Marine Science and Engineering Guangdong Laboratory (Guangzhou);
Guangdong Provincial Key Laboratory of Water Quality Improvement and
Ecological Restoration for Watersheds, School of Ecology, Environment and
Resources, Guangdong University of Technology, Guangzhou; Simon F.S. Li
Marine Science Laboratory, School of Life Sciences, The Chinese University of
Hong Kong, Hong Kong Special Administrative Region, China

Susan Guldberg Graungård Petersen
Department of Biology, University of Southern Denmark, Odense, Denmark

Cintia Organo Quintana
Department of Biology, University of Southern Denmark, Odense, Denmark

Judith A. Rosentreter
Yale School of the Environment, Yale University, New Haven, CT, United States;
Centre for Coastal Biogeochemistry, School of Environment, Science and
Engineering, Southern Cross University, Lismore, NSW, Australia

Karina V.R. Schäfer
Earth and Environmental Science Department, Rutgers University Newark,
Newark, NJ, United States

Nora F.Y. Tam
School of Science and Technology, The Open University of Hong Kong;
Department of Chemistry, City University of Hong Kong, Hong Kong Special
Administrative Region, China

Faming Wang
Xiaoliang Research Station for Tropical Coastal Ecosystems, Key Laboratory of
Vegetation Restoration and Management of Degraded Ecosystems, and the CAS
engineering Laboratory for Ecological Restoration of Island and Coastal
Ecosystems, South China Botanical Garden, Chinese Academy of Sciences,
Guangzhou, China

Yong Ye
Key Laboratory of the Ministry of Education for Coastal and Wetland Ecosystems,
College of the Environment and Ecology, Xiamen University, Xiamen, China

Acknowledgments

We cordially thank the reviewers who review the chapters of our book, including a few anonymous reviewers. Their comments and ideas improve the quality and expand the outlook of our book.

We appreciate the help of Elsevier and the different Publishing Editors who advance the progress of our book at different stages from proposal drafting, book preparation to production. In particular, Louisa Hutchins invited the first author to contribute a book in his research area. Andrea Dulberger, the Editorial Project Manager, helps to negotiate book production and keep our book writing and production well on track and are patient to communicate the progress of our book with our editorial team during the past 2 years. We also thank Mohan Raj Rajendran for reminding us of obtaining permission files used in each chapter, and Vijay Bharath Rajan for assistance in the book production process.

Xiaoguang Ouyang is supported by an Impact Postdoctoral Fellowship from The Chinese University of Hong Kong, Key Special Project for Introduced Talents Team of Southern Marine Science and Engineering Guangdong Laboratory (Guangzhou) (GML2019ZD0403) and Guangdong Provincial Key Laboratory Project (2019B121203011). Faming Wang is funded by the Guangdong Basic and Applied Basic Research Foundation (2021B1515020011), the CAS Youth Innovation Promotion Association (2021347), the Key Special Project for Introduced Talents Team of Southern Marine Science and Engineering Guangdong Laboratory (Guangzhou, China) (GML2019ZD0408), the National Forestry and Grassland Administration Youth Talent Support Program (2020BJ003), and the R & D program of Guangdong Provincial Department of Science and Technology (2018B030324003). Judith A. Rosentreter would like to acknowledge funding from the Yale Institute for Biospheric Studies..

Any findings, opinions, conclusions, and recommendations presented in this book are those of the authors.

CHAPTER

Introduction

1

Xiaoguang Ouyang[a,b,c], Derrick Y.F. Lai[d], Cyril Marchand[e], and Shing Yip Lee[f]

[a]*Southern Marine Science and Engineering Guangdong Laboratory (Guangzhou), Guangzhou, China,*

[b]*Guangdong Provincial Key Laboratory of Water Quality Improvement and Ecological Restoration for Watersheds, School of Ecology, Environment and Resources, Guangdong University of Technology, Guangzhou, China,*

[c]*Simon F.S. Li Marine Science Laboratory, School of Life Sciences, The Chinese University of Hong Kong, Hong Kong Special Administrative Region, China,*

[d]*Department of Geography and Resource Management, The Chinese University of Hong Kong, Hong Kong Special Administrative Region, China,*

[e]*University of New Caledonia, ISEA, Noumea, New Caledonia,*

[f]*Institute of Environment, Energy and Sustainability, Simon F.S. Li Marine Science Laboratory, School of Life Sciences, The Chinese University of Hong Kong, Hong Kong Special Administrative Region, China*

1.1 Concepts and background

Carbon mineralization is defined as the conversion of organic to inorganic carbon and occurs mainly during organic matter decomposition (Molles, 2015; Purnobasuki and Suzuki, 2005). Carbon mineralization is a key process of carbon cycling in ecosystems, during which large organic polymers are decomposed into monomers, and ultimately transformed into inorganic carbon. This has significant implications for the studies of food webs, carbon accumulation, and carbon gas fluxes and thus on climate changes mitigation programs. There are several general definitions of wetlands or coastal wetlands. We adopt the general approach rather than a precise definition for coastal wetlands: coastal wetlands are ecosystems that develop within an elevation gradient that ranges between subtidal depths, where light penetrates to support photosynthesis of benthic plants, to the landward edge, where the sea passes its hydrologic influence to groundwater and atmospheric processes (Perillo et al., 2018). Coastal wetlands, lying between land and sea, are the hotspot of carbon cycling since carbon comes from both autochthonous sources (such as carbon fixed by vegetation and algae via photosynthesis) and allochthonous inputs, including riverine transport of terrestrial carbon and tidal transport of marine carbon.

Carbon Mineralization in Coastal Wetlands. https://doi.org/10.1016/B978-0-12-819220-7.00007-8

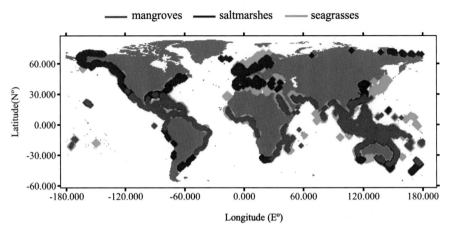

FIG. 1.1

Global distribution of "blue carbon" ecosystems. The distribution maps of mangroves, saltmarshes, and seagrasses come from 2010 Global Baseline of Mangrove Extent provided by Global Mangrove Watch (Bunting et al. 2018), global distribution of saltmarsh (ver. 4.0) (Mcowen et al. 2017), and global distribution of seagrasses (version 6.0) (UNEP-WCMC, 2018).

Vegetated coastal wetlands (including mangroves, tidal marshes, and seagrasses) have been coined as "blue carbon" ecosystems (Nellemann et al. 2009) since these ecosystems have a disproportionately important role in carbon sequestration (Breithaupt et al., 2012; Duarte et al., 2013; Ouyang and Lee, 2014). Despite the fact that blue carbon ecosystems only cover approximately 49 million hectares (Pendleton et al., 2012), their carbon accumulation capacity in sediments is one or two orders of magnitude higher than terrestrial ecosystems (Ouyang and Lee, 2014). Fig. 1.1 shows the global distribution of "blue carbon" ecosystems.

Coastal wetlands are unique habitats that are distinct from both terrestrial and oceanic ecosystems, which means the patterns and well-established carbon cycling models in latter ecosystems cannot be directly applied in coastal wetlands, especially blue carbon ecosystems. Oceanic sediments are saturated with relatively constant temperature, light, and physico-chemical conditions. Terrestrial soils are usually aerobic and well drained with lower water content except during flooding and heavy precipitation. By contrast, characteristics of coastal wetlands sediments are highly variable and depend notably on tides, season, position along the tidal zone, and plant species. Coastal wetlands experience periodical tidal flooding, importing, and exporting carbon from marine and terrestrial sources, but also electron acceptors influencing decay processes. These variabilities make carbon cycling be everchanging in coastal wetlands, while carbon cycling in terrestrial and oceanic ecosystems are less variable and probably easier to constrain and assess. Better integration of carbon mineralization (especially decomposition) mechanisms into blue-carbon models may improve predictions of sediment organic carbon stocks and facilitate incorporation of coastal vegetated wetlands into global carbon budgets and management tools (Spivak et al., 2019). Further, vegetated coastal wetlands are habitats for vascular plants, including eudicot herbs in saltmarshes, eudicot trees and palms in mangroves, and alismatid moncots in seagrasses. Decomposition processes of lignin and cellulose from vascular plant in waterlogged and saline sediments depend on the redox conditions of the environment (Marchand et al., 2005). If carbohydrates are rapidly degraded, lignin-derived phenols are more refractory, and lost at lower rates than total bulk organic carbon, facilitating, on the one hand, their long-term storage in sediments (Lallier-Vergès et al., 2008), and on the other hand, their export to adjacent ecosystem and eventually to the open ocean (Dittmar et al., 2001).

Twelve years ago, a review synthesized data on mangrove primary production and carbon sinks and revealed that over 50% of the carbon fixed by mangroves was unaccounted for, notably because carbon mineralization was severely underestimated and quantified (Bouillon et al., 2008). This highlights the importance of carbon mineralization in constraining carbon budgets in global mangroves. After this review, there was a surge in the number of studies on carbon mineralization in mangroves. A search in Web of Science with combined topics including carbon mineralization, decomposition, and coastal wetlands as well as various substitutes for the terms show 15, 389 publications relevant to this topic in the past 10 years. This number is quite inspiring given the limited fundings sponsoring biogeochemical, environmental, and ecological studies on coastal wetlands worldwide. There are studies that try to better constrain the different components of the unaccounted carbon with more case studies. In terms of CO_2 flux rate, Alongi (2014) and Rosentreter et al. (2018) estimated slightly lower rates (43 and 56.8 mmol m^{-2} day^{-1}, respectively) in comparison with the previous review (59 ± 52 mmol m^{-2} day^{-1}). In terms of sediment organic carbon accumulation rate, Breithaupt et al. (2012) updated the rate to reach 163 g C m^{-2} year^{-1} and revised the unaccounted carbon to be 104.3 g C m^{-2} year^{-1}, which accounts for around 48% of carbon fixed by mangroves. However, there are few studies providing a global perspective on carbon cycling in other coastal wetlands.

Despite the importance of quantifying carbon mineralization to constrain carbon budgets, carbon mineralization is the engine underlying benthic dynamics. Deciphering the influence of forces, driving this process is critical in determining the evolution of coastal wetlands. On top of that, the lignin component of halophytes is less readily depolymerized, and detritus becomes lignin enriched in anoxic sediments (e.g., in coastal wetlands) and forms a major carbon sink in blue carbon ecosystems or is carried into the oceans (Cragg et al., 2019). This further strengthens the earlier justification on the role of coastal wetlands (e.g., mangroves and saltmarshes) in supporting nearshore communities by exporting carbon under favourable tidal conditions (Dame et al., 1986; Lee, 1995). If substantial amount of particulate organic carbon is exported during each ebb tide, the nutritional quality of mangrove-derived organic matter is low, being tannin-rich, challenging the processes, allowing mangrove-derived organic matter utilization by consumers. However, mangrove-derived carbon can be exported through different forms, porewater seepage induces dissolved inorganic carbon and CO_2 export to tidal creeks and further (Maher et al., 2013); subsequently, the CO_2 provided by the tidal pumping of mangrove pore water can be used by the phytoplankton for its development, and thus sustaining trophic chain in coastal waters (David et al., 2018). Studying organic carbon mineralization in coastal wetlands and its evolution with global changes will also help understanding the future modification of food chain and biodiversity along coastlines.

1.2 Potential climatic and anthropogenic drivers for carbon mineralization

Coastal wetlands, especially blue carbon ecosystems, are hotspots of carbon sequestration, but the sequestered carbon tends to be mineralized, producing greenhouse gas (GHG, including CO_2, CH_4) subsequently emitted toward the atmosphere. Production and emissions of GHG may be enhanced by anthropogenic disturbance and climate change.

Blue carbon ecosystems are highly efficient in carbon sequestration and accumulation due to their high primary production, anoxic sediments, and efficiency in trapping suspended matter and associated inorganic/organic carbon during tidal inundation, and lignin-enriched detritus forming a major carbon sink (Bouillon et al., 2008; Cragg et al., 2019; Kristensen et al., 2008; Mcleod et al., 2011; Ouyang and Lee, 2020). However, anthropogenic disturbance and climate change may pose threats to carbon sequestered in blue carbon ecosystems and other coastal wetlands. In particular, sites with high sediment carbon stocks and where sediment undergoes physical disturbance have the high risk of CO_2 emissions, and this risk is accentuated in settings that favor the decomposition of organic substance to CO_2 (Lovelock et al., 2017) due to rapid exposure of organic carbon to fundamentally different biogeochemical environments and dispersal of organic carbon. Different coastal wetlands may lose at different rates subject to geographical regions and may confront with different disturbance activities; tidal marshes and seagrasses witness conversion rates of 1%–2% and 0.4%–2.6% (Pendleton et al., 2012), while recent loss rates of mangroves have declined to 0.16%–0.39% (Hamilton and Friess, 2018) due to mangrove restoration in some countries and regions (Lee et al., 2019; Vaiphasa et al., 2007). South Asia, Southeast Asia and Asia-Pacific contain around 46% of the mangrove forests but also demonstrate the highest global rates of mangrove loss (Gandhi and Jones, 2019), due to mangrove transformation to other land uses, e.g., rice and palm oil fields (Friess et al., 2020). Human impacts are documented to directly modify or destroy saltmarshes at the local scale (e.g., alteration of coastal hydrology, salt hay farming, and metal and nutrient pollution), the regional scale (e.g., subsurface withdrawal of groundwater, oil and gas), and even the global scale (e.g., human-induced climate change) (Gedan et al., 2009; Kennish, 2001). Saltmarshes are particularly susceptible to coastal eutrophication. Current nutrient loading rates to coastal ecosystems have overwhelmed the capacity of marshes to remove nitrogen without deleterious effects in many places (Deegan et al., 2012). Seagrass meadows are also increasingly stressed by human activities such as coastal construction, degradation of water quality and eutrophication-fuelled bloom-forming macroalgae (Orth et al., 2006; Waycott et al., 2009). Table 1.1 summarizes

Table 1.1 Coastal wetlands confronting with different anthropogenic disturbance.

Anthropogenic disturbance	Type of influence	Impact on physico-chemical and biological processes related to carbon mineralization in coastal wetlands	References
Vegetation removal	Deforestation	Deforestation results in the loss of aboveground biomass. Biomass carbon stocks recovered over the first 40 years after regeneration, but there is nonsignificant recovery of sediment bulk density despite an obvious increase in carbon content. Plantation after vegetation removal may alter the partitioning of CO_2 from difference sources	Sasmito et al. (2019) and Ouyang et al. (2018a)
Transformation to other land uses (e.g., aquaculture ponds, salt works, urban lands, rice and oil palm fields)	Land use change	Land use change causes substantial reduction in biomass and soil carbon stocks. Sediment carbon is exposed to destabilization or exposed to oxygen, resulting in increased microbial activities. Aquaculture can reduce mangrove growth rates, increase mangrove mortality rates, and lead to ecological degradation through excess pond waste materials in sediments discharged from adjacent aquaculture ponds. Urbanization has profound effect on the hydrological and sedimentary regimes and the dynamics of nutrients and contaminants	Friess et al. (2020), Vaiphasa et al. (2007), Lee et al. (2006), and Sasmito et al. (2019)
Eutrophication, metal pollution, waste dumping, Sewage discharge	Pollution	Eutrophication results in restructured plant zonation. Relief of nitrogen limitation for nitrogen-limited species causes an increase in aboveground plant height and biomass and often accompanied reduction in belowground biomass. Nitrogen and phosphorus enhance microbial activity, carbon assimilation, and accumulation. The continuous availability of high nitrate in the water and more decomposable detritus increase decomposition rates. Heavy metal uptake by wetland vascular plants inhibits photosynthetic activity. Sewage pollution may also constrain bioturbation activities. Plastic pollution increases CO_2 emission and leafdrop.	Lee et al. (2006), Deegan et al. (2012), Orth et al. (2006), Waycott et al. (2009), Bartolini et al. (2011), Herbert et al. (2020), and Ouyang et al. (2022)

Continued

Table 1.1 Coastal wetlands confronting with different anthropogenic disturbance—cont'd

Anthropogenic disturbance	Type of influence	Impact on physico-chemical and biological processes related to carbon mineralization in coastal wetlands	References
Introduction of nonnative species	Species invasion	Plant invasion affects plant and animal communities and density, reduces aquatic habitat quality by infilling nearby water bodies, has effects on physiochemical and hydrodynamic environment, detritus quality and availability. Introduced species may decay at lower rates in the invaded sites than its native habitats	Siple and Donahue (2013)
Runaway consumer effects	Changes in consumer control	Consumer control triggered by human disturbances has catastrophic consequences on animal populations and structure	Silliman et al. (2005)
Ditching, tidal restriction, damming, dredging, drainage, leveeing	Hydrologic alteration	Impoundments lower inorganic sediment inputs from land drainage. Dredging alters salinity, water levels, and the input and dispersal of sediments across marsh surfaces. Canals and associated levees modify hydrologic processes and greatly reduce the mineral sediment supply. Altered hydrology related to levees also impacts sediment chemistry. Damming causes a sharp drop in sediment loading. Diking and accompanied drainage disrupt salinity gradients and cause an abrupt transition from halophytic vegetation to brackish and freshwater species. Diked and drained sediments become acidified, and organic matter is more rapidly oxidized	Gedan et al. (2009), Spivak et al. (2019), Ezcurra et al. (2019), and Kennish (2001)
Salt production, pasturelands for livestock, harvest for timber, and others	Resources extraction	Animal grazing removes biomass, affects plant species composition, and compacts the sediments, resulting in higher sediment salinities. Drainage and peat removal for salt collection in wetlands result in wetland degradation	Gedan et al. (2009)
Concrete and earthen seawall and road construction on the migration pathway of mangroves	Breakdown of landward margins	Seawall blocks the path of plant migration and creates a narrowing shoreline—a phenomenon known as "Coastal squeeze." This results in the higher ratio of endangered species and destroys sea-to-land connectivity by disrupting hydrodynamic patterns, organism movements, nutrient flows, and modifying sediment balance	Wang et al. (2020) and Doody (2004)

Dredging   Ditching

Spartina invasion into mangroves   Transformation to aquaculture ponds

Mangrove deforestation   Transformation to oil palm field

FIG. 1.2

Anthropogenic disturbance on coastal wetlands.

Pictures are provided by Xiaoguang Ouyang and Shing Yip Lee.

the impact of anthropogenic disturbance on biogeochemical processes related to carbon mineralization in coastal wetlands. Fig. 1.2 shows the anthropogenic disturbance on coastal wetlands. Land use change stemmed from anthropogenic disturbance was claimed to result in significant reduction in biomass and sediment organic carbon stocks but insignificant changes in GHG emissions (Sasmito et al., 2019). In contrast, restoration resulted in significant increase in carbon stock in comparison with degraded sites after 4 years and reach parity with natural sites after 7–17 years, while restored sites emit less than natural sites but more than degraded sites (O'Connor et al., 2020). Increased CH_4 production in reforested mangrove sites was attributed to high capacity of CH_4 production and decreased CH_4 consumption resulting from the shift of sediment CH_4-cycling microbial communities (Yu et al., 2020).

Climate change has profound influence on coastal wetlands, in particular their carbon cycling and related processes. Increasing atmospheric CO_2 concentrations were observed to lengthen the growing season of coastal marshes, the phenological changes of which may increase CO_2 uptake (Mo et al., 2019). Sea level rise was projected to result in the submersion of most of the Indo-Pacific mangroves by the end of the 21st century without considering inland migration (Lovelock et al., 2015). However, with nature-based adaptation solutions, coastal wetlands were estimated to gain area with rising sea levels which create accommodation space for inland migration (Schuerch et al., 2018). This kind of solutions can be implemented in coastal regions where seawalls and other hurdles prohibiting landward migration of wetland plants.

Coastal wetlands mitigate the negative impact of extreme weather events (Ouyang et al., 2018b), such as cyclones, droughts, and heatwaves, on coastal communities. However, these events may restructure ecosystems (Stuart-Smith et al., 2018). For example, extreme climatic conditions including high temperature, drought, and EI-Niño-Southern Oscillation-induced low sea level resulted in severe mangrove mortality along around 1000 km of Australian coastline (Duke et al., 2017). Extreme weather events may have profound impact on coastal wetlands and some impacts are long-lasting and irreversible (Harris et al., 2018). For example, hurricanes may result in the loss and conversion of mangroves to mudflat or saltmarshes due to the long-term consequences of sediment deposition or smothering rather than the immediate effects of winds or waves arising from hurricanes (Paling et al., 2008). Table 1.2 summarizes the impact of climate change on biogeochemical processes related to carbon mineralization in coastal wetlands.

1.3 Conceptual model on carbon mineralization and related processes

Carbon mineralization is a complex process from litter decomposition to greenhouse gas production and is related to other carbon cycling processes and sedimentology in coastal wetlands. Carbon dioxide emission from the water column or the sediment is a component of ecosystem respiration that also includes carbon assimilated or expired by halophytes.

The waterlogged and anoxic conditions prevailing in coastal wetland sediments induce slow decomposition rates. Accordingly, CO_2 releases at the sediment-air interface in these wetlands are lower than those measured both in temperate and tropical terrestrial environments. Nevertheless recent studies showed that GHG emission are highly variable and depend on ecosystems productivity, position along the tidal elevation gradient, anthropogenic impact, and seasons (Kristensen et al., 2017). In addition, high rates of methanogenesis can also occur in coastal wetland

Table 1.2 Coastal wetlands confronting with climate change.

Climate change	Type of influence	Impact on physico-chemical and biological processes related to carbon mineralization in coastal wetlands	References
Rising air temperature	Ecosystem conversion	Nonlinear temperature thresholds regulate the potential for marsh-to-mangrove conversion. Tropical seagrasses are expanding and replacing temperate seagrasses	Gabler et al. (2017) and He and Silliman (2019)
	Ecosystem function	Mangrove encroachment in salt marshes under warming can promote wetland carbon stock	Saintilan and Rogers (2015)
	Phenology	Rising air temperature increases the growing season of wetland plants and thus carbon production in biomass	Mo et al. (2019)
	Physiological/ metabolic processes	Photosynthesis and respiration increase with temperature within the range of species tolerance (before the tolerance threshold is reached and acute thermal death occurs). Rising temperature may increase root decomposition rates in mangroves and saltmarshes	He and Silliman (2019) and Ouyang et al. (2017)
Precipitation	Foundation species	Precipitation thresholds exist for dominance by various functional groups, including succulent plants and unvegetated mudflats	Gabler et al. (2017)
	Physiology	Precipitation may regulate root decay processes by influencing oxygen supply to, and thus the redox potential of sediments, as well as their salinity	Ouyang et al. (2017)
Sea level rise	Ecosystem conversion	Rising sea levels create accommodation space that allows saline ecosystems to migrate to brackish, freshwater, or even upland ecosystems	Schuerch et al. (2018)
	Ecosystem function	Conversion of freshwater wetlands or upland ecosystems to coastal wetlands results in the increase of carbon stock. On the other hand, the stored carbon deposits in submerged mangroves may be eroded by wave action and oxidized back to CO_2. By contrast, flooding of a marsh or mangrove may also permanently bury the accumulated peat and inhibit its decomposition. Rising sea levels may reduce organic decomposition rates, and thus increase the carbon storage capacity of intertidal sediments	He and Silliman (2019) and Mcleod et al. (2011)
	Physiology	Rising sea levels can increase inundation stress, reducing the photosynthesis and growth of salt marshes and mangroves on their seaward edge	He and Silliman (2019)

Continued

Table 1.2 Coastal wetlands confronting with climate change—cont'd

Climate change	Type of influence	Impact on physico-chemical and biological processes related to carbon mineralization in coastal wetlands	References
	Sediment biogeochemistry	Sea level rise induces saltwater intrusion and increases inundation, which reduce the pool of the organic carbon substrate but may expand that of microbes with strong carbon metabolism capacities. Sulfate availability also increases with increasing salinity, while availability of other electron acceptors could transiently increase but would finally decline with increasing salinity and inundation periods	Luo et al. (2019)
Extreme weather events	Ecosystem health and conversion	Extreme high water levels and storms may affect the position and health of coastal wetlands through altered sediment elevation and sulfide sediment toxicity. Hurricanes may result in massive mangrove dieback and conversion to tidal mudflats or other ecosystems, such as tidal marshes	Gilman et al. (2008), Smith et al. (2009), and Paling et al. (2008)
	Physiology	Hurricanes result in plant mortality, decrease in leaf area index and living tree densities, and crown damage. Canopy loss decreases with distance from the hurricane eyewall. Few saplings survive hurricanes	Sherman et al. (2001), Milbrandt et al. (2006), Li et al. (2007), and Salmo et al. (2014)
	Sediment biogeochemistry	Microbial biomass and extracellular enzyme activities would be lower in the storm sediment layer arising from hurricane input. Allochthonous mineral inputs from hurricanes represent a significant source of nutrient resources	Breithaupt et al. (2020) and Castañeda-Moya et al. (2010)

sediments influenced by freshwater, even in sulfate-rich environment, methanogenic bacteria can coexist with sulfate-reducing ones producing high amount of CH_4 (Chauhan et al., 2015). Carbon gases produced during carbon mineralization, either CO_2 or CH_4, emit from the sediment- or water-air interfaces through or accelerated by an array of pathways, including (1) diffusion, (2) ebullition, (3) aerenchyma tissues, (4) xylem and tap tissues, and (5) animal bioturbation.

Decomposition fuels carbon mineralization and has an important role in carbon cycling in coastal wetlands. Plant litter in coastal wetlands contains lignocellulose—an energy-rich polymer consisting of cellulose, hemicelluloses, and lignin that is resistant to enzymatic decomposition (Cragg et al., 2019). This results in the high proportion of autochthonous carbon stored in sediments and contributes to high carbon accumulation capacity of coastal wetlands (Ouyang et al., 2017). Nonetheless, macroinvertebrate (e.g., crabs) can breakdown leaf litter and thus increase the surface area to volume ratio of litter, facilitating microbial decomposition and physical leaching of feeding deterrents (Lee, 1997; Werry and Lee, 2005). Decomposition of sediment organic matter is another source of carbon input to carbon stock in coastal wetlands. The decomposed organic matter originates not only from autochthonous litter inputs but also from allochthonous organic matter inputs from riverine and marine sources. The rate of organic matter decomposition in marine sediments is donor controlled and closely associated with its production, transport, and alteration in the ocean (Arndt et al., 2013).

Organic carbon in coastal wetlands is mineralized in sediments and porewaters by microbial transformations. Much of the produced carbon gases dissolve in porewater under relatively high in situ pressure, which result in significant supersaturation, and finally part of the gases escape to the atmosphere through diffusion and direct ebullition (under submergence) (Dutta et al., 2013). King (1984) found that CH_4 ebullition is a significant process varying both seasonally and spatially in tidal marshes. Estimates of the magnitude of CH_4 inputs to the atmosphere by ebullition suggest that it may be as important as diffusional losses across the sediment-air interface (Bartlett et al., 1985).

Some coastal wetland plants have aerenchyma tissues (e.g., pneumatophores) that transport carbon gas produced underground to the atmosphere. The aerenchyma of young *Avicennia marina* horizontal roots and pneumatophores has a tubular, non-random structure, the tubes being relatively straight, with air spaces representing 70% of the total root volume (Curran, 1985). Kitaya et al. (2002) confirmed that gas movement through the lenticels and aerenchyma passages in the pneumatophores of *A. marina* and *Sonneratia alba* and prop roots of *Rhizophora stylosa*. Pneumatophores have open lenticels during emersion, permitting rapid diffusion of gases into (e.g., O_2) and from (e.g., CO_2 and CH_4) deep sediments through the air-filled aerenchyma tissue to the atmosphere (Purnobasuki and Suzuki, 2005) but can also stimulate sulfate reduction via root exudates (Kristensen and Alongi, 2006).

Bioturbation of macroinvertebrates (e.g., crabs) can increase the surface area of the sediment water/air interface (Lee, 2008) and re-distribute sediments. Mangrove crabs are coined as ecosystem engineer since their burrowing activities have considerable impact on ecosystem functioning (Kristensen, 2008). Both the sediment percolation rate and sediment hardness were demonstrated to be significantly higher outside than inside the crab inhabited areas (Bortolus and Iribarne, 1999). Burrowing activities of macroinvertebrates can alter fundamental biogeochemical processes, which may also be regulated through vertical re-distribution of sediments during burrow maintenance and construction (e.g., Botto and Iribarne, 2000). Ocypodid crabs, e.g., *Uca* spp. may oxygenate marsh sediments and promote sediment drainage via burrowing activities (Iribarne et al., 1997). Moreover, the tidal subsidy effect in coastal wetlands is apparently regulated by bioturbating activities of burrowing and deposit-feeding macrofauna such as crabs, whose burrows may remarkably give rise to porewater flow and aeration of the sediment (Lee, 1999).

There are emerging evidences of carbon gases emitted belowground to the air via other avenues. Methane emission from mangrove stems is attributed to the pressurized process (xylem and sap flow) in spite of the nonpressurized diffusion process, and the flux is higher from dead than from live threes since dead trees have an array of empty internal cavities that facilitate more upward diffusive flux than live trees do (Jeffrey et al., 2019). Fluxes of CO_2 from coarse mangrove woody debris $(2.34 \pm 0.23 \, \mu mol \, m^{-2} \, s^{-1})$ were reported to be even higher than those from prop roots $(1.94 \pm 0.45 \, \mu mol \, m^{-2} \, s^{-1})$ (Troxler et al., 2015).

Fig. 1.3 describes the carbon mineralization and related processes in coastal wetlands. Table 1.3 shows the relevant terms in each chapter of the book. The relevant processes are described as below.

(1) Ecosystem respiration

$$R_{ecosystem} = R_{diff} + R_{bioturb} \pm R_{biomass} \pm R_{MPB}$$
$$R_{diff} = R_{diff1} + R_{diff2} + R_{diff3}$$

where R_{diff} is the diffusive flux from sediment surfaces (R_{diff1}), and/or water surfaces, bubbling from the water column (R_{diff2}, only included when the ecosystem is inundated), and plant aerenchyma or xylem and tap tissues (R_{diff3}, e.g., stems/branches and pneumatophores), $R_{biomass}$ and R_{MPB} are CO_2 assimilated via photosynthesis or respired by plants and microphytobenthos and can be positive or negative depending on the balance between photosynthesis and respiration. $R_{bioturb}$ is the sum of epifauna respiration and increase in GHG emission from bioturbation activities.

(2) GHG emission from sediment-air interface (emerged)

$$R_{sed} = R_{diff1} + R_{diff3} + R_{bioturb} \pm R_{MPB}$$

where R_{sed} is GHG emission from sediment-air interface, including R_{diff1}, R_{diff3}, $R_{bioturb}$, and R_{MPB}.

(3) GHG emission from water-air interface (submerged)

$$R_{water} = R_{diff2} + R'_{bioturb} \pm R'_{MPB}$$

where R_{water} is GHG emission from water-air interface, including R_{diff2}, $R_{bioturb}'$, and R_{MPB}'.

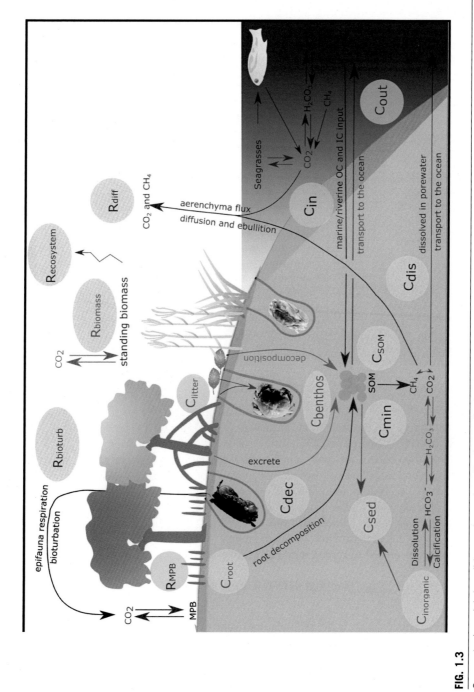

FIG. 1.3

Conceptual model describing carbon mineralization and related processes in coastal wetlands.

Table 1.3 A list of above terms and processes to be included in the various book chapters.

Term	Chapters
All	Chapter 1: Introduction
C_{root}, C_{litter}, C_{dec}, C_{min}, C_{dis}	Chapter 2: Decomposition of vascular plants and carbon mineralization in coastal wetlands
R_{diff} (R_{diff1}, R_{diff3}), $R_{bioturb}$, R_{MPB}, R_{sed}	Chapter 3: CO_2 and CH_4 emissions from coastal wetland soils
$R_{biomass}$, $R_{ecosystem}$	Chapter 4: Biosphere-atmosphere greenhouse gas fluxes at the ecosystem scale
$C_{benthos}$	Chapter 5: Macrofaunal consumption as a mineralization pathway
R_{diff} (R_{diff2}), $R_{bioturb}'$, R_{MPB}', R_{water}	Chapter 6: Water-air gas exchange of CO_2 and CH_4 in coastal wetlands
R_{sed} and others	Chapter 7: The impact of climate change on greenhouse gas emission
$R_{bioturb}$, R_{sed}	Chapter 8: The role of biogenic structures for greenhouse gas balance in vegetated intertidal wetlands
R_{sed} and others	Chapter 9: Greenhouse gas emission from anthropogenic disturbances
All	Chapter 10: Carbon storage and mineralization in coastal wetlands

(4) Belowground total carbon balance

$$C_{in} + C_{root} + C_{litter} = C_{sed} + C_{dec} + C_{out} + C_{benthos} \pm C_{inorganic}$$
$$C_{dec} = C_{min} + C_{dis}$$

where C_{in} is the rate of allochthonous carbon input rate from marine or riverine sources, C_{root} is the dead root C production, C_{litter} is the litter C production, C_{dec} is the rate of carbon from SOM, litter, root decomposition, and excretion, C_{sed} is the carbon buried rate (both organic and inorganic) in sediments, C_{min} is the respiration rate of carbon from the decomposition processes, C_{dis} is the rate of carbon dissolved in porewater and export via groundwater, C_{out} is the rate of carbon export via tides, $C_{benthos}$ is the rate of benthos-processed carbon from sources such as leaf litter and MPB, and $\triangle C_{inorganic}$ is the change rate in $C_{inorganic}$ due to dissolution or calcification.

(5) Processes related to carbon mineralization

    (1) $C_{benthos}$

        where $C_{benthos}$ is related to CO_2 respired by epifauna in $R_{bioturb}$.

    (2) $C_{inorganic}$

        where $C_{inorganic}$ is the sediment inorganic carbon from skeletons, shells, feces, and nonbiogenic carbonate minerals. It will be transformed to bicarbonate and then CO_2 and vice versa by dissolution or calcification processes. It is related to R_{diff}.

1.4 **Motivation for writing the book**

The book is the first synthesis of research on carbon mineralization in coastal wetlands, counter-balancing the bias toward carbon storage in these systems. Studies on blue carbon have strongly emphasized the storage potential of coastal wetlands. Currently, studies on carbon mineralization are patchy because they usually focus on either leaf litter/root decomposition or GHG emissions. Few studies systematically review the processes and principles of GHG production and emissions and the linkage between litter/root decomposition and GHG emissions.

There are a few recently published books on coastal wetlands, management on the ecosystem, and soil carbon dynamics in terrestrial ecosystems. Faridah-Hanum et al. (2014) describe the distribution, challenges, and opportunities of mangroves in Asia but mainly presents the distribution of mangrove species and fauna in Malaysia, Indonesia, Thailand only. The book examines a diverse range of topics in mangroves but are presented without emphasis or a clear theme and without coverage on other coastal wetlands. Mitra (2013) explores the response of mangrove ecosystems to climate change and clarifies the impact of climate change on aquaculture and fisheries, including freshwater and brackish waters. The response of aquaculture to climate change does not fall in the scope of this book, especially freshwater and brackish water aquaculture. Muñoz and Zornoza (2017) overview recent finding and future challenges with regard to factors controlling soil carbon and GHG emissions and integrates many case studies on soil carbon mineralization and GHG emissions as independent chapters. The scope of the book is confined to terrestrial ecosystems and do not cover coastal wetlands, and there is a lack of syntheses on the sources, processes, and mechanisms of carbon mineralization and/or GHG. This book overviews carbon mineralization at the global scale rather than the regional scale and includes other types of coastal wetlands and focuses on carbon mineralization. The book addresses the hotspot of research in mangroves, saltmarshes, and other coastal wetlands, such as litter decomposition, GHG emissions, and anthropogenic activities, including eutrophication and deforestation. This book fills the current knowledge gap in carbon mineralization and provides a balanced view of the carbon dynamics of coastal wetlands. Potential readers will get a holistic treatment of carbon mineralization, from litter/root decomposition pathways and their contribution to carbon mineralization, and the processes and sources of GHG production. The book compares carbon mineralization in mangroves and saltmarshes and highlight the differences in carbon dynamics.

The book aims to

- provide comprehensive perspectives on the processes and mechanisms of carbon mineralization in coastal wetlands,
- identify factors regulating organic matter decomposition and GHG emission,
- clarify the linkage between litter decomposition and GHG emission,
- unravel how GHG emissions are modified by anthropogenic activities, including eutrophication and deforestation, and climate changes.

- put forward suggestions on "blue carbon" management from the view of carbon mineralization by providing a holistic view of carbon storage and mineralization dynamics.

This book meets the need of researchers who study coastal wetlands and decision-makers on coastal ecosystems. This is fulfilled in a number of aspects

- integrating knowledge on both mangroves and saltmarshes, which are usually adjoining but different ecosystems
- covering two research areas, i.e., litter decomposition and GHG emissions, and improving current understanding on carbon mineralization in carbon cycling processes
- promoting effective "blue carbon" management, which is biased toward carbon storage and largely ignores the pathways of carbon mineralization, in particular, under anthropogenic disturbances and climate change

Chapter 2 discusses factors affecting processes related to litter decomposition in coastal wetlands, including biotic and abiotic factors as well as anthropogenic activities. Indicators related to litter decomposition are described. The authors examine the variability of leaf litter, wood, and root decomposition with a series of factors, such as media for decomposition experiment, stoichiometry, and ecosystem types. The authors analyze the contribution of mineralized carbon during litter decomposition to carbon budgets, including GHG production in "blue carbon" ecosystems where data are available. The authors describe and clarify the composition and the characteristics of sediment organic matter that shape the dynamics of decomposition processes in coastal wetlands. Sediment organic matter decomposition is elucidated in terms of decomposition pathways, processes, and energetics. The authors illustrate processes related to carbon mineralization, including inorganic carbon dynamics and sedimentation.

Chapter 3 synthesizes GHG emissions from vegetated coastal wetlands in comparison with other ecosystems and highlights data gaps in geographic regions. Then the chapter gives an overview of the physical and biological drivers of GHG emission from coastal wetlands, in relation to leaf litter, wood, and root decomposition (Chapter 2). The impacts of sediment and nutrient inputs from autochthonous and allochthonous sources are discussed and linked to suggestions on "blue carbon" management (Chapter 10).

Chapter 4 Coastal wetlands have the highest carbon sequestration rate among all natural ecosystems because of their comparatively high productivity and low ecosystem respiration caused by frequent dal flooding. Meanwhile, they have long been considered as minor CH_4 sources due to their saline environments. Therefore, coastal wetlands are among the best candidate ecosystems for rehabilitation or restoration projects to accomplish long-term greenhouse gas (GHG) sinks. The eddy covariance technique provides direct, long-term, and quasicontinuous measurements of the CO_2 and CH_4 exchange between coastal wetlands and the atmosphere. In this chapter, the authors briefly introduced the eddy covariance technique and then synthesized the ecosystem-scale CO_2 and CH_4 fluxes in mangroves and tidal marshes from the existing eddy covariance literature. The discussion focused on the magnitude and temporal variability of ecosystem-scale GHG fluxes and their major biophysical drivers in coastal wetlands.

Chapter 5 examines the consumption of organic production of coastal wetlands by animals and the implications for energy flow and carbon storage. Emphasis is put on the consumption of vascular plant detritus by the macrofauna. The historical development of the concept of detritivory and the significance of the detritus-based food chain are reviewed. The controversy of the trophic value of vascular plant production in coastal food webs is discussed in the light of recent evidence from field observations, chemical tracers, and molecular analyses. The fate of the carbon assimilated by macrofauna is also discussed in terms of the overall carbon budget of coastal wetlands.

Chapter 6 focuses on GHG fluxes at the water-air interface in coastal wetlands. In this chapter, the authors review the existing literature to report the magnitude of GHG fluxes in mangroves, salt-marshes, and seagrasses. The authors identify and explain the main environmental factors that drive water-air fluxes including tidal dynamics, salinity, rainfall, water temperature, and light availability. The authors then describe and discuss the advantages and disadvantages of the main methods used to estimate GHG fluxes at the water-air interface, which are the floating chamber method, the concentration gradient method, and the Eddy-covariance technique.

Chapter 7 reviews the latest data and analyses on the impact of climate changes on GHG production and emissions from coastal wetlands and related processes such as autotroph productivity in coastal wetlands and carbon burial in coastal wetland sediments. The influence of temperature on GHG production is examined associated with rates of organic matter decomposition. The influences of rainfall patterns and sea level rise on GHG production are revealed related to sediment water content and the renewal of electron acceptors in sediments. Elevated atmospheric CO_2 concentrations may affect GHG production from different interfaces by regulating stoichiometry of autotroph tissues and ocean acidification. Sea level rise, rainfall, and/ temperature may induce changes in sediment water content and biofilm development which affect GHG diffusion.

Chapter 8 elucidates in detail the current knowledge on the role of biogenic structures for greenhouse gas biogeochemistry and dynamics in coastal intertidal wetlands. It starts with a description of intertidal wetlands according to our definition in the present context. This is followed by an overview of the relevant biogeochemical processes affecting greenhouse gases in these wetlands. Then the major types of biogenic structures formed by plants and animals in intertidal wetlands are portrayed. The functioning of these biogenic structures for greenhouse gas biogeochemistry and exchange in intertidal wetlands are demonstrated by assessing and compiling the current knowledge from densely vegetated and bioturbated tropical mangrove forests, temperate *Spartina* marshes, and unvegetated intertidal flats. The stage is then set for a discussion of the climate perspective with respect to ecosystem functioning of coastal wetlands in a world with rising sea level and increasing temperatures.

Chapter 9 Mangrove soils have been recognized as potential sources of greenhouse gases (GHG), including carbon dioxide (CO_2), methane (CH_4), and nitrous oxide (N_2O), and anthropogenic activities influence the emission patterns of the GHGs through changing the environmental settings and substrates regulating the productions of the GHGs. In this chapter, the authors summarize the impacts of various anthropogenic activities, including (while not restricted to) the rewetting, revegetation, deforestation, and wastewater discharge, on the soil to atmosphere GHG fluxes in intertidal vegetated wetlands. The potential biogeochemical processes for the gas productions were also examined in the impacted wetland soils. Generally, the responses of GHG emissions from wetlands depend on the species of gases and the anthropogenic activity. The authors also propose future studies for better understanding the responses of gas emissions and mechanisms behind the gas productions in wetland soils subjected to anthropogenic activities.

Chapter 10 is a synthesis and conclusion chapter to compare the contrast processes and drivers in different coastal wetlands and gives suggestions on "blue carbon management" in view of GHG emission and/or litter decomposition of different coast wetlands. The authors discuss factors constraining different processes of carbon mineralization, including litter and live biomass decomposition, sediment organic matter decomposition, GHG production, and emission at sediment-air and water-air interfaces. The authors synthesize knowledge on how carbon mineralization in coastal wetlands is affected by climate change and anthropogenic activities, which help to enhance the understanding on the vulnerability of carbon stocks to negative global changes and to leverage carbon management by reducing carbon emissions. The authors highlight the past lessons learnt from past and current coastal wetland management associated with carbon budgets, which pave the way for future research and management initiatives and practices.

This book is expected to bridge the gap between carbon mineralization and carbon accumulation and broadens the horizon of current study on carbon cycling in coastal wetlands. The chapter authors of the book highlight the necessity to take into account of different biotic and abiotic aspects in relation to the different processes of carbon mineralization, from litter and organic matter decomposition to GHG emissions. This will be a holistic treatment for different potential readers, including researchers and decision-makers. We foresee that the book enhances effective management of carbon budgets in coastal wetlands.

References

Alongi, D.M., 2014. Carbon cycling and storage in mangrove forests. Ann. Rev. Mar. Sci. 6, 195–219.

Arndt, S., Jørgensen, B.B., LaRowe, D.E., Middelburg, J., Pancost, R., Regnier, P., 2013. Quantifying the degradation of organic matter in marine sediments: a review and synthesis. Earth Sci. Rev. 123, 53–86.

Bartlett, K.B., Harriss, R.C., Sebacher, D.I., 1985. Methane flux from coastal salt marshes. J. Geophys. Res. Atmos. 90 (D3), 5710–5720.

Bartolini, F., Cimò, F., Fusi, M., Dahdouh-Guebas, F., Lopes, G.P., Cannicci, S., 2011. The effect of sewage discharge on the ecosystem engineering activities of two East African fiddler crab species: consequences for mangrove ecosystem functioning. Mar. Environ. Res. 71 (1), 53–61.

Bortolus, A., Iribarne, O., 1999. Effects of the SW Atlantic burrowing crab *Chasmagnathus granulata* on a *Spartina* salt marsh. Mar. Ecol. Prog. Ser. 178, 79–88.

Botto, F., Iribarne, O., 2000. Contrasting effects of two burrowing crabs (*Chasmagnathus granulata and Uca uruguayensis*) on sediment composition and transport in estuarine environments. Estuar. Coast. Shelf Sci. 51, 141–151.

Bouillon, S., Borges, A.V., Castañeda-Moya, E., Diele, K., Dittmar, T., Duke, N.C., Kristensen, E., Lee, S.Y., Marchand, C., Middelburg, J.J., Rivera-Monroy, V.H., Smith III, T.J., Twilley, R.R., 2008. Mangrove production and carbon sinks: a revision of global budget estimates. Global Biogeochem. Cycles 22 (2), GB2013.

Breithaupt, J.L., Smoak, J.M., Smith, T.J., Sanders, C.J., Hoare, A., 2012. Organic carbon burial rates in mangrove sediments: strengthening the global budget. Global Biogeochem. Cycles 26 (3), GB3011. https://doi.org/10.1029/2012GB004375.

Breithaupt, J.L., Hurst, N., Steinmuller, H.E., Duga, E., Smoak, J.M., Kominoski, J.S., Chambers, L.G., 2020. Comparing the biogeochemistry of storm surge sediments and pre-storm soils in coastal wetlands: Hurricane Irma and the Florida Everglades. Estuar. Coasts 43, 1090–1103. https://doi.org/10.1007/s12237-019-00607-0.

Bunting, P., Rosenqvist, A., Lucas, R., Rebelo, L.-M., Hilarides, L., Thomas, N., Hardy, A., Itoh, T., Shimada, M., Finlayson, C., 2018. The global mangrove watch—a new 2010 global baseline of mangrove extent. Remote Sens. (Basel) 10 (10), 1669.

Castañeda-Moya, E., Twilley, R., Rivera-Monroy, V., Zhang, K., Davis III, S., Ross, M., 2010. Sediment and nutrient deposition associated with Hurricane Wilma in mangroves of the Florida Coastal Everglades. Estuar. Coasts 33 (1), 45–58. https://doi.org/10.1007/s12237-009-9242-0.

Chauhan, R., Datta, A., Ramanathan, A.L., Adhya, T.K., 2015. Factors influencing spatio-temporal variation of methane and nitrous oxide emission from a tropical mangrove of eastern coast of India. Atmos. Environ. 107, 95–106.

Cragg, S.M., Friess, D.A., Gillis, L.G., Trevathan-Tackett, S.M., Terrett, O.M., Watts, J.E., Distel, D.L., Dupree, P., 2019. Vascular plants are globally significant contributors to marine carbon fluxes and sinks. Ann. Rev. Mar. Sci. 12, 469–497.

Curran, M., 1985. Gas movements in the roots of *Avicennia marina (Forsk.) Vierh.* Funct. Plant Biol. 12 (2), 97–108.

Dame, R., Chrzanowski, T., Bildstein, K., Kjerfve, B., McKellar, H., Nelson, D., Spurrier, J., Stancyk, S., Stevenson, H., Vernberg, J., 1986. The outwelling hypothesis and North inlet, South Carolina. Mar. Ecol. Prog. Ser., 217–229.

David, F., Marchand, C., Taillardat, P., Nho, N.T., Meziane, T., 2018. Nutritional composition of suspended particulate matter in a tropical mangrove creek during a tidal cycle (Can Gio, Vietnam). Estuar. Coast. Shelf Sci. 200, 126–130.

Deegan, L.A., Johnson, D.S., Warren, R.S., Peterson, B.J., Fleeger, J.W., Fagherazzi, S., Wollheim, W.M., 2012. Coastal eutrophication as a driver of salt marsh loss. Nature 490 (7420), 388.

Dittmar, T., Lara, R.J., Kattner, G., 2001. River or mangrove? Tracing major organic matter sources in tropical Brazilian coastal waters. Mar. Chem. 73, 253–271.

Doody, J.P., 2004. 'Coastal squeeze'—an historical perspective. J. Coast. Conserv. 10 (1), 129–138.

Duarte, C.M., Losada, I.J., Hendriks, I.E., Mazarrasa, I., Marbà, N., 2013. The role of coastal plant communities for climate change mitigation and adaptation. Nat. Clim. Chang. 3 (11), 961–968.

Duke, N.C., Kovacs, J.M., Griffiths, A.D., Preece, L., Hill, D.J., Van Oosterzee, P., Mackenzie, J., Morning, H.S., Burrows, D., 2017. Large-scale dieback of mangroves in Australia's Gulf of Carpentaria: a severe ecosystem response, coincidental with an unusually extreme weather event. Mar. Freshw. Res. 68 (10), 1816–1829.

Dutta, M.K., Chowdhury, C., Jana, T.K., Mukhopadhyay, S.K., 2013. Dynamics and exchange fluxes of methane in the estuarine mangrove environment of the Sundarbans, NE coast of India. Atmos. Environ. 77 (Suppl. C), 631–639. https://doi.org/10.1016/j.atmosenv.2013.05.050.

Ezcurra, E., Barrios, E., Ezcurra, P., Ezcurra, A., Vanderplank, S., Vidal, O., Villanueva-Almanza, L., Aburto-Oropeza, O., 2019. A natural experiment reveals the impact of hydroelectric dams on the estuaries of tropical rivers. Sci. Adv. 5 (3), eaau9875.

Faridah-Hanum, I., Latiff, A., Hakeem, K.R., Özturk, M., 2014. Mangrove Ecosystems of Asia: Status, Challenges and Management Strategies. Springer.

Friess, D.A., Yando, E.S., Abuchahla, G.M.O., Adams, J.B., Cannicci, S., Canty, S.W.J., Cavanaugh, K.C., Connolly, R.M., Cormier, N., Dahdouh-Guebas, F., Diele, K., Feller, I.C., Fratini, S., Jennerjahn, T.C., Lee, S.Y., Ogurcak, D.E., Ouyang, X., Rogers, K., Rowntree, J.K., Sharma, S., Sloey, T.M., Wee, A.K.S., 2020. Mangroves give cause for conservation optimism, for now. Curr. Biol. 30 (4), R153–R154. https://doi.org/10.1016/j.cub.2019.12.054.

Gabler, C.A., Osland, M.J., Grace, J.B., Stagg, C.L., Day, R.H., Hartley, S.B., Enwright, N.M., From, A.S., McCoy, M.L., McLeod, J.L., 2017. Macroclimatic change expected to transform coastal wetland ecosystems this century. Nat. Clim. Chang. 7 (2), 142.

Gandhi, S., Jones, T.G., 2019. Identifying mangrove deforestation hotspots in South Asia, Southeast Asia and Asia-Pacific. Remote Sens. 11 (6), 728.

Gedan, K.B., Silliman, B.R., Bertness, M.D., 2009. Centuries of human-driven change in salt marsh ecosystems. Ann. Rev. Mar. Sci. 1, 117–141.

Gilman, E.L., Ellison, J., Duke, N.C., Field, C., 2008. Threats to mangroves from climate change and adaptation options: a review. Aquat. Bot. 89 (2), 237–250.

Hamilton, S.E., Friess, D.A., 2018. Global carbon stocks and potential emissions due to mangrove deforestation from 2000 to 2012. Nat. Clim. Chang. 8 (3), 240–244. https://doi.org/10.1038/s41558-018-0090-4.

Harris, R.M., Beaumont, L.J., Vance, T.R., Tozer, C.R., Remenyi, T.A., Perkins-Kirkpatrick, S.E., Mitchell, P.J., Nicotra, A., McGregor, S., Andrew, N., 2018. Biological responses to the press and pulse of climate trends and extreme events. Nat. Clim. Chang. 8 (7), 579.

He, Q., Silliman, B.R., 2019. Climate change, human impacts, and coastal ecosystems in the Anthropocene. Curr. Biol. 29 (19), R1021–R1035.

Herbert, E.R., Schubauer-Berigan, J.P., Craft, C.B., 2020. Effects of 10 yr of nitrogen and phosphorus fertilization on carbon and nutrient cycling in a tidal freshwater marsh. Limnol. Oceanogr. https://doi.org/10.1002/lno.11411.

Iribarne, O., Bortolus, A., Botto, F., 1997. Between-habitat differences in burrow characteristics and trophic modes in the southwestern Atlantic burrowing crab *Chasmagnathus granulata*. Mar. Ecol. Prog. Ser. 155, 137–145.

Jeffrey, L.C., Reithmaier, G., Sippo, J.Z., Johnston, S.G., Tait, D.R., Harada, Y., Maher, D.T., 2019. Are methane emissions from mangrove stems a cryptic carbon loss pathway? Insights from a catastrophic forest mortality. New Phytol. 224 (1), 146–154.

Kennish, M.J., 2001. Coastal salt marsh systems in the US: a review of anthropogenic impacts. J. Coast. Res. 17 (3), 731–748.

King, G.M., 1984. Utilization of hydrogen, acetate, and "noncompetitive"; substrates by methanogenic bacteria in marine sediments. Geomicrobiol J. 3 (4), 275–306.

Kitaya, Y., Yabuki, K., Kiyota, M., Tani, A., Hirano, T., Aiga, I., 2002. Gas exchange and oxygen concentration in pneumatophores and prop roots of four mangrove species. Trees 16 (2–3), 155–158. https://doi.org/10.1007/s00468-002-0167-5.

Kristensen, E., 2008. Mangrove crabs as ecosystem engineers; with emphasis on sediment processes. J. Sea Res. 59 (1–2), 30–43. https://doi.org/10.1016/j.seares.2007.05.004.

Kristensen, E., Alongi, D.M., 2006. Control by fiddler crabs (*Uca vocans*) and plant roots (*Avicennia marina*) on carbon, iron, and sulfur biogeochemistry in mangrove sediment. Limnol. Oceanogr. 51 (4), 1557–1571.

Kristensen, E., Bouillon, S., Dittmar, T., Marchand, C., 2008. Organic carbon dynamics in mangrove ecosystems: a review. Aquat. Bot. 89, 201–219.

Kristensen, E., Connolly, R., Ferreira, T.O., Marchand, C., Otero, X.L., Rivera-Monroy, V.H., 2017. Biogeochemical cycles; global approaches and perspectives. In: Mangrove Ecosystems: A Global Biogeographic Perspective Structure, Function and Services. Springer.

Lallier-Vergès, E., Marchand, C., Disnar, J.-R., Lottier, N., 2008. Origin and diagenesis of lignin and carbohydrates in mangrove sediments of Guadeloupe (French West Indies): evidence for a two-step evolution of organic deposits. Chem. Geol. 255, 388–398.

Lee, S.Y., 1995. Mangrove outwelling: a review. Hydrobiologia 295, 203–212.

Lee, S.Y., 1997. Potential trophic importance of the faecal material of the mangrove sesarmine crab *Sesarma messa*. Mar. Ecol. Prog. Ser. 159, 275–284.

Lee, S.Y., 1999. Tropical mangrove ecology: physical and biotic factors influencing ecosystem structure and function. Aust. J. Ecol. 24 (4), 355–366.

Lee, S.Y., 2008. Mangrove macrobenthos: assemblages, services, and linkages. J. Sea Res. 59 (1–2), 16–29. https://doi.org/10.1016/j.seares.2007.05.002.

Lee, S.Y., Dunn, R.J.K., Young, R.A., Connolly, R.M., Dale, P.E.R., Dehary, R., Lemckert, C. J., Mckinnon, S., Powell, B., Teasdale, P.R., Welsh, D.T., 2006. Impact of urbanization on coastal wetland structure and function. Austral Ecol. 31, 149–163.

Lee, S.Y., Hamilton, S., Barbier, E., Primavera, J.H., Lewis III, R.R., 2019. Better restoration policies are needed to conserve mangrove ecosystems. Nat. Ecol. Evol. 3, 870–872.

Li, J., Powell, T.L., Seiler, T.J., Johnson, D.P., Anderson, H.P., Bracho, R., Hungate, B.A., Hinkle, C.R., Drake, B.G., 2007. Impacts of Hurricane Frances on Florida scrub-oak ecosystem processes: defoliation, net CO_2 exchange and interactions with elevated CO_2. Glob. Chang. Biol. 13 (6), 1101–1113.

Lovelock, C.E., Cahoon, D.R., Friess, D.A., Guntenspergen, G.R., Krauss, K.W., Reef, R., Rogers, K., Saunders, M.L., Sidik, F., Swales, A., 2015. The vulnerability of indo-Pacific mangrove forests to sea-level rise. Nature 526 (7574), 559–563. https://doi.org/10.1038/nature15538.

Lovelock, C.E., Fourqurean, J.W., Morris, J.T., 2017. CO_2 emissions from coastal wetland transitions to other land uses: tidal marshes, mangrove forests, and seagrass beds. Front. Mar. Sci. 4, 143. https://doi.org/10.3389/fmars.2017.00143.

Luo, M., Huang, J.-F., Zhu, W.-F., Tong, C., 2019. Impacts of increasing salinity and inundation on rates and pathways of organic carbon mineralization in tidal wetlands: a review. Hydrobiologia 827 (1), 31–49.

Maher, D.T., Santos, I.R., Golsby-Smith, L., Gleeson, J., Eyre, B.D., 2013. Groundwater derived dissolved inorganic and organic carbon exports from a mangrove tidal creek:

the missing mangrove carbon sink? Limnol. Oceanogr. 58, 475e488. https://doi.org/10.4319/lo.2013.58.2.0475.

Marchand, C., Disnar, J.-R., Lallier-Vergès, E., Lottier, N., 2005. Early diagenesis of carbohydrates and lignin in mangrove sediments submitted to variable redox conditions (French Guiana). Geochim. Cosmochim. Acta 69, 131–142.

Mcleod, E., Chmura, G., Bouillon, S., Salm, R., Björk, M., Duarte, C., Lovelock, C., Schlesinger, W., Silliman, B., 2011. A blueprint for blue carbon: toward an improved understanding of the role of vegetated coastal habitats in sequestering CO_2. Front. Ecol. Environ. 9 (10), 552–560.

Mcowen, C.J., Weatherdon, L.V., Van Bochove, J.-W., Sullivan, E., Blyth, S., Zockler, C., Stanwell-Smith, D., Kingston, N., Martin, C.S., Spalding, M., 2017. A global map of saltmarshes. Biodivers. Data J. (5). https://doi.org/10.3897/BDJ.5.e11764, e11764.

Milbrandt, E., Greenawalt-Boswell, J., Sokoloff, P., Bortone, S., 2006. Impact and response of Southwest Florida mangroves to the 2004 hurricane season. Estuar. Coasts 29 (6), 979–984.

Mitra, A., 2013. Sensitivity of Mangrove Ecosystem to Changing Climate. Springer.

Mo, Y., Kearney, M.S., Turner, R.E., 2019. Feedback of coastal marshes to climate change: Long-term phenological shifts. Ecol. Evol. 9 (12), 6785–6797.

Molles, M., 2015. Ecology: Concepts and Applications, seventh ed. McGraw-Hill Education, New York, NY.

Muñoz, M.Á., Zornoza, R., 2017. Soil Management and Climate Change: Effects on Organic Carbon, Nitrogen Dynamics, and Greenhouse Gas Emissions. Academic Press.

Nellemann, C., Corcoran, E., Duarte, C., Valdés, L., De Young, C., Fonseca, L., Grimsditch, G., 2009. Blue Carbon: A Rapid Response Assessment. UNEP/Earthprint.

O'Connor, J.J., Fest, B.J., Sievers, M., Swearer, S.E., 2020. Impacts of land management practices on blue carbon stocks and greenhouse gas fluxes in coastal ecosystems–a meta-analysis. Glob. Chang. Biol. 26, 1354–1366. https://doi.org/10.1111/gcb.14946.

Orth, R.J., Carruthers, T.J., Dennison, W.C., Duarte, C.M., Fourqurean, J.W., Heck, K.L., Hughes, A.R., Kendrick, G.A., Kenworthy, W.J., Olyarnik, S., 2006. A global crisis for seagrass ecosystems. Bioscience 56 (12), 987–996.

Ouyang, X., Lee, S.Y., 2014. Updated estimates of carbon accumulation rates in coastal marsh sediments. Biogeosciences 11, 5057–5071. https://doi.org/10.5194/bg-11-5057-2014.

Ouyang, X., Lee, S.Y., 2020. Improved estimates on global carbon stock and carbon pools in tidal wetlands. Nat. Commun. 11, 317. https://doi.org/10.1038/s41467-019-14120-2.

Ouyang, X., Lee, S.Y., Connolly, R.M., 2017. The role of root decomposition in global mangrove and saltmarsh carbon budgets. Earth Sci. Rev. 166, 53–63. https://doi.org/10.1016/j.compchemeng.2016.09.009.

Ouyang, X., Lee, S.Y., Connolly, R.M., 2018a. Using isotope labeling to partition sources of CO_2 efflux in newly established mangrove seedlings. Limnol. Oceanogr. 63 (2), 731–740. https://doi.org/10.1002/lno.10663.

Ouyang, X., Lee, S.Y., Connolly, R.M., Kainz, M.J., 2018b. Spatially-explicit valuation of coastal wetlands for cyclone mitigation in Australia and China. Sci. Rep. 8 (1), 3035.

Ouyang, X., Duarte, C.M., Cheung, S.G., Tam, N.F.Y., Cannicci, S., Martin, C., Lo, H.S., Lee, S.Y., 2022. Fate and effects of macro-and microplastics in coastal wetlands. Environ. Sci. Technol. 56 (4), 2386–2397.

Paling, E., Kobryn, H., Humphreys, G., 2008. Assessing the extent of mangrove change caused by Cyclone Vance in the eastern Exmouth Gulf, northwestern Australia. Estuar. Coast. Shelf Sci. 77 (4), 603–613.

Pendleton, L., Donato, D.C., Murray, B.C., Crooks, S., Jenkins, W.A., Sifleet, S., Craft, C., Fourqurean, J.W., Kauffman, J.B., Marbà, N., 2012. Estimating global "blue carbon" emissions from conversion and degradation of vegetated coastal ecosystems. PLoS One 7 (9), e43542.

Perillo, G., Wolanski, E., Cahoon, D.R., Hopkinson, C.S., 2018. Coastal Wetlands: an Integrated Ecosystem Approach. Elsevier.

Purnobasuki, H., Suzuki, M., 2005. Aerenchyma tissue development and gas- pathway structure in root of *Avicennia marina* (Forsk.) Vierh. J. Plant Res. 118, 285–294.

Rosentreter, J.A., Maher, D., Erler, D., Murray, R., Eyre, B., 2018. Seasonal and temporal CO_2 dynamics in three tropical mangrove creeks–a revision of global mangrove CO_2 emissions. Geochim. Cosmochim. Acta 222, 729–745.

Saintilan, N., Rogers, K., 2015. Woody plant encroachment of grasslands: a comparison of terrestrial and wetland settings. New Phytol. 205 (3), 1062–1070.

Salmo, S.G., Lovelock, C.E., Duke, N.C., 2014. Assessment of vegetation and soil conditions in restored mangroves interrupted by severe tropical typhoon 'Chan-hom'in the Philippines. Hydrobiologia 733 (1), 85–102.

Sasmito, S.D., Taillardat, P., Clendenning, J.N., Cameron, C., Friess, D.A., Murdiyarso, D., Hutley, L.B., 2019. Effect of land-use and land-cover change on mangrove blue carbon: a systematic review. Glob. Chang. Biol. 25, 4291–4302.

Schuerch, M., Spencer, T., Temmerman, S., Kirwan, M.L., Wolff, C., Lincke, D., McOwen, C. J., Pickering, M.D., Reef, R., Vafeidis, A.T., 2018. Future response of global coastal wetlands to sea-level rise. Nature 561 (7722), 231–234.

Sherman, R.E., Fahey, T.J., Martinez, P., 2001. Hurricane impacts on a mangrove forest in the Dominican Republic: damage patterns and early recovery 1. Biotropica 33 (3), 393–408.

Silliman, B.R., Van De Koppel, J., Bertness, M.D., Stanton, L.E., Mendelssohn, I.A., 2005. Drought, snails, and large-scale die-off of southern US salt marshes. Science 310 (5755), 1803–1806.

Siple, M.C., Donahue, M.J., 2013. Invasive mangrove removal and recovery: food web effects across a chronosequence. J. Exp. Mar. Biol. Ecol. 448, 128–135.

Smith, T.J., Anderson, G.H., Balentine, K., Tiling, G., Ward, G.A., Whelan, K.R., 2009. Cumulative impacts of hurricanes on Florida mangrove ecosystems: sediment deposition, storm surges and vegetation. Wetlands 29 (1), 24–34.

Spivak, A.C., Sanderman, J., Bowen, J.L., Canuel, E.A., Hopkinson, C.S., 2019. Global-change controls on soil-carbon accumulation and loss in coastal vegetated ecosystems. Nat. Geosci. 12 (9), 685–692.

Stuart-Smith, R.D., Brown, C.J., Ceccarelli, D.M., Edgar, G.J., 2018. Ecosystem restructuring along the great barrier Reef following mass coral bleaching. Nature 560 (7716), 92–96.

Troxler, T.G., Barr, J.G., Fuentes, J.D., Engel, V., Anderson, G., Sanchez, C., Lagomasino, D., Price, R., Davis, S.E., 2015. Component-specific dynamics of riverine mangrove CO_2 efflux in the Florida coastal Everglades. Agric. For. Meteorol. 213, 273–282.

UNEP-WCMC, S. F, 2018. Global Distribution of Seagrasses (version 6.0). Sixth Update to the Data Layer Used in Green and Short (2003). UN Environment World Conservation Monitoring Centre, Cambridge. http://data.unep-wcmc.org/datasets/7.

Vaiphasa, C., De Boer, W., Skidmore, A., Panitchart, S., Vaiphasa, T., Bamrongrugsa, N., Santitamnont, P., 2007. Impact of solid shrimp pond waste materials on mangrove growth and mortality: a case study from Pak Phanang, Thailand. Hydrobiologia 591 (1), 47–57.

Wang, W., Fu, H., Lee, S.Y., Fan, H., Wang, M., 2020. Can strict protection stop the decline of mangrove ecosystems in China? From rapid destruction to rampant degradation. Forests 11 (1), 55.

Waycott, M., Duarte, C., Carruthers, T.J.B., Orth, R.J., Dennison, W.C, Olyarnik, S., Calladine, A., Fourqurean, J.W., Heck, K.L., Hughes, A., Kendrick, G.A., Kenworthy, W.J., Short, F.T., Williams, S.L., 2009. Accelerating loss of seagrass across the globe threatens coastal ecosystems. Proc. Natl. Acad. Sci. U. S. A 106 (30), 12377–12381.

Werry, J., Lee, S.Y., 2005. Grapsid crabs mediate link between mangrove litter production and estuarine planktonic food chains. Mar. Ecol. Prog. Ser. 293, 165–176.

Yu, X., Yang, X., Wu, Y., Peng, Y., Yang, T., Xiao, F., Zhong, Q., Xu, K., Shu, L., He, Q., Tian, Y., Yan, Q., Wang, C., Wu, B., He, Z., 2020. *Sonneratia apetala* introduction alters methane cycling microbial communities and increases methane emissions. Soil Biol. Biochem. https://doi.org/10.1016/j.soilbio.2020.107775, 107775.

Decomposition of vascular plants and carbon mineralization in coastal wetlands

Xiaoguang Ouyang[a,b,c] **and Shing Yip Lee**[d]

[a]*Southern Marine Science and Engineering Guangdong Laboratory (Guangzhou), Guangzhou, China,*
[b]*Guangdong Provincial Key Laboratory of Water Quality Improvement and Ecological Restoration for Watersheds, School of Ecology, Environment and Resources, Guangdong University of Technology, Guangzhou, China,*
[c]*Simon F.S. Li Marine Science Laboratory, School of Life Sciences, The Chinese University of Hong Kong, Hong Kong Special Administrative Region, China,*
[d]*Institute of Environment, Energy and Sustainability, Simon F.S. Li Marine Science Laboratory, School of Life Sciences, The Chinese University of Hong Kong, Hong Kong Special Administrative Region, China*

2.1 Introduction

Autotrophs, including vascular plants, macroalgae, and the microphytobenthos, in vegetated coastal wetlands convert inorganic carbon to organic carbon via photosynthesis. During photosynthesis, plant primary productivity accumulates in aboveground tissues (including leaves, wood, and reproductive tissues) and belowground roots. In mangroves and saltmarshes, only a small portion of aboveground vascular plant productivity is grazed by herbivores (e.g., insects) (Lee, 1991; Bosire et al., 2005; Cannicci et al., 2008; Teal, 1962), but the fate of this production is different in subtidal seagrasses, which are directly consumed by grazers up to at least 50%, considering that dynamic seagrass responses can lead to underestimates of grazing loss (Valentine and Duffy, 2007). Both live aboveground and belowground tissues turnover and form litter. The fate of aboveground litterfall includes shredding and consumption by macro- and meio-fauna (Lee, 1998; Bosire et al., 2005; Imgraben and Dittmann, 2008), decomposition by microbial communities (Rice and Tenore, 1981), export to nearshore ecosystems (Lee, 1995; Odum, 1968), and the storage of the remainder in sediments (Ouyang et al., 2017b). Belowground dead roots lack a direct export pathway. Tidal movement may help fragment aboveground litter (Fenchel, 1972) and also affect export. Litter

Carbon Mineralization in Coastal Wetlands. https://doi.org/10.1016/B978-0-12-819220-7.00002-9

may be shifted from vegetated coastal wetlands to tidal flats, but the importance of this linkage depends on wetland types. Mangrove leaf litter can be exported to tidal flats and beyond via tidal flushing, whereas the mineralization of saltmarsh-derived organic matter happens almost completely within marshes dominated by grasses, which do not abscise leaves (Lee, 1990; Bouchard and Lefeuvre, 2000). Litter decomposition is an important process of nutrient cycling in coastal wetlands. Fast decomposition allows recycling of carbon and other nutrients to fuel the nearshore food web, while slow decomposition facilitates their accumulation in sediments. When litter is decomposed, part of the nutrients are retained and support further primary production in the wetland (Imgraben and Dittmann, 2008).

Whole plant litter decomposition in coastal wetlands generally consists of three stages: (1) the initial stage of autolysis and leaching with loss of soluble organic matter and inorganic nutrients (Fell and Master, 1975); (2) the second stage of microbial degradation of labile organic matter by bacteria and fungi (Swift, 1976) and the leaching of hydrolyzed substances (Valiela et al., 1985); and (3) the third stage of slower decomposition of recalcitrant organic matter, e.g., lignin (Jorgensen, 1980). In addition to whole plant litter, there are fragmented litter pieces processed by shredders. Werry and Lee (2005) observed that shredding by the mangrove crab *Parasesarma erythodactyla* resulted in reduction of mangrove leaf litter to fragments around 200 µm in their faecal material, which had surface bacteria density of around 70 times higher than that of whole leaf litter after 28 days of decomposition.

Currently, there is a lack of systematic synthesis on factors regulating vascular plant litter decomposition in coastal wetlands. This chapter describes kinetics of litter decomposition. It is followed by an overview of the abiotic, biotic, and anthropogenic factors influencing processes related to litter decomposition in coastal wetlands. Then the results of a metaanalysis on litter decomposition patterns of coastal vascular plants are presented, with a focus on carbon mineralization. Decomposition pathways of recalcitrant compounds are discussed in the context of greenhouse gas production. The last section explores the sinks of mineral carbon in coastal wetlands (Supply.

2.2 Kinetics of litter decomposition

Various indicators are used to describe litter disappearance. These indicators involve residence time (T_r), litter decomposition rate (R_d), mass loss percentage (M_L), and litter half-life time $(t_{1/2})$. While residence time may be associated not only with litter decomposition but also macro-/meio-faunal consumption and other processes (e.g., export), others are exclusive indicators of litter decomposition, over either the whole or part of the decomposition period.

2.2.1 Residence time (T_r)

Litter residence time (T_r) describes the time taken between aboveground litter production and disappearance. It is determined from the ratio of standing crop biomass to litter production:

$$T_r = \frac{B_L}{GPP_L}$$

where B_L is the steady-state litter biomass (in mass m^{-2}) on the substrate, and GPP$_L$ is the aboveground litter production rate (mass m^{-2} year^{-1}), i.e., litterfall rate of one component or the combination of leaf, wood, and reproductive parts.

2.2.2 Decomposition rate (R_d)

Decomposition rate describes the rate of litter mass loss during leaching and decomposition of labile and recalcitrant materials, usually expressed in the form of a decomposition rate constant k (day^{-1}). The commonly used models include linear and negative exponential models. Both models estimate litter decomposition rates from litter mass remaining (% original mass) and time of decomposition.

(1) Linear model

$$M_r = M_0 - kt$$

(2) Exponential model

$$M_r = M_0 \, e^{-kt}$$

where M_r is the percentage remaining of initial litter dry mass. k is the decomposition rate constant (day^{-1}). t is the decomposition time (day). M_0 is the initial litter dry mass (100%).

However, these models have recently been criticized for simplifying the decomposition processes in both terrestrial and marine ecosystems, and there are a few attempts to tackle the complexity of organic matter decomposition in coastal wetlands. Some studies tried to use more complex models to simulate the dynamics of litter decomposition in coastal wetlands, e.g., the double exponential model and the asymptotic model (Siple and Donahue, 2013). Nonetheless, application of these complex models depends on differentiating the decomposition of labile and recalcitrant organic matter, which is unlikely without priori knowledge of litter quality changes during decomposition. Therefore, the linear and single exponential models are convenient to estimate litter decomposition rates. For exponential models, higher sampling frequencies during the initial fast leaching phase may contribute to better estimates of the litter decomposition rates.

(3) Double exponential model

$$M_r = M_0 \left[ce^{-k1t} + (1-c) \, e^{-k2t} \right]$$

This model assumes that a percentage $(1-c)$ of the material is less labile and has a smaller decomposition rate constant k_2, while the more labile material decomposes with a higher rate constant k_1. c is the percentage of the labile material.

(4) Asymptotic model

$$M_r = M_0 \left[ce^{-kt} + (1-c) \right]$$

This model assumes that a percentage $(1-c)$ of the material is recalcitrant and will decompose only over extremely long time frames.

2.2.3 Half-life ($t_{1/2}$)

Half-life ($t_{1/2}$) describes the time taken to lose half of the initial litter mass through decomposition. It is determined from the litter decomposition rate constant k for the respective models.

(1) Linear model

$$t_{1/2} = \frac{1}{2k}$$

(2) Single exponential model

$$t_{1/2} = \frac{\ln 2}{k}$$

2.3 Factors affecting litter decomposition in coastal wetlands

2.3.1 Priori knowledge of influential factors

Litter decomposition in coastal wetlands is affected by a combination of biotic and abiotic factors, as well as anthropogenic activities (Fig. 2.1). Biotic influence comes from plants, animals, and microbes. Abiotic factors include but are not limited to climate, environmental conditions, tidal positions and inundation, physical fragmentation, seasons, and media. Different management regimes also affect litter decomposition in coastal wetlands, e.g., eutrophication, vegetation clearing, and replanting.

(1) Biotic factors

Faunal influence on litter decomposition consists of two major aspects: fragmentation and colonization. These may have contrasting effects on litter decomposition. Litter fragmentation by detritivores may increase the surface area of litter, which facilitate the loss of soluble material as well as colonization by microbial communities (Werry and Lee, 2005). However, meiofaunal and microbial colonization during litter decomposition can increase the litter mass (Gwyther, 2003; Newell, 1984) and bias the estimate of litter decomposition rates. In addition, some macro-detritivores have been proposed to move leaf litter to burrows to enhance in situ decomposition, e.g., the leaf-aging hypothesis (Giddins et al., 1986), but this behavior is not supported by recent evidence. In essence, microbes and macro-/meio-fauna can maintain positive interactions during litter decomposition. Litter fragmentation by macro-/meio-fauna facilitates microbial decomposition. In return, conditioned litter via microbial colonization enhances the nutritional quality of leaf litter for macrofauna, which may also feed on the biofilm, cannibalize, or scavenge

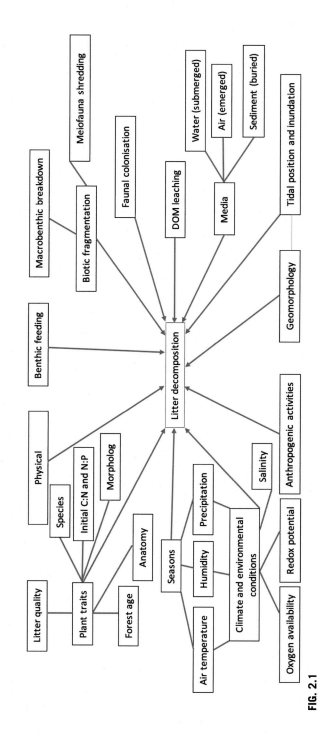

FIG. 2.1

Factors affecting processes of litter decomposition in coastal wetlands. *Arrows* represent influential factors or related processes associated with litter decomposition. *Straight lines* represent elements related to the factors and processes.

animal carcasses to meet nutrients, especially nitrogen, demands (Lee, 1998; Kristensen et al., 2017).

Floral influence is related to different plant traits, including litter quality, forest/stand age, and species identity. The fast decomposition rate of some mangrove species (e.g., *Avicennia marina*) was attributed to the high initial nitrogen concentration (low initial C:N) and low condensed tannin, characteristics that encourage macrofaunal consumption and microbial colonization (Lee, 1993; Mfilinge et al., 2002; Zhou et al., 2010). Leaf litter of the saltmarsh grass *Spartina alterniflora* decomposed faster than that of *Spartina patens*, since stem and leaf sheaths of *S. alterniflora* are largely aerenchyma tissues that provide great surface area for microbial colonization, while *S. patens* has few aerenchyma tissues (Frasco and Good, 1982). There are also differences in litter decomposition among different coastal wetlands. In general, mangrove leaves are more resistant to decay than seagrass blades (Holmer and Olsen, 2002). Differences in the chemistry (e.g., wax and fiber contents) of mangrove and seagrass leaves, combined with the higher surface area to volume ratio of thinner seagrass blades in comparison with thicker mangrove leaves, typically lead to significantly higher decomposition rates of seagrass compared to mangrove litter (Ainley and Bishop, 2015; Fourqurean and Schrlau, 2003). In particular, seagrass blades have a more obvious leaching phase with higher mass loss than mangrove leaves (Holmer and Olsen, 2002; Ainley and Bishop, 2015).

Different indices of litter quality, including initial litter nitrogen content and litter stoichiometry, may have a distinct impact on litter decomposition. Higher litter nitrogen content generally results in higher decomposition rates (Pelegraí et al., 1997), due to increased microbial decomposition activities stimulated by higher nutrient supply. Low litter C/N ratios would also lead to high microbial assimilation and mineralization efficiencies, and thus high decomposition rates. Differences may also result from the macro-molecular composition of litter. The lignocellulose content of mangrove wood (82.6 ± 4.7%) is almost twice that of mangrove leaves (48.4 ± 0.2%) for *Rhizophora mangle* (Benner and Hodson, 1985), likely resulting in slower decomposition of recalcitrant organic matter in the later stage. Mangrove leaf litter was also found to decompose faster in older than in younger forests, attributed to faster leaching of soluble organic matter and nutrient compounds in the former (Li and Ye, 2014), but differences in microhabitat condition, e.g., humidity, may have a significant influence (see section below). The impact of plant morphology on litter decomposition is inconclusive. Tall form of *Spartina* decomposed more slowly than the medium and short forms in a North Carolina saltmarsh (McKee and Seneca, 1982), while nonsignificant differences in decomposition rates were reported for tall versus short forms of *Spartina* in another saltmarsh (Reice and Stiven, 1983).

(2) Abiotic factors

Abiotic factors usually interact with biotic factors to influence litter decomposition in coastal wetlands. Seasonal differences in decomposition were measured for mangrove leaf litter, with faster decomposition during the wet seasons (Bosire et al., 2005), likely due to differences in litter stoichiometry, moisture, and microbial as well as meio-fauna colonization. Precipitation may regulate litter decomposition by modifying the microbial substrate and enhancing microbial decomposition activities (Mackey and Smail, 1996). Lee (1989) observed that cumulative day-degree and immersion time served as good indicators of decomposition rates for mangrove leaf litter, suggesting air temperature and moisture availability are drivers of litter decomposition rates. A similar result was reported for the decomposition of the brackish reed *Phragmites australis* (identified as *P. communis*) (Lee, 1990).

The availability of oxygen generally enhances litter decomposition. In other words, the lack of oxygen suppresses litter decomposition, but the impact of aerobic/anaerobic conditions on litter decomposition rates is sometimes confounded with tidal flooding that promotes organic matter leaching from litter at low tidal positions where anoxic conditions prevail. Anaerobic mineralization rates of the leachable and lignocellulosic components of mangrove leaves and wood were 10–30 times lower than the corresponding aerobic mineralization rates (Benner and Hodson, 1985). Low marshes are exposed to longer tidal inundation, which hampers litter decomposition compared with high marshes. Nevertheless, saltmarsh (*S. alterniflora*) leaf litter decomposed faster in low marshes than in high marshes, due to longer inundation periods in low tidal positions (Montemayor et al., 2011), resulting in more leaching of labile material (Chale, 1993). This is consistent with results of leaf litter decomposition at different tidal positions for other saltmarsh species (e.g., *Atriplex portulacoides* and *S. patens*) (Bouchard et al., 1998; Frasco and Good, 1982). Mangroves in different geomorphological settings differ in root decomposition rates, with higher decomposition rates in riverine and fringe mangroves than those in basin and overwash mangroves (Fig. 2.2), due to differences in plant morphology (e.g., height form) and tidal inundation frequency (e.g., higher inundation frequency in fringe than basin mangroves) (Ouyang et al., 2017a).

(3) Anthropogenic activities

Anthropogenic activities may regulate decomposition through modifying the combined impact of abiotic and/or biotic factors. Different management regimes may have distinct effects on litter decomposition. Mangrove leaf litter decomposition was slower in cleared than intact sites, owing to lower redox potentials and less litterfall in cleared sites (Ashton et al., 1999). Faster evaporation resulting in generally less humid conditions in cleared sites may also have contributed to the lower decomposition rate. Fertilization experiments showed that direct effects on nutrient availability to decomposers might be insignificant, whereas the indirect effects through enhanced litter quality might be substantial in mangroves (Keuskamp et al., 2015). Yang et al. (2018) also observed that nitrogen addition significantly enhanced the dry mass loss rates of *Bruguiera gymnorrhiza* litter by 52%. Mass loss of mangrove leaves was greater

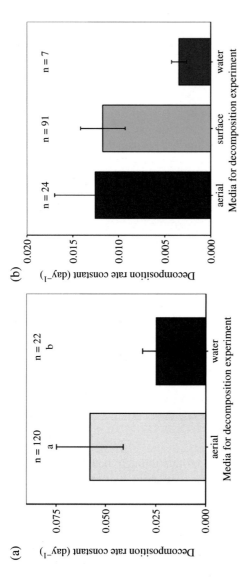

FIG. 2.2

Variation of mangrove (A) and saltmarsh (B) leaf litter decomposition rate constant with media for decomposition experiments.

Data from Rubio, G., Childers, D.L., 2006. Controls on herbaceous litter decomposition in the estuarine ecotones of the Florida Everglades. Estuar. Coasts 29, 257–268, Chen, H., Li, B., Fang, C., Chen, J., Wu, J., 2007. Exotic plant influences soil nematode communities litter input. Soil Biol. Biochem. 39, 1782–1793, Palomo, L., Niell, F.X., 2009. Primary production and nutrient budgets of Sarcocornia perennis ssp alpini (Lag.) Castroviejo in the salt marsh of the Palmones River estuary (Southern Spain). Aquat. Bot. 91, 130–136, Montemayor, D.I., Addino, M., Fanjul, E., Escapa, M., Alvarez, M.F., Botto, F., Iribarne, O.O., 2011. Effect of dominant Spartina species on salt marsh detritus production in SW Atlantic estuaries. J. Sea Res. 66, 104–110, McKee, K.L., Seneca, E.D., 1982. The influence of morphology in determining the decomposition of two salt marsh macrophytes. Estuaries 5, 302–309. https://doi.org/10.2307/1351753, Simões, M.P., Calado, M.D.L., Madeira, M., Gazarini, L.C., 2011. Decomposition and nutrient release in halophytes of a Mediterranean salt marsh. Aquat. Bot. 94, 119–126, Li, H., Liu, Y., Li, J., Zhou, X., Li, B., 2016. Sterols in decomposing Spartina alterniflora and the use of ergosterol in estimating the contribution of fungi to detrital nitrogen. Limnol. Oceanogr. 25, 290–303, de la Cruz, A.A., Gabriel, B.C., 1974. Elemental, and nutritive changes in Phragmites in Spartina-invaded salt marshes. Ecol. Eng. 90, 459–465, Lee, C., Howarth, R.W., Howes, B.L., 1980. Decomposition of marsh grass by aerobic marine bacteria. Bull. decomposing Juncus Roemerianus leaves. Ecology 55, 882–886, Burkholder, P.R., Bornside, G.H., 1957. Decomposition of Spartina anglica, Elytrigia pungens and Halimione portulacoides in a Dutch salt marsh in Torrey. Bot. Club, 84, 366–383, Buth, G.J.C., de Wolf, L., 1985. Decomposition of Spartina anglica, Elytrigia pungens and Halimione portulacoides in a Dutch salt marsh in association with faunal and habitat influences. Vegetatio 62, 337–355, Sun, Z., Mou, X., Sun, W., 2016. Decomposition and heavy metal variations of the typical halophyte litters in coastal marshes of the Yellow River estuary, China. Chemosphere, 147, 163–172, Liao, C.Z., Luo, Y.Q., Fang, C.M., Chen, J.K., Li, B., 2008. Litter pool sizes, decomposition, and nitrogen dynamics in Spartina alterniflora-invaded and native coastal marshlands of the Yangtze Estuary. Oecologia, 156, 589–600, Opsahl, S., Benner, R., 1995. Early diagenesis of vascubr phmt tisswes:Lignin and cutin decompo&ion and biogeochemicai impkations. Geochim. Cosmochim. Acta 59, 4889–4904, Windham, L., 2001. Comparison of biomass production and decomposition between Phragmites australis (common reed) and Spartina patens (salt hay) in brackish tidal marsh of New Jersey. Wetlands 21, 179–188, Rice, D.L., Tenore, K.R., 1981. Dynamics of carbon and nitrogen during the decomposition of detritus derived from estuarine macrophytes. Estuar. Coast. Shelf Sci. 13, 681–690, Woodroffe, C.D., 1982. Litter production and decomposition in the New Zealand mangrove, Avicennia marina var. resinifera. N.Z. J. Mar. Freshw. Res. 16, 179–188, Woodroffe, C.D., 1984. Litter fall beneath Rhizophora stylosa griff., Vaitupu Tuvalu, South Pacific. Aquat. Bot. 18, 249–255, Albright, L.J., 1976.

In situ degradation of mangrove tissues (note). N. Z. J. Mar. Freshw. Res. 10, 385–389, Lee, S.Y., 1989. The importance of sesarminae crabs Chiromanthes spp. and inundation frequency on mangrove (Kandelia candel (L.) Druce) leaf litter turnover in a Hong Kong tidal shrimp pond. J. Exp. Mar. Biol. Ecol. 131, 23–43, Tam, N.F.Y., Wong, Y.S., Lan, C.Y., Wang, L.N., 1998. Litter production and decomposition in a subtropical mangrove swamp receiving wastewater. J. Exp. Mar. Biol. Ecol. 226, 1–18, Lu, C.Y., Lin, P., 1990. Study on litter fall and decomposition of Bruguiera sexangula (Lour.) Poir, community on Hainan Island, China. Bull. Mar. Sci. 47, 139–148, Steinke, T.D., Ward, C.J., 1987. Degradation of mangrove leaf litter in the St Lucia estuary as influenced by season and exposure. S. Afr. J. Bot. 53, 323–328, Chale, F., 1993. Degradation of mangrove leaf litter under aerobic conditions. Hydrobiologia 257, 177–183, Bosire, J.O., Dahdouh-Guebas, F., Kairo, J.G., Kazungu, J., Dehairs, F., Koedam, N. 2005. Litter degradation and CN dynamics in reforested mangrove plantations at Gazi Bay, Kenya. Biol. Conserv. 126, 287–295, Imgraben, S., Dittmann, S., 2008. Leaf litter dynamics and litter consumption in two temperate South Australian mangrove forests. J. Sea Res. 59, 83–93, Li, T., Ye, Y., 2014. Dynamics of decomposition and nutrient release of leaf litter in Kandelia obovata mangrove forests with different ages in Jiulongjiang Estuary, China. Ecol. Eng. 73, 454–460, Wafar, S., Untawale, A., Wafar, M., 1997. Litter fall and energy flux in a mangrove ecosystem. Estuar. Coast. Shelf Sci. 44, 111–124, Mfilinge, P., Atta, N., Tsuchiya, M., 2002. Nutrient dynamics and leaf litter decomposition in a subtropical mangrove forest at Oura Bay, Okinawa, Japan. Trees 16, 172–180. https://doi.org/10.1007/s00468-001-0156-0, Twilley, R.R., Pozo, M., Garcia, V.H., Rivera-Monroy, V.H., Zambrano, R., Bodero, A., 1997. Litter dynamics in riverine mangrove forests in the Guayas River estuary, Ecuador. Oecologia 111, 109–122. https://doi.org/10.1007/s004420050214, Middleton, B.A., McKee, K.L., 2001. Degradation of mangrove tissues and implications for peat formation in Belizean island forests. J. Ecol. 89, 818–828. https://doi.org/10.1046/j.0022-0477.2001.00602.x, Mackey, A., Smail, G., 1996. The decomposition of mangrove litter in a subtropical mangrove forest. Hydrobiologia 332, 93–98, Sessegolo, G., Lana, P., 1991. Decomposition of Rhizophora mangle, Avicennia schaueriana and Laguncularia racemosa leaves in a mangrove of Paranagua Bay (southeastern Brazil). Bot. Mar. 34, 285–290. https://doi.org/10.1515/botm.1991.34.4.285, Woitchik, A., Ohowa, B., Kazungu, J., Rao, R., Goeyens, L., Dehairs, F., 1997. Nitrogen enrichment during decomposition of mangrove leaf litter in an east African coastal lagoon (Kenya): relative importance of biological nitrogen fixation. Biogeochemistry 39, 15–35, Ashton, E., Hogarth, P., Ormond, R., 1999. Breakdown of mangrove leaf litter in a managed mangrove forest in Peninsular Malaysia. In: Diversity and Function in Mangrove Ecosystems, Springer, pp. 77–88, Shafique, S., Siddiqui, P.J., Aziz, R., Shaukat, S., Farooqui, Z., 2015. Decomposition of Avicennia marina (Forsk.) Vierh. Foliage under field and laboratory conditions in the backwaters of Karachi, Pakistan. Bangladesh J. Bot. 44, 1–7, Barroso-Matos, T., Bernini, E., Rezende, C.E., 2012. Decomposition of mangrove leaves in the estuary of Paraiba do Sul River Rio de Janeiro, Brazil. Lat. Am. J. Aquat. Res. 40, 398–407, Aké-Castillo, J.A., Vazquez, G., Lopez-Portillo, J., 2006. Litterfall and decomposition of Rhizophora mangle L. in a coastal lagoon in the southern Gulf of Mexico. Hydrobiologia 559, 101–111, Menezes, G.V., Schaeffer-Novelli, Y., 2000. Produção e decomposição em bosques de mangue na ilha do Cardoso, Cananéia, SP. Proceedings of the 15th Simpósio de ecossistemas brasileiros. ACIESP, São Paulo, pp. 349–356., Oliveira, A.B.D., Rizzo, A.E., Couto, E.D.C.A.O.G., 2013. Assessing decomposition rates of Rhizophora mangle and Laguncularia racemosa leaves in a Tropical Mangrove. Estuar. Coasts, 36, 1354–1362, Zhou, H.-C., Tam, N.F.-Y., Lin, Y.-M., Wei, S.-D., Li, Y.-Y., 2012. Changes of condensed tannins during decomposition of leaves of Kandelia obovata in a subtropical mangrove swamp in China. Soil Biol. Biochem. 44, 113–121, Ainley, L.B., Bishop, M.J., 2015. Relationships between estuarine modification and leaf litter decomposition vary with latitude. Estuar. Coast. Shelf Sci. 164, 244–252, Gatune, C., Vanreusel, A., Cnudde, C., Ruwa, R., Bossier, P., De Troch, M., 2012. Decomposing mangrove litter supports a microbial biofilm with potential nutritive value to penaeid shrimp post larvae. J. Exp. Mar. Biol. Ecol. 426, 28–38, d'Croz, L., Del Rosario, J., Holness, R., 1989. Degradation of red mangrove (Rhizophora mangle L.) leaves in the Bay of Panama. Rev. Biol. Trop. 37, 101–103, eSilva, C.A.R., Oliveira, S.R., Rêgo, R.D., Mozeto, A.A., 2007. Dynamics of phosphorus and nitrogen through litter fall and decomposition in a tropical mangrove forest. Mar. Environ. Res. 64, 524–534, Davis Iii, S.E. Corronado-Molina, C., Childers, D.L., Day Jr. J.W., 2003. Temporally dependent C, N, and P dynamics associated with the decay of Rhizophora mangle L. leaf litter in oligotrophic mangrove wetlands of the Southern Everglades. Aquat. Bot. 75, 199–215. https://doi.org/10.1016/S0304-3770(02)00176-6, Dick, T.M. Osunkoya, O.O., 2000. Influence of tidal restriction floodgates on decomposition of mangrove litter. Aquat. Bot. 68, 273–280, Robertson, A.I., 1988. Decomposition of mangrove leaf litter in tropical Australia. J. Exp. Mar. Biol. Ecol. 116, 235–247, Vinh, T.V., Allenbach, M., Linh, K.T.V., Marchand, C., 2020. Changes in leaf litter quality during its decomposition in a tropical planted mangrove Forest (can Gio, Vietnam). Front. Environ. Sci. 8. https://doi.org/10.3389/fenvs.2020.00010.

in nutrient enriched estuaries than in the largely unmodified estuaries in Australia (Ainley and Bishop, 2015). With the hydrological regime of coastal wetlands extensively modified and managed along populated coastlines, e.g., impoundments for aquaculture production, created wetlands for wildlife and water treatment, decomposition rates of wetland organic carbon production may be significantly managed to optimize ecosystem services related to carbon dynamics, e.g., offsetting urban carbon emission through reduced decomposition (Lee et al., 2014; Macreadie et al., 2017). For example, Dick and Osunkoya (2000) found that decomposition rate of mangrove leaf litter was much greater on the tidal side of floodgates than those on the landward side of impoundments.

2.3.2 A metaanalysis of litter decomposition patterns

This section examines litter decomposition patterns via metaanalysis of available data. We collected data from 50 studies on mangrove and saltmarsh leaf litter decomposition in different media. Decomposition at the aerial, sediment, and aquatic environments means positioning the litter bags above the sediments (aerial), on the sediment surface and in the water (aquatic), respectively. There are significant differences in mangrove leaf litter decomposition rate constants between the aerial (mean ± standard error: $0.058 \pm 0.017 \, day^{-1}$) and aquatic environments ($0.025 \pm 0.007 \, day^{-1}$) (Wilcoxon rank sum test, $W = 1844$, $P < .05$). No significant differences are found among saltmarsh leaf litter decomposition rate constants among the aerial ($0.0125 \pm 0.0044 \, day^{-1}$), surface sediments ($0.0117 \pm 0.0024 \, day^{-1}$), and water environments ($0.0035 \pm 0.0008 \, day^{-1}$) (Kruskal-Wallis test, $\chi^2(2) = 4.4$, $P > .11$). Generally, average leaf litter decomposition rate constants at the aerial and surface sediment environments are more than double those at the aquatic environment, and the lack of significant differences may be attributed to the small sample size of measurements under submerged conditions. The overall higher leaf litter decomposition rate constants at the aerial and/or surface sediment environments may be attributed to aerobic conditions, which favor microbial decomposition.

We collected data from 11 studies on saltmarsh leaf litter decomposition and initial leaf litter C/N. There is a significant negative relationship between log-transformed leaf litter decomposition rate constant and log-transformed initial leaf litter C/N in saltmarshes ($R^2 = 0.41$, $P < .001$, Fig. 2.3). This relationship is consistent with the linear relationship found for mangroves (Kristensen et al., 2008). These relationships reflect that leaf litter decomposition is promoted by the nitrogen content of leaf litter, which determines nutrient availability for microbes participating in decomposition in tidal wetlands.

There are significant differences in root decomposition rates among different mangrove geomorphological and physiognomic settings (Fig. 2.4A, Ouyang et al., 2017a). Decomposition rates were highest in riverine mangroves, intermediate in fringe and scrub, and lowest in overwash and basin mangroves. Root decomposition rate is lowest for overwash mangroves, which accumulate substrate slowly and only via autochthonous input (Middleton and McKee, 2001). The low sediment supply

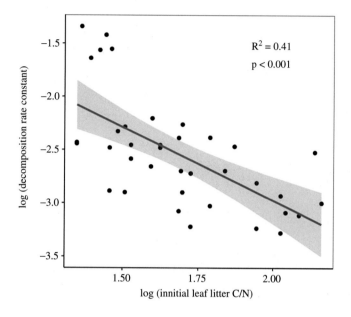

FIG. 2.3

Relationship between decomposition rate constant of tidal marshes and initial leaf litter C/N. The relationship is described by a linear relationship after log-transformation: log (decomposition rate constant) $= -1.37\log(C/N) - 0.23$.

Data from Liao, C.Z., Luo, Y.Q., Fang, C.M., Chen, J.K., Li, B., 2008. Litter pool sizes, decomposition, and nitrogen dynamics in Spartina alterniflora-invaded and native coastal marshlands of the Yangtze Estuary. Oecologia 156, 589–600, Chen, H., Li, B., Fang, C., Chen, J., Wu, J., 2007. Exotic plant influences soil nematode communities through litter input. Soil Biol. Biochem. 39, 1782–1793, de la Cruz, A.A., Gabriel, B.C., 1974. Elemental, and nutritive changes in decomposing Juncus Roemerianus leaves. Ecology 55, 882–886, Sun, Z., Mou, X., Sun, W., 2016. Decomposition and heavy metal variations of the typical halophyte litters in coastal marshes of the Yellow River estuary, China. Chemosphere 147, 163–172, Lee, C., Howarth, R.W., Howes, B.L., 1980. Sterols in decomposing Spartina alterniflora and the use of ergosterol in estimating the contribution of fungi to detrital nitrogen. Limnol. Oceanogr. 25, 290–303, Twilley, R.W., Lugo, A.E., Patterson-Zucca, C., 1986. Litter production and turnover in basin mangrove forests in Southwest Florida. Ecology 67, 670–683. https:/doi.org/10.2307/1937691, Palomo, L., Niell, F.X., 2009. Primary production and nutrient budgets of Sarcocornia perennis ssp alpini (Lag.) Castroviejo in the salt marsh of the Palmones River estuary (Southern Spain). Aquat. Bot. 91, 130–136, Buth, G.J.C., de Wolf, L., 1985. Decomposition of Spartina anglica, Elytrigia pungens and Halimione portulacoides in a Dutch salt marsh in association with faunal and habitat influences. Vegetatio 62, 337–355, Li, H., Liu, Y., Li, J., Zhou, X., Li, B., 2016. Dynamics of litter decomposition of dieback Phragmites in Spartina-invaded salt marshes. Ecol. Eng. 90, 459–465, Frasco, B.A., Good, R.E., 1982. Decomposition dynamics of Spartina alterniflora and Spartina patens in a New Jersey salt marsh. Am. J. Bot. 69, 402–406, Bouchard, V., Lefeuvre, J.-C., 2000. Primary production and macro-detritus dynamics in a European salt marsh: carbon and nitrogen budgets. Aquat. Bot. 67, 23–42.

and thus allochthonous nutrient limitation may account for the lower root decomposition rates in these mangroves. Among the other mangrove geomorphological settings, riverine mangroves show the highest rates of root decomposition. These mangroves are dominant along river and creek drainages and experience regular freshwater dilution, thereby alleviated salinity stress. These conditions enhance root

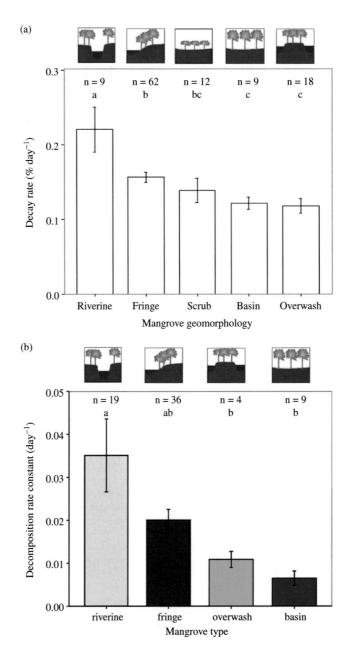

FIG. 2.4

Variation of root (A) and leaf litter (B) decomposition rates among different mangroves. The letters above the error bars indicate significance levels.

(A) From Ouyang, X., Lee, S. Y., Connolly, R.M., 2017. The role of root decomposition in global mangrove and saltmarsh carbon budgets. Earth Sci. Rev. 166, 53–63. https://doi.org/10.1016/j.earscirev.2017.01.004.

(Continued)

decomposition. Additionally, sediments of riverine mangroves may have higher nutrient levels than scrub mangroves (Castañeda-Moya et al., 2011). The higher nutrient supply may also result in higher root decomposition rates.

We collected data from 18 references on leaf litter decomposition in mangroves for which mangrove types are known. Again, there are significant differences in leaf litter decomposition rate constants among different mangrove types (Kruskal-Wallis test, $\chi^2(3) = 18.9$, $P < .001$, Fig. 2.4B). Leaf litter decomposition rate constant is the lowest for overwash ($0.0109 \pm 0.0019\,\mathrm{day^{-1}}$) and basin ($0.0065 \pm 0.0017\,\mathrm{day^{-1}}$) and the highest for riverine ($0.0351 \pm 0.0085\,\mathrm{day^{-1}}$) and fringe mangroves ($0.0201 \pm 0.0025\,\mathrm{day^{-1}}$). This pattern is in agreement with that for root decomposition rate and is probably underpinned by the same biotic and hydrological conditions.

The decomposed root carbon can contribute to porewater dissolved inorganic (DIC) and organic carbon (DOC), then potentially exported to other nearshore environments via both superficial as well as subterranean flow. Some carbon may be released as CO_2 or CH_4 gases, and the balance of the two being strongly influenced by salinity, oxygen availability, and the dominance of sulfate reduction. Further, global sediment carbon gases (including CO_2 and CH_4) emitted are 38 Tg C year^{-1},

FIG. 2.4, cont'd *(B) Data from Woodroffe, C.D., 1982. Litter production and decomposition in the New Zealand mangrove, Avicennia marina var. resinifera. N. Z. J. Mar. Freshw. Res. 16, 179–188, Woodroffe, C.D., 1984. Litter fall beneath Rhizophora stylosa griff., Vaitupu Tuvalu, South Pacific. Aquat. Bot. 18, 249–255, Tam, N.F.Y., Vrijmoed, L.L.P., Wong, S.Y., 1990. Nutrient dynamics associated leaf decomposition in a small subtropical mangrove community in Hong Kong. Bull. Mar. Sci. 47, 68–78, Tam, N.F.Y., Wong, Y.S., Lan, C.Y., Wang, L.N., 1998. Litter production and decomposition in a subtropical mangrove swamp receiving wastewater. J. Exp. Mar. Biol. Ecol. 226, 1–18, Lu, C.Y., Lin, P., 1990. Study on litter fall and decomposition of Bruguiera sexangula (Lour.) Poir, community on Hainan Island, China. Bull. Mar. Sci. 47, 139–148, Steinke, T.D. Ward, C.J., 1987. Degradation of mangrove leaf litter in the St Lucia estuary as influenced by season and exposure. S. Afr. J. Bot. 53, 323–328, Van der Valk, A.G., Attiwill, P.M., 1984. Decomposition of leaf and root litter of Avicennia marina at Westernport Bay, Victoria, Australia. Aquat. Bot. 18, 205–221, Hegazy, A.K., 1998. Perspectives on survival, phenology, litter fall and decomposition, and caloric content of Avicennia marina in the Arabian Gulf region. J. Arid Environ. 40, 417–429, Imgraben, S., Dittmann, S., 2008. Leaf litter dynamics and litter consumption in two temperate South Australian mangrove forests. J. Sea Res. 59, 83–93, Li, T., Ye, Y. 2014. Dynamics of decomposition and nutrient release of leaf litter in Kandelia obovata mangrove forests with different ages in Jiulongjiang Estuary, China. Ecol. Eng. 73, 454–460, Wafar, S., Untawale, A., Wafar, M., 1997. Litter fall and energy flux in a mangrove ecosystem. Estuar. Coast. Shelf Sci. 44, 111–124, Mfilinge, P., Atta, N., Tsuchiya, M., 2002. Nutrient dynamics and leaf litter decomposition in a subtropical mangrove forest at Oura Bay, Okinawa, Japan. Trees, 16, 172–180. https://doi.org/10.1007/s00468-001-0156-0, Twilley, R.R., Pozo, M., Garcia, V.H., Rivera-Monroy, V.H., Zambrano, R., Bodero, A., 1997. Litter dynamics in riverine mangrove forests in the Guayas River estuary, Ecuador. Oecologia 111, 109–122. https://doi.org/10.1007/s004420050214, Middleton, B.A., McKee, K.L., 2001. Degradation of mangrove tissues and implications for peat formation in Belizean island forests. J. Ecol. 89, 818–828. https://doi.org/10.1046/j.0022-0477.2001.00602.x, Mackey, A., Smail, G., 1996. The decomposition of mangrove litter in a subtropical mangrove forest. Hydrobiologia 332, 93–98, Shafique, S., Siddiqui, P.J., Aziz, R., Shaukat, S., Farooqui, Z., 2015. Decomposition of Avicennia marina (Forsk.) Vierh. Foliage under field and laboratory conditions in the backwaters of Karachi, Pakistan. Bangladesh J. Bot. 44, 1–7, Oliveira, A.B.D., Rizzo, A.E., Couto, E.D.C.A.O. G., 2013. Assessing decomposition rates of Rhizophora mangle and Laguncularia racemosa leaves in a Tropical Mangrove. Estuar. Coasts 36, 1354–1362, Zhou, H.-C., Tam, N.F.-Y., Lin, Y.-M., Wei, S.-D., Li, Y.-Y., 2012. Changes of condensed tannins during decomposition of leaves of Kandelia obovata in a subtropical mangrove swamp in China. Soil Biol. Biochem. 44, 113–121.*

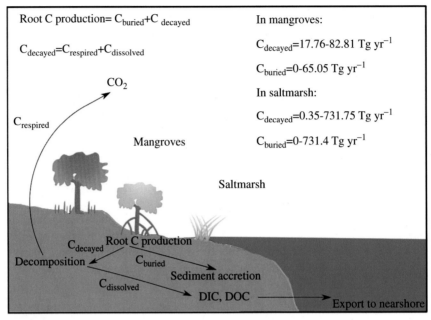

Root C production= C_{buried}+$C_{decayed}$

$C_{decayed}$=$C_{respired}$+$C_{dissolved}$

CO_2

$C_{respired}$

Mangroves

Saltmarsh

$C_{decayed}$ Root C production

Decomposition

C_{buried}

Sediment accretion

$C_{dissolved}$

DIC, DOC

Export to nearshore

In mangroves:

$C_{decayed}$=17.76-82.81 Tg yr^{-1}

C_{buried}=0-65.05 Tg yr^{-1}

In saltmarsh:

$C_{decayed}$=0.35-731.75 Tg yr^{-1}

C_{buried}=0-731.4 Tg yr^{-1}

FIG. 2.5

Global fate of root C in mangroves and saltmarsh. Data outside and inside brackets were calculated from the median and area-averaged root decomposition rates in mangroves, respectively.

From *Ouyang, X., Lee, S.Y., Connolly, R.M., 2017. The role of root decomposition in global mangrove and saltmarsh carbon budgets. Earth Sci. Rev. 166, 53–63. https://doi.org/10.1016/j.earscirev.2017.01.004.*

while DIC and DOC export rates are 86 and 15 Tg C year^{-1}, respectively, in mangroves (Alongi, 2014). As a result, emitted carbon gases, DIC, and DOC account for 27.3%, 61.9%, and 10.8% of mangrove belowground C mineralization. The dissolved carbon (DIC+DOC) forms are significantly higher (10×) than C emissions in their contribution to the products of mangrove root decomposition. The remaining root carbon in mangroves and saltmarshes, manifest as standing dead belowground biomass, is estimated as the difference between total dead root C production and decayed root C. This part contributes to sediment C burial (Fig. 2.5).

2.4 Refractory compounds

Compared to leaves and roots, few studies examined the decomposition of woody tissues in coastal wetlands. In coastal wetlands, woody tissues usually relate to mangrove forests as tidal marshes mainly, and seagrass meadows exclusively, support herbaceous plants. Lignocellulose from mangrove (*R. mangle*) wood decomposed 1.6 times slower than lignocellulose from mangrove leaves (Benner and Hodson,

1985). For bulk organic matter, $t_{1/2}$ of *Avicennia germinans* wood was ~1.4 years while only 77 days for leaves. Mass remaining of woody tissues is an order of magnitude higher than that of leaf tissues after 4 years (Opsahl and Benner, 1995). This means mangrove wood is more recalcitrant than leaf tissues. There are different patterns in wood decomposition in standing dead trees and downed trees, but this may be species specific. Twigs of *R. mangle* and *Avicennia germinans* decomposed faster on the sediment surface than being aerial, while those of *Laguncularia racemosa* were highly resistant to degradation regardless their position (Middleton and McKee, 2001). This pattern for *Avicennia germinans* is corroborated by Romero et al. (2005), who also showed that wood decomposition rate under buried conditions was intermediate between aerial and on the sediment surface.

Sediments of coastal wetlands support a diversity of organic compounds, including carbohydrates, proteins (e.g., amino acids), lipids (e.g., fatty acids and n-alkanes), and phenols (e.g., lignin-derived phenols and tannins) (Kristensen et al., 2008). There are exceptions to the above-mentioned main components of sediment organic matter. For example, microbes are not only decomposers of sediment organic matter but also a component of the sediment organic matter pool. Nucleic acids (i.e., deoxyribonucleic acid (DNA) and ribonucleic acid (RNA)) make up 23% of the dry body mass of prokaryotes (Canfield et al., 2005). The sources of sediment organic matter in coastal wetlands are diverse, owing to the mixing of fresh and oceanic waters, both of which carry allochthonous organic matter, respectively, from terrestrial sources via riverine flow or from marine sources via tidal exchange. These sources, combined with autochthonous primary production, determine the quantity and quality of sediment organic matter in coastal wetlands.

Different halophytes in coastal wetlands may have different compositions of organic compounds, thus affecting the composition of sediment organic carbon. Mangrove leaves may contain around 1 mmol of phenolic carbon derived from lignin per 100 mmol of total organic carbon (Lallier-Vergès et al., 2008; Dittmar and Lara, 2001), while the woody tissues are 2.7–4.8 times richer in lignin than leaves (Lallier-Vergès et al., 2008; Marchand et al. 2005). Similarly, mangrove woody tissues are 1.6–2.9 times richer in carbohydrates than leaves, mostly being structural compounds in the former.

Sediment organic compounds have different chemical characteristics, which shape the dynamic decomposition processes. Detrital sediment organic matter typically contains similar proportions of identifiable compounds: 10%–20% carbohydrates, 10% nitrogenous compounds (mostly amino acids), and 5%–15% lipids, with the rest of the uncharacterized organic matter collectively grouped as humic substances (Arndt et al., 2013). Humic substances (i.e., humic and fulvic acids) were found to account for 6.9% of the carbon pool in sediments of a saltmarsh (Alberts et al., 1988). Humic substances are relatively resistant to mirobial degradation and can be formed by abiotic polymerizaton of simple aliphatic and aromatic molecules (e.g., carbohydrates and proteins) and by microbial synthesis (Filip and Alberts, 1989).

Carbohydrates are the most abundant constituents of vascular plants and assume a major role as storage and structural components (Marchand et al., 2005). Different groups of neutral carbohydrates show different degradation patterns. In coastal wetland sediments, refractory carbohydrates such as arabinose, rhamnose, fucose, hemicellulosic glucose, and cellulose are common, with the latter two showing different resistance to microbial activities. Hemicellulosic glucose tends to be removed by microbial hydrolysis, while cellulose is more refractory to microbial degradation (Stout et al., 1988). Lignin present in vascular plants is a nitrogen-free copolymer of various phenylpropenyl alcohols. Lignin oxidation products are decomposed at a lower rate than carbohydrates. Different groups of lignin-derived phenols or indicators were used to characterize vascular plant lignin sources. Lamda-6 ($\Lambda 6$) is defined as the sum of vanillyl (vanillin, acetovanillone, and vanillic acid) and syringyl (syringaldehyde, acetosyringone, and syringic acid) phenols. Lamda-8 ($\Lambda 8$) is defined as the cinnamyl (p-coumaric and ferulic acid) phenols. They are described, along with the derived indicators, by the equations below:

$$\Lambda 6 = V + S$$
$$\Lambda 8 = C$$

where V is vanillyl phenols (including V_1 and V_2, i.e., vanillic acid and vanillin, respectively), S is syringyl phenols, C is cinnamyl phenols, and C/V and S/V are indicators of vascular plant lignin sources. These indicators may reflect differences in plant age and morphology. For example, C/V is around 0.1 for young *Avicennia germinans* plantings, due to the organic matter input from cable roots. S/V is slightly lower in sediments than in wood of young mangrove trees while significantly lower in sediments than mature mangrove plants. $(Ad/Al)_V$, i.e., the ratio of vanillic acid to vanillin, shows rather stable profiles in different sediment depths, suggesting that aromatic ring cleavage is the primary means by which lignin is degraded (Bianchi et al., 2013).

Despite the traditional attribution of organic carbon preservation capacity of coastal wetlands to intrinsically recalcitrant macro-molecules (e.g., lignin), recent evidence in terrestrial and oceanic ecosystems suggests that microbes can extensively decompose complex plant macro-molecules and their degradation products under suitable environmental conditions (Arndt et al., 2013; Schmidt et al., 2011; Lehmann and Kleber, 2015). In addition, the ability to exploit vascular plant detritus and to overcome the recalcitrance of lignocellulose is found in a range of crustaceans and other wetland macrofauna (see Chapter 5 for a more detailed account). Some mangrove detritivores (e.g., sesarmid crabs) produce endogenous enzymes capable of breaking down the molecular chains of cellulose (Bui and Lee, 2015). Limnoriid isopods (*Limnoria* spp.) have a gut that is devoid of resident microbiota but has the ability to digest crystalline cellulose in lignocellulose, due to the activities of proteins secreted by its hepatopancreas (King et al., 2010). Lignocellulosic detritus breakdown is achieved by a combination of microbial action and detritivore activity that vary in relative rates and sequence according to the nature of the substrate and the environment in which the breakdown takes place (Cragg et al., 2020).

Long-chain n-alkanes are characteristic components of epicuticular leaf waxes of mangroves and saltmarshes, with chain length mainly ranging from C_{23} to C_{35} (Kristensen et al., 2008; Tanner et al., 2010). For sediments of coastal wetlands, peak abundances of long-chain n-alkanes can mainly occur in the lower range of C_{21} to C_{33}, with strong odd-to-even carbon number predominance (Wang et al., 2003). The lowest peak abundance of long-chain n-alkanes for halophytes in coastal wetlands was found in marsh plants with Crassulacean acid metabolism (CAM) photosynthesis pathway (e.g., *Salicornia depressa*) and nonemergent plants (e.g., *Ruppia maritima*) (Tanner et al., 2010). Distinct seasonal variability in the $\delta^{13}C$ values of long-chain n-alkanes has also been reported for C_3 and C_4 halophytes in coastal wetlands. $\delta^{13}C$ values of C_4 marsh halophytes show lower variability (<2%), while C_3 halophytes show higher variability (8%–10%) (Eley et al., 2016). In addition to fatty acids and n-alkanes, there are other types of lipids in the marine environment such as alkanols and alkanoic acids, which can indicate past changes in organic matter sources (Ouyang et al., 2015).

The mineralization of organic carbon in coastal wetlands involves the fermentation and hydrolysis of large biopolymers to produce low molecular weight compounds and ultimately carbon gases (i.e., CO_2 and CH_4) via various aerobic and anaerobic pathways (Fig. 2.6). Generally, oxygen penetration and aerobic decomposition in coastal wetlands are limited to the shallow aerobic zone (typically the top few millimeters), except where microbenthic bioturbation and plant root activity allow oxygen to penetrate to deeper layers below the sediment surface (Holmboe et al., 2001). Below the aerobic zone, different anaerobic pathways dominate the decomposition of organic carbon. CO_2 is mainly produced by sulfate, iron, and manganese reduction. Sulfate reduction is usually the dominant pathway of CO_2 production in coastal wetlands due to the high abundance of sulfate in saline environments (Skyring, 1987). Nonetheless, other pathways may dominate CO_2 production, e.g., iron reduction where iron is abundant (Kristensen et al., 2000). Methane is the reduction product during methanogenesis. Generally, methanogenesis is thought to be suppressed by sulfate reduction since methanogens may be outcompeted by sulfate-reducing bacteria for common substrates, such as H_2 and acetate (Abram and Nedwell, 1978). However, methane formation from CO_2 but not from acetate was detected within the same horizon of saltmarsh sediments where sulfate reduction was most active (Senior et al., 1982). On the other hand, methane production can be restrained by the activities of methanotrophs, as has been observed in a Great Sippewissett saltmarsh (Buckley et al., 2008). Sulfate-reduction bacteria may also inhibit the microphytobenthos when they compete for common substrates (Winfrey and Ward, 1983). The fermentation of low molecular weight organic compounds and the pathways of CH_4 and CO_2 production are described in Fig. 2.6.

Positive values of the free energy yields (ΔG^0) indicate that the reaction is endergonic and energy from adenosine triphosphate (ATP) or an accompanying process is needed to drive the reaction, while negative values indicate that the reaction is exergonic and can proceed spontaneously or be biologically catalyzed. The lower the ΔG^0, the easier the reaction would happen. For the above reactions, ΔG^0 of methanal aerobic decomposition ($-479\,kJ\,mol^{-1}$) is lower than anaerobic decomposition pathways, e.g., iron reduction with 2-line ferrihydrite as electron acceptor ($-114\,kJ\,mol^{-1}$) and manganese reduction with pyrolusite as electron acceptor

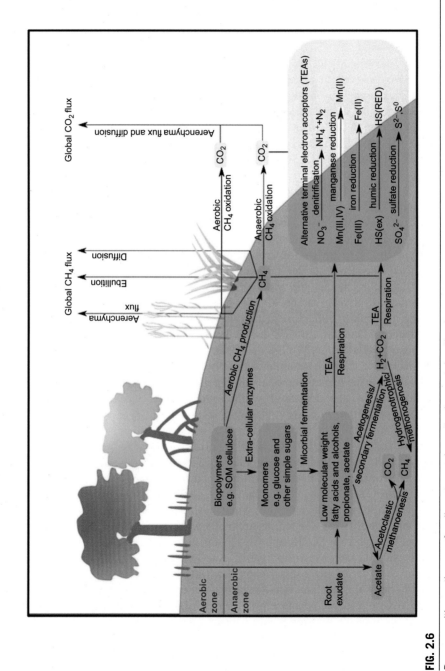

FIG. 2.6

Decomposition pathways of sediment organic matter in coastal wetlands.

representative value of 34.7 Mg ha^{-1} (Ouyang and Lee, 2020). Contents of sediment inorganic carbon in mangroves can reach 549.5 Mg ha^{-1} in karstic regions (Adame et al., 2013) and 93.4 Mg ha^{-1} in mangroves adjacent to coral reefs (Guan et al., 2018). However, mangrove environments do not promote calcification compared to seagrasses, due to the generally lower pH (Camp et al., 2016), which facilitates calcium carbonate dissolution and suppresses calcification. Likewise, the inorganic carbon process of saltmarshes in Chenier plain, China, is dominated by carbonate dissolution and has nearly no carbonate production in situ (Lu et al., 2019). Inorganic carbon stocks in global seagrasses were estimated to be 590 Mg ha^{-1} (median) in the top-meter sediments based on the result of a recent study (Saderne et al., 2019). Aside from the lithogenic origin and input from coral reefs, seagrasses may provide relatively good conditions (i.e., higher pH) for calcification by shellfish, zooplankton, and pteropods (Howard et al., 2018). Nevertheless, with the accelerating trend of ocean acidification due to anthropogenic CO_2 emissions (Burnell et al., 2013; Ouyang and Guo, 2016), pH of seawater has declined 0.1 units since the 1800s and the continuing decline in pH may promote more carbonate dissolution in seagrass beds in the future. Aside from the above linkages, dissolved organic carbon in surface waters can be consumed by wetland autotrophs to produce organic carbon (Liu et al., 2018), such as acetate. Some macrofauna are also known to be able to assimilate DOC directly. Fig. 2.8 presents the processes of inorganic carbon cycling linked to organic carbon in coastal wetlands.

Inorganic carbon dynamics (including calcification and dissolution) are closely linked to carbon remineralization in coastal wetlands. Some amount of CO_2 produced by remineralizing stored organic carbon could be consumed by the dissolution of $CaCO_3$ sediments, evidenced by the correlation between dissolution rates of sediment $CaCO_3$ and sulfide oxidation in porewater beneath seagrasses (Ku et al., 1999). Nonetheless, the ability of $CaCO_3$ dissolution to buffer the release of CO_2 from organic carbon remineralization is claimed to be limited to areas where calcification commonly occurs and $CaCO_3$ is present (Howard et al., 2018). Further, photosynthesis decreases the CO_2 concentration in surface waters as dissolved organic carbon is incorporated into organic carbon (Macreadie et al., 2019) of underwater vegetation (e.g., seagrasses), while respiration and remineralisation increase the CO_2 concentration. This means the underwater balance between inorganic and organic carbon can be directly altered by photosynthesis and mineralization processes. Calcification and $CaCO_3$ deposition may enhance organic carbon preservation by accelerating its burial in anoxic sediment layers where remineralization is considered to be low (Arndt et al., 2013).

Changes in inorganic carbon dynamics are subject to global environmental changes. Fig. 2.9 describes the changes in carbonate species and K_H with pH and temperature. With rising air temperature, the Henry constant K_H declines linearly, suggesting less CO_2 dissolved in water and thus less carbon sink in coastal wetlands under inundated conditions. Furthermore, with the overwhelming trend of ocean acidification, water pH declines. This results in the formation of carbonic acid and simultaneous transformation of carbonate, while the pattern of bicarbonate is

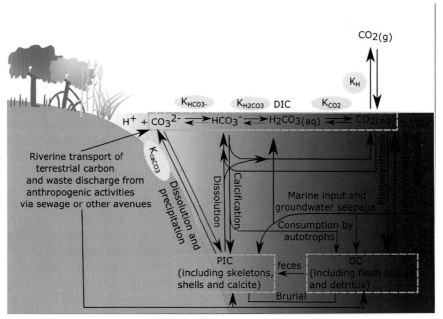

FIG. 2.8

Conceptual model describing inorganic carbon dynamics and linkage to organic carbon in coastal wetlands. K_H is the constant of Henry's law. K_{CO2}, K_{H2CO3}, $K_{HCO_3^-}$, and K_{CaCO3} are the equilibrium constant of different chemical reactions, respectively. *DIC*, dissolved inorganic carbon. Constants are described in the Supplementary material in the online version at https://doi.org/10.1016/B978-0-12-819220-7.00002-9. *PIC*, particulate inorganic carbon; *OC*, organic carbon.

patchy. Overall, inorganic carbon dynamics are sensitive to global changes subject to anthropogenic disturbances, e.g., rising temperatures resulting from anthropogenic emissions and deforestation.

Sedimentation is another indirect factor that may affect carbon mineralization aside from the dynamics of inorganic carbon. Generally, carbon mineralization rates (both sulfate reduction and total carbon oxidation) show linear increase with increasing sedimentation rates (Canfield et al., 2005). The significance of sulfate reduction declines when sediment deposition rate decreases (Ferdelman et al., 1999).

2.6 Conclusions

Decomposition of coastal vascular plants (including leaf litter, wood, and roots) is an important component of carbon mineralization in coastal wetlands. It is affected by biotic factors, including faunal and floral influence. The former facilitates microbial decomposition and thus enhances vascular plant decomposition. The latter

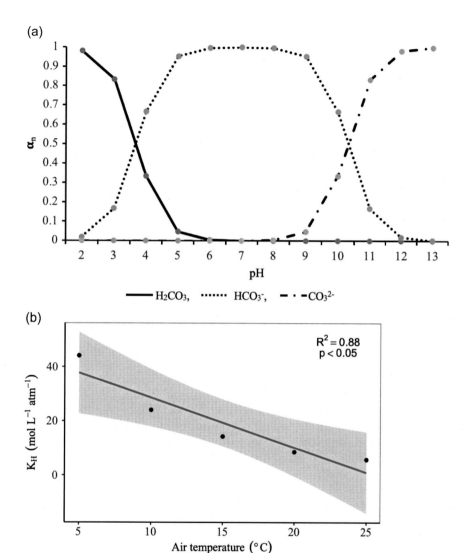

FIG. 2.9

Change in carbonate species and K_H with pH and temperature. (A) distribution of carbonate species as a portion of dissolved carbonate related to water pH; (B) relationship between K_H and air temperature: $K_H = 47 - 1.8T$. $\alpha_n = H_{2-n}CO_3/C_T$. C_T is the sum of all carbonate species concentration.

Data from Leng, C.B., Roberts, J.E., Zeng, G., Zhang, Y.H., Liu, Y., 2015. Effects of temperature, pH, and ionic strength on the Henry's law constant of triethylamine. Geophys. Res. Lett. 42, 3569–3575.

modulates decomposition via different plant traits such as litter quality. The influence of biotic factors on vascular plant decomposition can be enhanced by interacting with abiotic factors such as seasons and tidal positions. Anthropogenic activities influence litter decomposition via the combined impact on biotic and abiotic factors.

The decomposition of vascular plants is quantified in relation to environmental media, stoichiometry, and ecosystem types. The decomposition of leaf litter of vascular plants occurs under different environmental media, including aerial, sediment, and submerged conditions. Significant differences are found for mangrove leaf litter decomposing at aerial and aquatic environments, whereas no significant differences are found for saltmarshes. Similar to mangroves, there is a significant relationship between the decomposition rate constant of saltmarsh leaf litter and initial leaf litter C/N. Decomposition rates of mangrove leaf litter vary among different mangrove geomorphology and physiognomic settings, corroborating the pattern found for mangrove roots.

Sediment organic matter of coastal wetlands consists of a variety of organic compounds characterized by labile or humic substances. The production of carbon greenhouse gases (CO_2 and CH_4) in coastal wetland sediments is determined by the decomposition pathways of different sediment organic matter, from large to small molecular weight organic compounds. Free energy yields indicate which decomposition reaction is easier to take place than others. The same decomposition pathway may show different reaction rates in different types of coastal wetlands.

We highlight the importance of vascular plant and sediment organic matter decomposition in further clarifying carbon dynamics in coastal wetlands. Alongi (2014) provided a perspective of the global carbon cycle for mangrove forests without differentiating emerged and submerged conditions. Futures global synthesis on carbon cycling in coastal wetlands should consider the difference in plant litter decomposition under different environmental media. Few studies estimate mangrove wood decomposition, which is a component of carbon mineralization and is largely ignored in the current studies on carbon cycling. Exclusion of this important albeit refractory component of the carbon pool in coastal wetland leaves a large gap in the perspective on carbon mineralization in these habitats. Carbon mineralization is related to carbonate calcification and dissolution. It implicates that organic and inorganic carbon dynamics are tightly linked and should be integrated into the total carbon cycle of coastal wetlands.

References

Abram, J.W., Nedwell, D.B., 1978. Hydrogen as a substrate for methanogenesis and sulphate reduction in anaerobic saltmarsh sediment. Arch. Microbiol. 117, 93–97.

Adame, M.F., Kauffman, J.B., Medina, I., Gamboa, J.N., Torres, O., Caamal, J.P., Reza, M., Herrera-Silveira, J.A., 2013. Carbon stocks of tropical coastal wetlands within the karstic landscape of the Mexican Caribbean. PLoS One 8, e56569.

Ainley, L.B., Bishop, M.J., 2015. Relationships between estuarine modification and leaf litter decomposition vary with latitude. Estuar. Coast. Shelf Sci. 164, 244–252.

Alberts, J.J., Filip, Z., Price, M.T., Williams, D.J., Williams, M.C., 1988. Elemental composition, stable carbon isotope ratios and spectrophotometric properties of humic substances occurring in a salt marsh estuary. Org. Geochem. 12 (5), 455–467.

Alongi, D.M., 2009. The Energetics of Mangrove Forests. Springer, New York, NY.

Alongi, D.M., 2014. Carbon cycling and storage in mangrove forests. Ann. Rev. Mar. Sci. 6, 195–219.

Arndt, S., Jørgensen, B.B., LaRowe, D.E., Middelburg, J., Pancost, R., Regnier, P., 2013. Quantifying the degradation of organic matter in marine sediments: a review and synthesis. Earth Sci. Rev. 123, 53–86.

Ashton, E., Hogarth, P., Ormond, R., 1999. Breakdown of mangrove leaf litter in a managed mangrove forest in Peninsular Malaysia. In: Diversity and Function in Mangrove Ecosystems. Springer, pp. 77–88.

Benner, R., Hodson, R.E., 1985. Microbial degradation of the leachable and lignocellulosic components of leaves and wood from *Rhizophora mangle* in a tropical mangrove swamp. Mar. Ecol. Prog. Ser. 23, 221–230.

Bianchi, T.S., Allison, M.A., Zhao, J., Li, X., Comeaux, R.S., Feagin, R.A., Kulawardhana, R. W., 2013. Historical reconstruction of mangrove expansion in the Gulf of Mexico: linking climate change with carbon sequestration in coastal wetlands. Estuar. Coast. Shelf Sci. 119, 7–16.

Bosire, J.O., Dahdouh-Guebas, F., Kairo, J.G., Kazungu, J., Dehairs, F., Koedam, N., 2005. Litter degradation and CN dynamics in reforested mangrove plantations at Gazi Bay, Kenya. Biol. Conserv. 126, 287–295.

Bouchard, V., Lefeuvre, J.-C., 2000. Primary production and macro-detritus dynamics in a European salt marsh: carbon and nitrogen budgets. Aquat. Bot. 67, 23–42.

Bouchard, V., Creach, V., Lefeuvre, J., Bertru, G., Mariotti, A., 1998. Fate of plant detritus in a European salt marsh dominated by *Atriplex portulacoides (L.)* Aellen. Hydrobiologia 373, 75–87.

Buckley, D.H., Baumgartner, L.K., Visscher, P.T., 2008. Vertical distribution of methane metabolism in microbial mats of the great Sippewissett salt marsh. Environ. Microbiol. 10, 967–977.

Bui, T.H., Lee, S.Y., 2015. Endogenous cellulase production in the leaf litter foraging mangrove crab *Parasesarma erythodactyla*. Comp. Biochem. Physiol. B Biochem. Mol. Biol. 179, 27–36.

Burnell, O.W., Russell, B.D., Irving, A.D., Connell, S.D., 2013. Eutrophication offsets increased sea urchin grazing on seagrass caused by ocean warming and acidification. Mar. Ecol. Prog. Ser. 485, 37–46.

Camp, E.F., Suggett, D.J., Gendron, G., Jompa, J., Manfrino, C., Smith, D.J., 2016. Mangrove and seagrass beds provide different biogeochemical aervices for corals threatened by climate change. Front. Mar. Sci. 3, 52. https://doi.org/10.3389/fmars.2016.00052.

Canfield, D.E., Kristensen, E., Thamdrup, B., 2005. Aquatic Geomicrobiology. Elsevier, Amsterdam.

Cannicci, S., Burrows, D., Fratini, S., Smith III, T.J., Offenberg, J., Dahdouh-Guebas, F., 2008. Faunal impact on vegetation structure and ecosystem function in mangrove forests: a review. Aquat. Bot. 89, 186–200.

Castañeda-Moya, E., Twilley, R.R., Rivera-Monroy, V.H., Marx, B.D., Coronado-Molina, C., Ewe, S.M., 2011. Patterns of root dynamics in mangrove forests along environmental gradients in the Florida Coastal Everglades, USA. Ecosystems 14, 1178–1195.

Chale, F., 1993. Degradation of mangrove leaf litter under aerobic conditions. Hydrobiologia 257, 177–183.

Cragg, S.M., Friess, D.A., Gillis, L.G., Trevathan-Tackett, S.M., Terrett, O.M., Watts, J.E., Distel, D.L., Dupree, P., 2020. Vascular plants are globally significant contributors to marine carbon fluxes and sinks. Ann. Rev. Mar. Sci. 12, 469–497.

Dick, T.M., Osunkoya, O.O., 2000. Influence of tidal restriction floodgates on decomposition of mangrove litter. Aquat. Bot. 68, 273–280.

Dittmar, T., Lara, R.J., 2001. Molecular evidence for lignin degradation in sulfate-reducing mangrove sediments (Amazonia, Brazil). Geochim. Cosmochim. Acta 65, 1417–1428.

Eley, Y., Dawson, L., Pedentchouk, N., 2016. Investigating the carbon isotope composition and leaf wax n-alkane concentration of C3 and C4 plants in Stiffkey saltmarsh, Norfolk, UK. Org. Geochem. 96, 28–42.

Fell, J.W., Master, I., 1975. Phycomycetes (*Phytophthora* spp. nov. and *Pythium* sp. nov.) associated with degrading mangrove (*Rhizophora mangle*) leaves. Can. J. Bot. 53, 2908–2922.

Fenchel, T., 1972. Aspects of decomposer food chains in marine benthos. Verh Deutsch Zool Ges 14, 14–22.

Ferdelman, T.G., Fossing, H., Neumann, K., Schulz, H.D., 1999. Sulfate reduction in surface sediments of the Southeast Atlantic continental margin between 15 38'S and 27 57'S (Angola and Namibia). Limnol. Oceanogr. 44, 650–661.

Filip, Z., Alberts, J.J., 1989. Humic substances isolated from *Spartina alterniflora* (Loisel.) following long-term decomposition in sea water. Sci. Total Environ. 83, 273–285.

Fourqurean, J.W., Schrlau, J.E., 2003. Changes in nutrient content and stable isotope ratios of C and N during decomposition of seagrasses and mangrove leaves along a nutrient availability gradient in Florida Bay, USA. Chem. Ecol. 19, 373–390.

Frasco, B.A., Good, R.E., 1982. Decomposition dynamics of *Spartina alterniflora* and *Spartina patens* in a New Jersey salt marsh. Am. J. Bot. 69, 402–406.

Giddins, R.L., Lucas, J.S., Neilson, M.J., Richards, G.N., 1986. Feeding ecology of the mangrove crab *Neosarmatium smithi* (Crustacea: Decapoda:Sesarmidae). Mar. Ecol. Prog. Ser. 33, 147–155.

Guan, W., Xiong, Y., Liao, B., 2018. Soil inorganic carbon in mangroves of tropical China: patterns and implications. Biol. Lett. 14, 20180483.

Gwyther, J., 2003. Nematode assemblages from Avicenniamarina leaf litter in a temperate mangrove forest in South-Eastern Australia. Mar. Biol. 142, 289–297.

Holmboe, N., Kristensen, E., Andersen, F.Ø., 2001. Anoxic decomposition in sediments from a tropical mangrove forest and the temperate Wadden Sea: implications of N and P addition experiments. Estuar. Coast. Shelf Sci. 53, 125–140. https://doi.org/10.1006/ecss.2000.0794.

Holmer, M., Olsen, A.B., 2002. Role of decomposition of mangrove and seagrass detritus in sediment carbon and nitrogen cycling in a tropical mangrove forest. Mar. Ecol. Prog. Ser. 230, 87–101.

Howard, J.L., Creed, J.C., Aguiar, M.V., Fouqurean, J.W., 2018. CO_2 released by carbonate sediment production in some coastal areas may offset the benefits of seagrass "Blue Carbon" storage. Limnol. Oceanogr. 63, 160–172.

Imgraben, S., Dittmann, S., 2008. Leaf litter dynamics and litter consumption in two temperate South Australian mangrove forests. J. Sea Res. 59, 83–93.

Jorgensen, B.B., 1980. Mineralization and the bacterial cycling of carbon, nitrogen and sulphur in marine sediments. In: Ellwood, D.C., et al. (Eds.), Contemporary Microbial Ecology. Academic Press, London.

Keuskamp, J.A., Feller, I.C., Laanbroek, H.J., Verhoeven, J.T., Hefting, M.M., 2015. Short- and long-term effects of nutrient enrichment on microbial exoenzyme activity in mangrove peat. Soil Biol. Biochem. 81, 38–47.

King, A.J., Cragg, S.M., Li, Y., Dymond, J., Guille, M.J., Bowles, D.J., Bruce, N.C., Graham, I.A., McQueen-Mason, S.J., 2010. Molecular insight into lignocellulose digestion by a marine isopod in the absence of gut microbes. Proc. Natl. Acad. Sci. 107, 5345–5350.

Kristensen, E., Andersen, F.Ø., Holmboe, N., Holmer, M., Thongtham, N., 2000. Carbon and nitrogen mineralization in sediments of the Bangrong mangrove area, Phuket, Thailand. Aquat. Microb. Ecol. 22, 199–213.

Kristensen, E., Bouillon, S., Dittmar, T., Marchand, C., 2008. Organic carbon dynamics in mangrove ecosystems: a review. Aquat. Bot. 89, 201–219.

Kristensen, E., Lee, S.Y., Mangion, P., Quintana, C.O., Valdemarsen, T., 2017. Trophic discrimination of stable isotopes and potential food source partitioning by leaf-eating crabs in mangrove environments. Limnol. Oceanogr. 62, 2097–2112.

Ku, T., Walter, L., Coleman, M., Blake, R., Martini, A., 1999. Coupling between sulfur recycling and syndepositional carbonate dissolution: evidence from oxygen and sulfur isotope composition of pore water sulfate, South Florida Platform, USA. Geochim. Cosmochim. Acta 63, 2529–2546.

Lallier-Vergès, E., Marchand, C., Disnar, J.-R., Lottier, N., 2008. Origin and diagenesis of lignin and carbohydrates in mangrove sediments of Guadeloupe (French West Indies): evidence for a two-step evolution of organic deposits. Chem. Geol. 255, 388–398.

Lee, S.Y., 1989. The importance of sesarminae crabs *Chiromanthes* spp. and inundation frequency on mangrove (*Kandelia candel* (L.) Druce) leaf litter turnover in a Hong Kong tidal shrimp pond. J. Exp. Mar. Biol. Ecol. 131, 23–43.

Lee, S.Y., 1990. Net aerial primary productivity, litter production and decomposition of the reed *Phragmites communis* (L.) in a nature reserve in Hong Kong: management implications. Mar. Ecol. Prog. Ser. 66, 161–173.

Lee, S.Y., 1991. Herbivory as an ecological process in a *Kandelia candel* (Rhizophoraceae) mangal in Hong Kong. J. Trop. Ecol. 7 (337), 348.

Lee, S.Y., 1993. Leaf choice of the sesarmid crabs *Chiromanthes bidens* and *C. plicata* in a Hong Kong mangal. In: Morton, B. (Ed.), Proceedings of the International Conference on Marine Biology of Hong Kong and the South China Sea, University of Hong Kong, October 1990. Hong Kong University Press, Hong Kong, pp. 597–604.

Lee, S.Y., 1995. Mangrove outwelling: a review. Hydrobiologia 295, 203–212.

Lee, S.Y., 1998. Ecological role of grapsid crabs in mangrove ecosystems: a review. Mar. Freshw. Res. 49, 335–343.

Lee, S.Y., Primavera, J.H., Dahdouh-Guebas, F., McKee, K., Bosire, J.O., Cannicci, S., Diele, K., Fromard, F., Koedam, N., Marchand, C., Mendelssohn, I., Mukherjee, N., Record, S., 2014. Ecological role and services of tropical mangrove ecosystems: a reassessment. Glob. Ecol. Biogeogr. 23, 726–743. https://doi.org/10.1111/geb.12155.

Lehmann, J., Kleber, M., 2015. The contentious nature of soil organic matter. Nature 528, 60–68.

Li, T., Ye, Y., 2014. Dynamics of decomposition and nutrient release of leaf litter in *Kandelia obovata* mangrove forests with different ages in Jiulongjiang Estuary, China. Ecol. Eng. 73, 454–460.

Liu, Z., Macpherson, G.L., Groves, C., Martin, J.B., Yuan, D., Zeng, S., 2018. Large and active CO_2 uptake by coupled carbonate weathering. Earth Sci. Rev. 182, 42–49. https://doi.org/10.1016/j.earscirev.2018.05.007.

Lu, W., Liu, C.A., Zhang, Y., Yu, C., Cong, P., Ma, J., Xiao, J., 2019. Carbon fluxes and stocks in a carbonate-rich chenier plain. Agric. For. Meteorol. 275, 159–169. https://doi.org/10.1016/j.agrformet.2019.05.023.

Mackey, A., Smail, G., 1996. The decomposition of mangrove litter in a subtropical mangrove forest. Hydrobiologia 332, 93–98.

Macreadie, P.I., Nielsen, D.A., Kelleway, J.J., Atwood, T.B., Seymour, J.R., Petrou, K., Connolly, R.M., Thomson, A.C.G., Trevathan-Tackett, S.M., Ralph, P.J., 2017. Can we manage coastal ecosystems to sequester more blue carbon? Front. Ecol. Environ. 15, 206–213.

Macreadie, P.I., Anton, A., Raven, J.A., Beaumont, N., Connolly, R.M., Friess, D.A., Kelleway, J.J., Kennedy, H., Kuwae, T., Lavery, P.S., Lovelock, C.E., Smale, D.A., Apostolaki, E.T., Atwood, T.B., Baldock, J., Bianchi, T.S., Chmura, G.L., Eyre, B.D., Fourqurean, J.W., Hall-Spencer, J.M., Huxham, M., Hendriks, I.E., Krause-Jensen, D., Laffoley, D., Luisetti, T., Marbà, N., Masque, P., McGlathery, K.J., Megonigal, J.P., Murdiyarso, D., Russell, B.D., Santos, R., Serrano, O., Silliman, B.R., Watanabe, K., Duarte, C.M., 2019. The future of blue carbon science. Nat. Commun. 10, 3998. https://doi.org/10.1038/s41467-019-11693-w.

Marchand, C., Disnar, J.R., Lallier-Vergès, E., Lottier, N., 2005. Early diagenesis of carbohydrates and lignin in mangrove sediments subject to variable redox conditions (French Guiana). Geochim. Cosmochim. Acta 69, 131–142. https://doi.org/10.1016/j.gca.2004.06.016.

McKee, K.L., Seneca, E.D., 1982. The influence of morphology in determining the decomposition of two salt marsh macrophytes. Estuaries 5, 302–309. https://doi.org/10.2307/1351753.

Mfilinge, P., Atta, N., Tsuchiya, M., 2002. Nutrient dynamics and leaf litter decomposition in a subtropical mangrove forest at Oura Bay, Okinawa, Japan. Trees 16, 172–180. https://doi.org/10.1007/s00468-001-0156-0.

Middleton, B.A., McKee, K.L., 2001. Degradation of mangrove tissues and implications for peat formation in Belizean island forests. J. Ecol. 89, 818–828. https://doi.org/10.1046/j.0022-0477.2001.00602.x.

Montemayor, D.I., Addino, M., Fanjul, E., Escapa, M., Alvarez, M.F., Botto, F., Iribarne, O.O., 2011. Effect of dominant *Spartina* species on salt marsh detritus production in SW Atlantic estuaries. J. Sea Res. 66, 104–110.

Newell, S.Y., 1984. Carbon and nitrogen dynamics in decomposing leaves of three coastal marine vascular plants of the subtropics. Aquat. Bot. 19, 183–192.

Odum, E.P., 1968. A research challenge: evaluating the productivity of coastal and estuarine water. In: Proceedings of the Second Sea Grant Conference. vol. 63, p. 64.

Opsahl, S., Benner, R., 1995. Early diagenesis of vascubr phmt tisswes:Lignin and cutin decompo&ion and biogeochemicai impkations. Geochim. Cosmochim. Acta 59, 4889–4904.

Ouyang, X., Guo, F., 2016. Paradigms of mangroves in treatment of anthropogenic wastewater pollution. Sci. Total Environ. 544, 971–979. https://doi.org/10.1016/j.scitotenv.2015.12.013.

Ouyang, X., Lee, S.Y., 2020. Improved estimates on global carbon stock and carbon pools in tidal wetlands. Nat. Commun. 11, 317. https://doi.org/10.1038/s41467-019-14120-2.

Ouyang, X., Guo, F., Bu, H., 2015. Lipid biomarkers and pertinent indices from aquatic environment record paleoclimate and paleoenvironment changes. Quat. Sci. Rev. 123, 180–192. https://doi.org/10.1016/j.quascirev.2015.06.029.

Ouyang, X., Lee, S.Y., Connolly, R.M., 2017a. The role of root decomposition in global mangrove and saltmarsh carbon budgets. Earth Sci. Rev. 166, 53–63. https://doi.org/10.1016/j.earscirev.2017.01.004.

Ouyang, X., Lee, S.Y., Connolly, R.M., 2017b. Structural equation modelling reveals factors regulating surface sediment organic carbon content and CO_2 efflux in a subtropical mangrove. Sci. Total Environ. 578, 513–522. https://doi.org/10.1016/j.scitotenv.2016.10.218.

Pelegraí, S.P., Rivera-Monroy, V.H., Twilley, R.R., 1997. A comparison of nitrogen fixation (acetylene reduction) among three species of mangrove litter, sediments, and pneumatophores in South Florida, USA. Hydrobiologia 356, 73–79.

Reice, S.R., Stiven, A.E., 1983. Environmental patchiness, litter decomposition and associated faunal patterns in a *Spartina alterniflora* marsh. Estuar. Coast. Shelf Sci. 16, 559–571.

Rice, D.L., Tenore, K.R., 1981. Dynamics of carbon and nitrogen during the decomposition of detritus derived from estuarine macrophytes. Estuar. Coast. Shelf Sci. 13, 681–690.

Romero, L.M., Smith III, T.J., Fourqurean, J.W., 2005. Changes in mass and nutrient content of wood during decomposition in a South Florida mangrove forest. J. Ecol. 93, 618–631.

Saderne, V., Geraldi, N.R., Macreadie, P.I., Maher, D.T., Middelburg, J.J., Serrano, O., Almahasheer, H., Arias-Ortiz, A., Cusack, M., Eyre, B.D., Fourqurean, J.W., Kennedy, H., Krause-Jensen, D., Kuwae, T., Lavery, P.S., Lovelock, C.E., Marba, N., Masqué, P., Mateo, M.A., Mazarrasa, I., McGlathery, K.J., Oreska, M.P.J., Sanders, C.J., Santos, I.R., Smoak, J.M., Tanaya, T., Watanabe, K., Duarte, C.M., 2019. Role of carbonate burial in blue carbon budgets. Nat. Commun. 10, 1106. https://doi.org/10.1038/s41467-019-08842-6.

Schmidt, M.W., Torn, M.S., Abiven, S., Dittmar, T., Guggenberger, G., Janssens, I.A., Kleber, M., Kögel-Knabner, I., Lehmann, J., Manning, D.A., 2011. Persistence of soil organic matter as an ecosystem property. Nature 478, 49–56.

Senior, E., Lindström, E.B., Banat, I.M., Nedwell, D.B., 1982. Sulfate reduction and methanogenesis in the sediment of a saltmarsh on the east coast of the United Kingdom. Appl. Environ. Microbiol. 43, 987–996.

Siple, M.C., Donahue, M.J., 2013. Invasive mangrove removal and recovery: food web effects across a chronosequence. J. Exp. Mar. Biol. Ecol. 448, 128–135.

Skyring, G., 1987. Sulfate reduction in coastal ecosystems. Geomicrobiol J. 5, 295–374.

Stout, S.A., Boon, J.J., Spackman, W., 1988. Molecular aspects of the peatification and early coalification of angiosperm and gymnosperm woods. Geochim. Cosmochim. Acta 52, 405–414.

Swift, M., 1976. Species diversity and the structure of microbial communities in terrestrial habitats. In: Paper presented at the Symposium of the British Ecological Society, UK.

Tanner, B.R., Uhle, M.E., Mora, C.I., Kelley, J.T., Schuneman, P.J., Lane, C.S., Allen, E.S., 2010. Comparison of bulk and compound-specific $\delta^{13}C$ analyses and determination of carbon sources to salt marsh sediments using n-alkane distributions (Maine, USA). Estuar. Coast. Shelf Sci. 86, 283–291.

Teal, J.M., 1962. Energy flow in the salt marsh ecosystem of Georgia. Ecology 43, 614–624.

Valentine, J.F., Duffy, J.E., 2007. The central role of grazing in seagrass ecology. In: Seagrasses: Biology, Ecologyand Conservation. Springer, pp. 463–501.

Valiela, I., Teal, J.M., Allen, S.D., Van Etten, R., Goehringer, D., Volkmann, S., 1985. Decomposition in salt marsh ecosystems: the phases and major factors affecting disappearance of above-ground organic matter. J. Exp. Mar. Biol. Ecol. 89, 29–54.

Wang, X.-C., Chen, R., Berry, A., 2003. Sources and preservation of organic matter in Plum Island salt marsh sediments (MA, USA): long-chain n-alkanes and stable carbon isotope compositions. Estuar. Coast. Shelf Sci. 58, 917–928.

Werry, J., Lee, S.Y., 2005. Grapsid crabs mediate link between mangrove litter production and estuarine planktonic food chains. Mar. Ecol. Prog. Ser. 293, 165–176.

Winfrey, M.R., Ward, D.M., 1983. Substrates for sulfate reduction and methane production in intertidal sediments. Appl. Environ. Microbiol. 45, 193–199.

Yang, Z., Song, W., Zhao, Y., Zhou, J., Wang, Z., Luo, Y., Li, Y., Lin, G., 2018. Differential responses of litter decomposition to regional excessive nitrogen input and global warming between two mangrove species. Estuar. Coast. Shelf Sci. 214, 141–148.

Zhou, H.-C., Wei, S.-D., Zeng, Q., Zhang, L.-H., Tam, N.F.-Y., Lin, Y.-M., 2010. Nutrient and caloric dynamics in *Avicennia marina* leaves at different developmental and decay stages in Zhangjiang River estuary, China. Estuar. Coast. Shelf Sci. 87, 21–26.

CO$_2$ and CH$_4$ emissions from coastal wetland soils

Cyril Marchand[a,*], Frank David[b,*], Adrien Jacotot[c,*], Audrey Leopold[d,*], and Xiaoguang Ouyang[e,f,g,*]

[a]*University of New Caledonia, ISEA, Noumea, New Caledonia,*
[b]*Muséum National d'Histoire Naturelle, Station Marine de Concarneau, Concarneau, France,*
[c]*ISTO, Université d'Orléans, CNRS, BRGM, Orléans, France,*
[d]*Institut Agronomique néo-Calédonien, SolVeg, Noumea, New Caledonia,*
[e]*Southern Marine Science and Engineering Guangdong Laboratory (Guangzhou), Guangzhou, China,*
[f]*Guangdong Provincial Key Laboratory of Water Quality Improvement and Ecological Restoration for Watersheds, School of Ecology, Environment and Resources, Guangdong University of Technology, Guangzhou, China,*
[g]*Simon F.S. Li Marine Science Laboratory, School of Life Sciences, The Chinese University of Hong Kong, Hong Kong Special Administrative Region, China*

3.1 Introduction

Coastal wetlands are known for their great ability to fix carbon dioxide (CO$_2$) and store organic carbon (OC) in their biomass and soils, leading to the concept of blue carbon sequestration (Bouillon et al., 2008; Mcleod et al., 2011; Ouyang and Lee, 2020). Recently, Alongi (2020a) proposed that mangroves sequester globally 15 Tg OC year^{-1}, with mangrove net ecosystem production (628 g OC m^{-2} year^{-1}) being greater than that in saltmarshes (382 g OC m^{-2} year^{-1}) (Alongi, 2020b). High OC burial rates are characteristics of coastal wetland ecosystems, with Breithaupt et al. (2012) estimating a rate of 163 OC m^{-2} year^{-1} for mangroves, and Ouyang and Lee (2014) estimating a rate of 245 g OC m^{-2} year^{-1} for saltmarshes. Good management practices may thus lead these ecosystems to play a key role in climate change mitigation (Taillardat et al., 2018a; Alongi, 2020a). In order to assess accurately the carbon (C) storage capacity of ecosystems, one has to determine not only the C stocks but also the incoming and outgoing C fluxes. Incoming C fluxes encompass net primary productivity (NPP) and allochthonous inputs (Kristensen et al., 2008a). NPP determination has recently been improved, notably because of the set-up of eddy covariance towers in coastal wetlands that measure the net ecosystem CO$_2$ exchanges

---

[*]All authors have contributed equally to this chapter.

Carbon Mineralization in Coastal Wetlands. https://doi.org/10.1016/B978-0-12-819220-7.00006-6

(Barr et al., 2010). Quantifying the allochthonous inputs is more complex and requires the determination of the molecular and isotopic composition of soil organic matter. Outgoing C fluxes are diverse and include leaf litter tidal flushing (Lee, 1995), porewater seepage (Maher et al., 2013), and gaseous loss at different interfaces, notably from soils (e.g., Allen et al., 2007; Chen et al., 2010; Chauhan et al., 2015).

In soils, greenhouse gas (GHG) production results from autotrophic respiration of roots and heterotrophic respiration of soil fauna and microorganisms. Coastal wetlands are ecosystems that develop on very specific soils: waterlogged, mostly anoxic, and with variable salinity. These conditions limit OC mineralization and thus GHG production and emissions (Kristensen et al., 2017), which in turn are beneficial for climate change mitigation. In addition, methanogenesis, which occurs when all electron acceptors have been exhausted, is considered low or nonexistent in coastal wetlands since under the influence of sulfate-rich tidal water, sulfate-reducing bacteria can outcompete methanogens for the common carbon substrates, and thus inhibit methanogenesis (Alongi et al., 2001). However, Lyimo et al. (2002) demonstrated that sulfate reduction and methanogenesis can coexist in mangrove soils due to the utilization of other noncompetitive substrates by methanogens. Consequently, CH_4 emissions may have been underestimated, which may be of major concern considering its high global warming potential (Rosentreter et al., 2021).

While the soils in coastal wetland ecosystems generally exhibit low GHG emissions, the variability of emission rates is driven by different biotic and abiotic parameters. Temperature, soil water content (SWC), nutrient inputs, and pH control OC mineralization in coastal wetland soils (e.g., Chen et al., 2012; Jacotot et al., 2019). All these parameters vary notably with seasons, latitudes, anthropogenic pressures, and positions along the intertidal zone. In addition, mineralization depends on the bacterial communities in the soil and benthic photosynthetic microorganisms developing at the soil surface (Lovelock, 2008; Leopold et al., 2015). Furthermore, crab burrows and aerial roots can modify the soil redox conditions as well as act as a conduit for the transport of GHG produced in the soil (Kristensen et al., 2008b), which may influence GHG emissions.

The objective of this chapter is to review the latest published data and analyses regarding CO_2 and CH_4 emissions from coastal wetland soils. We reviewed the different methods that have been used to measure GHG fluxes at the soil-air interface in these ecosystems. We compared the magnitude of CO_2 and CH_4 emissions from coastal wetland soils with that from terrestrial ecosystems. Lastly, we reviewed in situ studies that have quantified the variability of CO_2 and CH_4 emissions to highlight the main biotic and abiotic drivers.

3.2 Methods to quantify GHG emissions from coastal wetland soils

The most common method of quantifying GHG emissions at the soil-air interface in coastal wetlands is the flux chamber. The chamber is placed on the studied soil, and the GHG concentration inside the chamber is measured during the deployment period. A GHG analyzer can be directly connected to the chamber to obtain continuous measurements of GHG concentrations, or discrete gas samples can be collected during chamber closure and further analyzed by gas chromatography. GHG fluxes are then calculated from linear regression of the GHG concentrations inside the chamber over time.

Either manual or automated chambers can be used in flux measurements. Some are commercially available, but many are custom-built systems. Their use in terrestrial ecosystems has been clearly described (e.g., Davidson et al., 2002); however, the specificities of coastal wetlands (including tides, burrowing activities, and aerial roots) must be considered for their application in such ecosystems. Manual chamber measurements are labor-intensive, but these chambers are not costly and do not require power supply to operate, which are ideal in remote areas such as coastal wetlands. Automated chambers can make continuous measurements, but a steady power supply for their operation is a huge challenge. In addition, tidal movement limits their long-term deployment in the intertidal zone. Consequently, manual chambers are used in coastal wetlands for most of the time. To be able to determine the fluxes, the volume and the inner diameter of the chamber must be known. The influence of the size of the chamber on the measured fluxes has been described in Davidson et al. (2002). In coastal wetlands, it can be difficult to set up chambers over aerial roots. Consequently, the ideal chamber size is highly site-specific, often requiring corrections be applied or errors be determined (Davidson et al., 2002). In coastal wetlands, the size of the chambers used is highly variable, with a surface area ranging from 78.5 to $1000\,cm^2$ and a volume ranging from 172 to $5700\,cm^3$ (Chen et al., 2010, 2012; Leopold et al., 2013, 2015; Bulmer et al., 2017; Ouyang et al., 2017; Jacotot et al., 2019). The base of the chamber can be directly inserted into the soil or installed on soil collars (Heinemeyer et al., 2011). It is recommended to install collars weeks before the actual flux measurement to let the soils stabilize. In some studies, chambers or collars were introduced a few centimeters into the soil (Chen et al., 2010; Troxler et al., 2015). However, chamber interference with natural soil gas flux must be minimized for getting accurate and consistent measurements. We recommend inserting the chamber edges only a few millimeters into the soil just to avoid gas leakage, which can be detected through the observation of GHG concentration variations within the chamber. This precaution will prevent soil disturbance and root sectioning (Heinemeyer et al., 2011) and minimize artificial CO_2 production (Bulmer et al., 2017). It is also recommended to install the chamber in locations without any mangrove seedling, aboveground root, and leaf litter to prevent their influence on gas fluxes (Chen et al., 2012). However, if the objective is to assess the influence of aerial roots or crab burrows on gas fluxes, then the number of these features within the chamber and their spatial dimensions must be determined. Since microtopography can exert an influence on gas fluxes, we suggest choosing a position in the middle of the wetland stand as walking with all the equipment from the lowest to the highest elevation of the site may be time consuming and the measurement conditions will be different. We also suggest avoiding placing the chambers in a micro-depression where water can accumulate, or on mounds created by crabs.

Headspace air mixing and unaltered diffusion gradient are critical for accurate flux measurements. Chanda et al. (2013) used fans to homogenize air within the chamber. However, fans can create pressure gradients within the chamber that may suppress or enhance flux. To achieve an optimal air mixing, we suggest using bowl-shaped chambers with specific air inlet/outlet positioning, with the outlet and inlet being located at the top and the base of the bowl, respectively. The duration of chamber closure should be minimized to avoid over accumulation of GHG within the chamber headspace, which may lead to a reduction of GHG flux by altering the GHG concentration gradient (Drewitt et al., 2002; Chanda et al., 2013). In addition, short deployment periods avoid excessive changes in microclimate (e.g., moisture and temperature) within the chamber headspace, which may affect gas diffusivity (Rochette et al., 1992; Jensen et al., 1996; Kabwe et al., 2002; Leopold et al., 2013). Leopold et al. (2013, 2015) performed flux measurement over $\sim$3 min, with the initial 30 s after chamber installation being used to allow equilibrium to be reached inside the chamber, and the CO_2 concentration within the chamber being subsequently measured every second for 2.5 min. However, depending on the size of the chamber and the prevailing soil properties (e.g., soil water content, grain size, GHG concentrations, etc.), this equilibration duration may vary. Bulmer et al. (2017) measured CO_2 concentration in the chamber at 5-s intervals over a period of 90 s. Jacotot et al. (2019) and Vinh et al. (2019) measured CO_2 concentrations every second over a 5-min period and did not take into account the data in the first 30 s in the calculation similar to the approach of Leopold et al. (2013, 2015). These latter studies measured CO_2 concentrations in the field using gas analyzers connected to the chambers. Some other studies measured GHG concentrations in the lab by gas chromatography, which used longer chamber deployment periods because of the smaller number of gas samples included. For instance, Chen et al. (2012) deployed the chamber for 30–45 min with sampling at 10- to 15-min intervals. Considering the microclimatic variability within the chamber and the tidal movement, we recommend an equilibration time of 30 s and a sampling duration of not longer than 5 min for soil CO_2 and CH_4 flux measurements.

Chambers can be opaque or transparent. Using opaque chambers is the most common method. Soil shading before installing the chamber was performed by some researchers because some organisms composing the biofilm may still perform photosynthesis under dark conditions for a few minutes (Kristensen et al., 2008b; Jacotot et al., 2019). However, the usefulness of soil-shading is still under discussion (Bulmer et al., 2017). Transparent chambers allow solar radiation to reach the soil surface and thus microphytobenthic photosynthesis can occur. To assess the influence of the biofilm, GHG emissions measured with transparent and opaque chambers must be compared. In addition, in order to accurately discuss the results obtained, it is suggested to equip the transparent chambers with a photosynthetically active radiation (PAR) sensor and collect the top few mm of soils to determine its pigment contents, especially chlorophyll-a (chl-a), after the measurement. Some researchers performed flux measurements after the removal of the upper few mm of the soil to assess the influence of biofilm on GHG emissions from coastal wetland soils

(Leopold et al., 2013, 2015; Bulmer et al., 2017). However, Jacotot et al. (2019) demonstrated that doing this resulted only in a partial removal of the biofilm, and higher CO_2 emissions are characterized by a depletion in ^{13}C isotopic value due to biofilm deterioration.

Coastal wetland soils are heterogeneous with highly variable properties over space and time. GHG production depends notably on OC content, redox conditions, soil water content, and temperature. Additionally, GHG emissions are controlled by the soil water table level as well as a series of biological parameters (pneumatophore density, crab burrowing activity, microphytobenthos developing at the soil surface). We suggest all these parameters to be determined for accurate GHG flux measurements and interpretation, as what has already been done for terrestrial ecosystems (Bond-Lamberty et al., 2020; Jian et al., 2020). Chen et al. (2014) suggested that the tidal range, the tidal flooding and exposure duration, as well as the meteorological conditions must be relatively similar among sampling days and site for comparing the flux results. Leopold et al. (2013, 2015) performed all their measurements at low tide and during the first steps of flood between 11 a.m. and 1 p.m. in order to limit differences between campaigns and stands due to daily variability, notably in solar radiation and soil water saturation. Jacotot et al. (2019) and Vinh et al. (2019) used approximately the same approach, starting measurements 1–2 h before low tide and finishing them 1–2 h after low tide. Concerning the number of replicates, researchers usually included 3–5 replicates for one measurement and chose 3–5 positions within a stand to obtain a mean flux value representative of the area (Leopold et al., 2013; Jacotot et al., 2019; Vinh et al., 2019). We suggest a minimum of 3 replicates per site and 3 replicate sites per coastal wetland stands (with homogenous properties) for flux measurement. If it takes more than 30 min to complete these replicate measurements, the tidal movement may modify the extent of the saturation zone and thus influence GHG fluxes, impairing reliable comparisons between replicates.

Different kinds of GHG analyzers can be coupled to the chambers to measure the changes of in situ GHG concentrations. Concerning CO_2, the nondispersive infrared gas analyzers (IRGA) are the most frequently used. Lovelock (2008) measured soil respiration using a LI-COR 6400 portable photosynthesis system configured with the LI-COR soil respiration chamber (LI-COR Biosciences, Lincoln, NE, United States). Leopold et al. (2013, 2015) and Vinh et al. (2019) used a LI-820A model from LI-COR Biosciences (Lincoln, NE, United States), which has an operating range of 0–20,000 ppm and an accuracy of better than 3% of the reading. To use this analyzer, a pump must be added to the circuit and the air must be dried prior to entering the analyzer to avoid CO_2 dilution by water vapor, which may increase with soil evapotranspiration within the chamber. Chanda et al. (2013) also measured CO_2 concentration with the help of an infrared gas analyzer, the LI-840A CO_2/H_2O (LI-COR Biosciences, Lincoln, NE, United States). This model also measures CO_2 and H_2O concentrations concurrently and thus can correct for the water vapor dilution effect on measured CO_2 concentrations. Troxler et al. (2015) used another model of IRGA, which is the LI-COR 8100 soil respiration system (LI-COR Biosciences, Lincoln, NE, United States) equipped with pumps and filters. The inconvenience of all these analyzers is that they need daily calibrations during a field campaign. Usually, the calibration is done using pure nitrogen gas (0 ppm CO$_2$) and one to two certified CO$_2$ spans with concentrations within the range of the expected measurements. Various other brands and models of IRGA also exist. For instance, Ouyang et al. (2017) used the SBA-5 gas analyzer, while Bulmer et al. (2017) used the Environmental Gas Monitor (EGM-4), both IRGAs produced by PP Systems (United States).

To perform simultaneous measurements of both CO$_2$ and CH$_4$, other techniques are available, such as (i) cavity ring-down spectroscopy (CRDS) developed by Picarro, (ii) high-resolution cavity enhanced direct-absorption spectroscopy developed by Los Gatos Research (LGR), and (iii) the optical feedback-cavity enhanced absorption spectroscopy of LI-COR. These analyzers have better performances and are more precise, but are more expensive. In addition, these instruments offer negligible zero and span drift, and thus do not require regular calibration. Some of these analyzers can even perform isotopic measurements, which can help to identify the sources of the measured GHG flux, and determine the partitioning between heterotrophic and autotrophic respiration using ad hoc mathematical models.

If the analyzer is able to perform isotopic measurements, a Keeling plot approach can be used to discriminate the isotopic value of the CO$_2$ (δ^{13}C-CO$_2$) released from soil, root respiration, and leaf litter decomposition from the background atmosphere trapped inside the chamber (Keeling, 1958, 1961; Pataki et al., 2003). This approach is based upon mass conservation during CO$_2$ exchange between the two compartments. The intercept of the linear regression of δ^{13}C-CO$_2$ against the inverse of CO$_2$ concentration corresponds to the δ^{13}C-CO$_2$ value of the CO$_2$ added to the chamber. This method allows separating the autotrophic and heterotrophic components of respiration. In projects that are interested in the impacts of land use change on coastal wetland ecosystems, heterotrophic respiration will be the most important pathway to quantify.

Discrete samples of GHG can be analyzed using a gas chromatograph equipped with a Porapak-Q column. CH$_4$ and CO$_2$ are detected by a flame ionization detector (FID) after prior methanation for CO$_2$. The recommended temperatures of the injector, column, and FID are 100°C, 60°C, and 250°C, respectively. Gas concentrations are quantified by comparing the peak areas of samples against certified references at different concentrations (Chen et al., 2010, 2012, 2014).

When the variations of GHG concentrations within the chamber headspace over time are known, the following equation is used to calculate the flux rate:

$$F = (\delta pGHG/\delta t)*V/(R*T*S)$$

where F is the GHG flux from the soil (μmol m^{-2} s^{-1}); $\delta pGHG/\delta t$ is the change of GHG concentrations as a function of time (ppm s^{-1}); V is the total volume of the system (chamber and analyzer) (m^3); R is the ideal gas constant of $8.20528 * 10^{-5}$ (atm m^3 K^{-1} mol^{-1}); T is the absolute air temperature (K); and S is the surface area covered by the flux chamber (m^2).

3.3 Magnitude of GHG emissions from coastal wetland soils

Carbon sources in the coastal wetlands include the carbon sequestered by vascular plants/microphytobenthos and the carbon inputs from riverine or marine sources. Carbon is stored in above- and below-ground biomass and soils and emitted via diffusion, bioturbation, as well as respiration by vascular plants and microphytobenthos. A detailed description of carbon cycling, remineralization, and ecosystem respiration is provided in Chapter 1 of this book.

In this section, we compare the magnitude of GHG emissions at the soil-air interface in coastal wetlands with that in terrestrial ecosystems and from the relative contributions of heterotrophic vs. autotrophic components to ecosystem respiration (Table 3.1). Soil-air CO_2 emissions from mangroves were reported to be 267.2 ± 201.5 and $-65.7 \pm 236.5\,g\,C\,m^{-2}\,year^{-1}$ (mean $\pm$ 1 standard deviation) under dark and light conditions, respectively, with a global mean of $100.7\,g\,C\,m^{-2}\,year^{-1}$ (Bouillon et al., 2008). However, these emissions are highly variable, as evidenced by the high standard deviation that approximate or even more than double the absolute values of mean fluxes. This variability results from the influence of biotic and abiotic parameters that are discussed in the following sections of this chapter.

Alongi (2014) updated the estimate of global mean CO_2 emissions from mangrove soils to $257.1\,g\,C\,m^{-2}\,year^{-1}$. The global estimate of saltmarsh soil respiration ($564 \pm 63\,g\,C\,m^{-2}\,year^{-1}$) includes the rates of oxygen consumption, as well as dissolved inorganic carbon and CO_2 emissions (Alongi, 2020b). These fluxes are estimated by different methods, including the chamber technique, and the measurements of oxygen consumption and dissolved inorganic carbon release. Considering that different techniques were used to determine these values, comparisons with soil respiration from global mangrove and other ecosystems measured by the chamber method are limited. Soil CO_2 emission rates from terrestrial ecosystems (e.g., $745.1\,g\,C\,m^{-2}\,year^{-1}$ from terrestrial forests) are more than twice of those from mangroves (area-weighted mean: $592.2\,g\,C\,m^{-2}\,year^{-1}$) (Chen et al., 2014, Warner et al., 2019; Tang et al., 2020). The methods used to quantify soil CO_2 emissions from terrestrial ecosystems include the chamber, eddy covariance, and gradient methods (Bond-Lamberty and Thomson, 2010). The low soil CO_2 emissions from mangroves can be attributed to the anaerobic, waterlogged, and saline conditions that are prevalent in coastal ecosystems, which inhibit carbon mineralization and thus CO_2 production (Muzuka and Shunula, 2006; Kristensen et al., 2008b, 2017; Ouyang et al., 2017).

Soil respiration from ecosystems can be partitioned into autotrophic and heterotrophic respiration. Autotrophic respiration involves the release of CO_2 by the respiration of belowground roots, while heterotrophic respiration involves the production of CO_2 by soil microbial activity. In terrestrial forests, autotrophic respiration ranges from $226 \pm 170\,g\,C\,m^{-2}\,year^{-1}$ in deciduous needle-leaf forests to $633 \pm 397\,g\,C\,m^{-2}\,year^{-1}$ in evergreen broad-leaf forests (Tang et al., 2020). It accounts for 39.9%–65.7% of the total soil respiration depending on the type of terrestrial forest considered (Warner et al., 2019). However, data on autotrophic

Table 3.1 Soil and/or ecosystem CO_2 and CH_4 emissions from coastal wetlands in comparison with other ecosystems.

CO_2 emissions

Ecosystems	R_h (gC m⁻² year⁻¹)	R_s (gC m⁻² year⁻¹)	R_a (gC m⁻² year⁻¹)	R_e (gC m⁻² year⁻¹)	References
Mangroves	254.1[a]	257.1 (average) 612.8[b] 267.2±201.5 (dark, n=82) −65.7±236.5 (light, n=14) 100.7 (average)	NA	3333.1[a] 1319.7±130.3	Alongi (2014), data in Chapter 4 Bouillon et al. (2008)
Saltmarshes	NA	564±63 (n=46)[b]	NA	2010	Alongi (2020b) and Duarte (2005)
Humid forests	877±96	3061±162	2323±144	3200	Luyssaert et al. (2007) and Malhi (2012)
Terrestrial ecosystems	NA	745.1 (terrestrial forests) 592.2±368.9 (area-weighted mean)	633±397 (EBF, n=82) 368±278 (GL, n=41) 353±182 (DBF, n=144) 226±170 (DNF, n=27)	NA	Chen et al. (2014), Tang et al. (2020), and Warner et al. (2019)

CH_4 emissions

Ecosystems	Global CH_4 emissions (gCH$_4$ m⁻² year⁻¹)	Global CH_4 emissions (gC m⁻² year⁻¹)	Global CH_4 (CO_2 eq.) emissions, GWP100 (gC m⁻² year⁻¹)	References
Mangroves	2.25±0.88 (n=14)	1.69±0.66 (n=14)	20.86±8.14 (n=14)	Rosentreter et al. (2018)
Saltmarshes	18.9±2.67 (n=46)	14.2±2 (n=46)	175.27±24.67 (n=46)	Alongi (2020b)
Terrestrial forests	0.29±0.12 (emission, n=68) 12.21±0.51 (uptake, n=68)	0.22±0.09 (emission, n=68) 9.16±0.38 (uptake, n=68)	2.021±0.84 (emission, n=68) 84.94±3.56 (uptake, n=68)	Yu et al. (2017)
Freshwater wetlands	27.5 (3.5–70.1)	20.6 (area-weighted mean) (2.6–52.5)	339.9 (43.3–866.2)	Roehm (2005)

DBF, deciduous broad-leaf forest; DNF, deciduous needle-leaf forest; EBF, evergreen broad-leaf forest; GL, grassland; GWP100, global warming potential for the 100-year time horizons; NA, not available; R_a, rhizosphere autotrophic respiration; R_e, ecosystem respiration; R_h, heterotrophic respiration; R_s, soil respiration.

[a]Estimated from the fluxes excluding pelagic respiration from tidal creeks and other waterways in the reference.

[b]Including dissolved and gaseous oxygen consumption and dissolved inorganic carbon production in addition to CO_2 emission measured from the chamber method.

respiration from mangroves and other coastal wetlands are too scarce to allow a reliable global synthesis. Among individual studies, Ouyang et al. (2018) investigated the sources of soil respiration for *Avicennia marina* (gray mangrove) seedlings and found that roots contributed to $31.8\% \pm 9.7\%$ of soil respiration, while their contribution may be higher for mature mangroves (Troxler et al., 2015). Alongi (2014) estimated that heterotrophic respiration from mangroves could reach $1101\,\mathrm{g\,C\,m^{-2}\,year^{-1}}$. Nonetheless, this estimate includes CO_2 emissions from adjacent waterways and subsurface CO_2 production that may lead to the production of dissolved inorganic carbon from dissolved CO_2. Therefore, heterotrophic respiration from mangrove soils should be lower than the estimate of $257.1\,\mathrm{g\,C\,m^{-2}\,year^{-1}}$ by Alongi (2014) and far lower than the rate of heterotrophic respiration in humid forests (Luyssaert et al., 2007; Malhi, 2012).

Wetlands, including coastal wetlands, are ecosystems that emit large quantities of methane into the atmosphere, accounting for 20%–39% of the global CH_4 emissions (Hoehler and Alperin, 2014). However, soil-air CH_4 emissions from global mangroves are one order of magnitude lower than those from freshwater wetlands ($20.6\,\mathrm{g\,C\,m^{-2}\,year^{-1}}$, Roehm, 2005) and saltmarshes ($14.2 \pm 2\,\mathrm{g\,C\,m^{-2}\,year^{-1}}$, Alongi, 2020b). Compared with saltmarshes, the low CH_4 emissions from mangrove soils may be partially due to the transport of CH_4 produced in soils by mangrove tree trunks (Jeffrey et al., 2019), and pneumatophores of many mangrove species, such as *A. marina* (Ouyang et al., 2017) (see Section 3.5 of this chapter). Additionally, being in intertidal areas, mangroves generally exhibit higher salinity than freshwater wetlands. Methane emissions have a significant negative relationship with salinity in tidal marshes (Poffenbarger et al., 2011), notably leading support to the influence of sulfate inputs in suppressing CH_4 emissions. Tidal pumping and porewater seepage (Maher et al., 2013) can export part of the GHG produced within coastal wetland soils to tidal creeks (Call et al., 2015; David et al., 2018, Taillardat et al., 2018b), resulting in lower emissions from soils. A recent study estimated that mean soil CH_4 emissions from global mangroves reach $1.69 \pm 0.66\,\mathrm{g\,C\,m^{-2}\,year^{-1}}$ (Rosentreter et al., 2018), which is higher than those from terrestrial forests ($0.22 \pm 0.09\,\mathrm{g\,C\,m^{-2}\,year^{-1}}$) (Yu et al., 2017). Terrestrial forests can consume CH_4 due to oxidation by methanotrophs with oxidation rates varying with soil conditions (e.g., Guckland et al., 2009). Intensive CH_4 oxidation may occur in the subsurface at 5–10 cm depth (Hütsch, 1998; Priemé and Christensen, 1997), where anaerobic conditions prevail and hinder CH_4 oxidation. Oxidation is a major consumption process of methane in coastal wetlands, substantially reducing CH_4 emission to the atmosphere. Methane oxidation is exerted following two major steps. First, anaerobic oxidation occurs during the diffusion of CH_4 in the deep anoxic layers of the soil. CH_4 is transformed into CO_2 by sulfate-reducing bacteria and archaea (Boetius et al., 2000) using sulfates as electron acceptors. Recent studies have estimated that anaerobic oxidation can account for 5%–20% of the CH_4 produced in the soil (Das et al., 2018). Then, around 60%–90% of CH_4 is oxidized by methanotrophs in the oxic surface layers (Le Mer and Roger, 2001), which ultimately results in low emissions to the atmosphere. Therefore, net CH_4 emissions at the soil surface mainly depend on the balance between production and oxidation, which is in turn controlled by the population of methanogens and methanotrophs. In the Sundarban mangroves, Das et al. (2018) calculated a CH_4 production rate that is 1.66 times faster than the rate of CH_4 oxidation, while the methanotrophs were 1.17 times more abundant than methanogens, suggesting that methanogenesis is more efficient than CH_4 oxidation.

The climatic impact of CO_2 and CH_4 emissions from soil carbon mineralization can be estimated considering the global warming potential of CH_4. The global warming potential of CH_4 over the 100-year time horizon presented in Rosentreter et al. (2018) was used to convert CH_4 emissions to CO_2 equivalent emissions for terrestrial forests in Table 3.1. Bouillon et al. (2008) and Alongi (2014) estimated mean CO_2 emission from mangrove soils of 100.7 and 257.1 $g\,C\,m^{-2}\,year^{-1}$, respectively, while that from terrestrial forests is greater by more than double ($745.1\,g\,C\,m^{-2}\,year^{-1}$) (Table 3.1).

3.4 Influence of physical parameters on GHG emissions from coastal wetland soils

3.4.1 Temperature

Temperature is a major factor driving GHG emissions to the atmosphere in coastal wetlands (Sun et al., 2013; Wang et al., 2015; Huertas et al., 2019). Temperature accelerates the decomposition of organic matter (Davidson and Janssens, 2006) by stimulating microbial activity in the soil and the associated CO_2 and CH_4 production. In addition, higher temperatures also increase plant productivity (Tong et al., 2011), which can increase the supply of labile carbon through root exudation and subsequently stimulate microbial activity (Olsson et al., 2015; Yang et al., 2019). Concerning CO_2, it is commonly accepted that temperature stimulates heterotrophic respiration exponentially (e.g., Fang and Moncrieff, 2001). This positive effect of temperature on CO_2 emissions from soils has been evidenced in many studies in coastal wetlands (e.g., Kirwan et al., 2014; Gebremichael et al., 2017; Cui et al., 2018; Huertas et al., 2019; among many others). Some of the highest CO_2 emissions observed from mangrove soils, reaching up to $800\,mmol\,CO_2\,m^{-2}\,day^{-1}$, could be attributed to the high temperatures combined with the first rainfall pulse of the monsoon season in low-latitude mangroves (Vinh et al., 2019; Fig. 3.1).

Regarding CH_4, higher temperatures may increase both methanogenesis and methanotrophy (Inglett et al., 2012; Tong et al., 2013; Liu et al., 2017; Martin and Moseman-Valtierra, 2017; Welti et al., 2017; Abdul-Aziz et al., 2018; Huertas et al., 2019). However, methanogenesis is more sensitive to temperature than methanotrophy, resulting in an increase in net CH_4 emission to the atmosphere (Born et al., 1990; Dunfield et al., 1993; Cai et al., 2013). The higher decay rates resulting from increased temperature can lead to soil acidification, which in turn affect

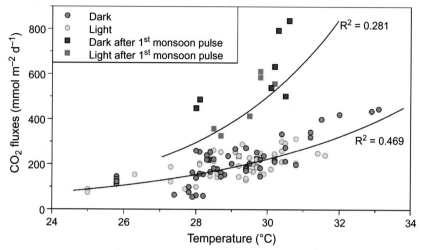

FIG. 3.1

Influence of temperature and of the first rainfall pulse of the monsoon on CO_2 emissions at the soil-air interface in a tropical *Rhizophora* mangrove forest.

Data from Vinh, T.V., Allenbach, M., Aimée, J., Marchand, C., 2019. Seasonal variability of CO_2 fluxes at different interfaces and vertical CO_2 concentrations profiles in a Rhizophora mangrove stand (Can Gio, Viet Nam). Atmos. Environ. https:/doi.org/10.1016/j.atmosenv.2018.12.049.

CH_4 mediation. Indeed, past studies showed that the optimum pH for methanogenesis lies between 7.0 and 8.6 (O'Flaherty et al., 1998; Chang and Yang, 2003; Das et al., 2018), whereas that for methanotrophs lies between 4 and 8.2 (Saari et al., 2004; Das et al., 2018). This suggests that soil acidification may result in higher CH_4 oxidation rates. Higher temperatures are also known to facilitate plant-mediated CH_4 transport (Hosono and Nouchi, 1997), which is one of the major pathways of CH_4 emissions to the atmosphere (Smith et al., 2003).

The effect of temperature on GHG emissions is often described using the temperature sensitivity factor Q_{10} (Lloyd and Taylor, 1994), which denotes the factor that needs to be multiplied to the fluxes for a 10°C rise. Reported Q_{10} values in coastal wetlands are highly variable, highlighting the strong interdependence of various predictors. In coastal wetlands, a mean Q_{10} value of 2 is reported for CO_2 in the existing literature (range 1.3–2.6; Lovelock, 2008; Kirwan et al., 2014; Leopold et al., 2015; Gebremichael et al., 2017; Cui et al., 2018; Jacotot et al., 2019), while higher values are usually obtained for CH_4. For instance, Yang et al. (2019) determined a lower threshold value of 2.5 for CH_4 fluxes, while Wang et al. (2015) reported higher values ranging from 3 to 5. Although special attention should be paid to Q_{10} values greater than 3 (Baldocchi, 2020), those reported in the literature for coastal wetlands clearly show strong temperature control on CO_2 and CH_4 fluxes, with the latter being the most sensitive.

3.4.2 Soil water content

Soil water content (SWC) is one of the primary factors driving GHG fluxes in coastal wetlands, affecting their production, consumption, and transport to the soil surface. Soil water content influences microbial activity (Stark and Firestone, 1995), chemical reaction rates (Liu et al., 2017), and gas diffusion rates through the soil column (Hu et al., 2018). Previous studies have shown that GHG production is at its maximum when water saturation reaches 50% of the water-holding capacity of soils (Schaufler et al., 2010). In coastal wetland soils, SWC is mainly determined by the length and frequency of tidal inundation, which in turn depend on elevation. SWC is low in the upper intertidal sites, as these areas mainly receive water from rains or equinox tides only and are therefore subjected to strong evaporation. However, evaporation intensity will depend on the climate and rainfall patterns, being more intense in arid or semiarid climates than in tropical humid climate. Conversely, SWC is higher at lower elevation sites that are submerged for several hours a day by regular tides. This gradient in SWC has repercussions in the redox state of the soil by filling the pore spaces with water, thus reducing the availability of oxygen and affecting the distribution of salinities. As a result, the higher elevation sites have higher oxygen content and redox values and are therefore dominated by the aerobic production of CO$_2$ and aerobic oxidation of CH$_4$ (Hirota et al., 2007; Chen et al., 2012; Bonnett et al., 2013; Treat et al., 2014). In semiarid climate, these zones also have higher salinity, which is known to inhibit methanogenesis and thus CH$_4$ emissions (e.g., Denier van der Gon and Neue, 1995; Purvaja and Ramesh, 2001; Dalal et al., 2003; Poffenbarger et al., 2011; Tong et al., 2012; Vizza et al., 2017; Yang et al., 2019). Conversely, coastal wetlands with lower elevations present a more anoxic environment, which triggers methanogenesis and reduces the oxic layer where aerobic oxidation prevails, ultimately increasing the net CH$_4$ flux to the atmosphere (Allen et al., 2007; Tong et al., 2013; Huang et al., 2019).

In addition, salinity gradients cause vegetation to develop zonations along the topographic slope in accordance with their tolerance to soil salinity. As a result, vegetated coastal wetlands most often exhibit clear zonations of halophilic communities (e.g., Snedaker, 1982; Pennings and Callaway, 1992; Pennings et al., 2005; Marchand et al., 2011, 2012; Lee et al., 2016; Fariña et al., 2018; Bourgeois et al., 2019). Generally, species that settle in areas with low salinity are more productive. This results in a gradient of soil organic matter content (Marchand et al., 2012) that has direct consequences on GHG fluxes, with the highest fluxes being observed in areas with the highest soil organic matter concentrations (e.g., Sutton-Grier et al., 2011; Chen et al., 2012; Leopold et al., 2015; Xu et al., 2020). Additional water inputs such as rainfall can also alter the spatial pattern of SWC along the intertidal slope, which in turn reduce oxygen content and salinity with implications for GHG production as discussed above. However, there is currently no clear consensus on the effects of rainfall on GHG production, which have been shown to be either positive, negative, or even neutral (Charles and Dukes, 2009; Emery et al., 2019;

Watson et al., 2014). This varying response can be attributed to the peculiar situation of coastal wetlands, featuring diurnal tidal cycles that buffer the spatial distribution of SWC according to the local amplitude and phase of tides.

The variability of SWC is governed by the frequency of tidal immersion and precipitation. The latter, like temperature, can vary seasonally but also with latitude. Seasonal variations of GHG emissions have been observed in many studies, with higher emissions during the wet and warm months as compared to the cold and dry ones (see e.g., Allen et al., 2007, 2011; Kirui et al., 2009; Chen et al., 2012; Sun et al., 2013; Murray et al., 2015; Olsson et al., 2015; Welti et al., 2017; Gnanamoorthy et al., 2019; Yang et al., 2019; among many others). Eventually, Hu et al. (2020) in a recent review of 70 studies in coastal wetlands showed that CO_2 and CH_4 emissions decrease with latitude.

3.4.3 Nutrient inputs

Nutrient inputs from anthropogenic disturbances, such as discharges from shrimp ponds, sewage and livestock effluents, and fertilizer from runoff or groundwater discharge from adjacent areas, are known to alter GHG emissions from coastal wetland soils. The response of CO_2 emissions to nutrient loading is inconsistent based on the results reported in the literature. Nitrogen enrichment increases above-ground biomass production, which in turn increases both respiration and photosynthetic CO_2 uptake (Morris and Bradley, 1999; Geoghegan et al., 2018; Martin et al., 2018a). Consequently, if the ecosystem respiration is greater than primary productivity, there will be a net CO_2 loss by the ecosystem. In addition, nutrient enrichment promotes both microbial respiration and leaf litter decomposition, which result in higher CO_2 fluxes to the atmosphere (Wigand et al., 2009; Geoghegan et al., 2018; Martin et al., 2018a). Contrasting results have been reported by Lovelock et al. (2014) with a stimulation of soil respiration in only 9 out of 22 mangrove sites following the addition of nitrogen and phosphorus. They suggested that nutrient stimulation of CO_2 emissions occurs mainly in hydrologically isolated areas that are subject to extremely limited availability of nutrients. Conversely, areas with better hydrological connectivity benefit less from nutrient addition. Compiling the paired data on land use and land cover changes (LULCCs), the mean ratio of CO_2 efflux between LULCC and control plots is less than 1 with ranges from near 0 to >20 for different types of LULCCs (Sasmito et al., 2019). In particular, the ratio shows the largest variability for aquaculture ponds. This pinpoints the above observation that the effect of nutrients on soil CO_2 emissions is site-specific in mangroves.

Methane emissions can also be affected by nutrient loading in contrasting ways. Indeed, some studies reported higher CH_4 fluxes in coastal wetlands that were subjected to anthropogenic activities than in pristine areas (Sotomayor et al., 1994; Kreuzwieser et al., 2003; Cotovicz et al., 2016) (Fig. 3.2), mainly due to the increase in substrate availability for methanogenesis and intensified hypoxia generated by the

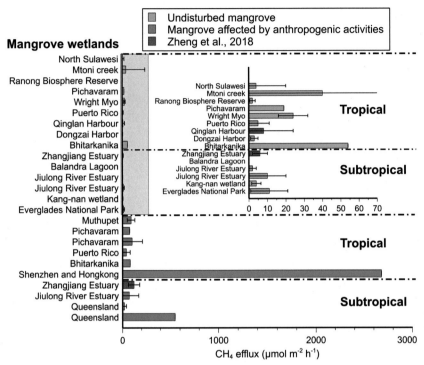

FIG. 3.2

Comparison of CH$_4$ emissions from 24 mangroves worldwide that are undisturbed or affected by anthropogenic activities.

From Zheng, X., Guo, J., Song, W., Feng, J., Lin, G., 2018. Methane emission from mangrove wetland soils is marginal but can be stimulated significantly by anthropogenic activities. Forests 9, 738. https:/doi.org/10.3390/ f9120738.

effluents (Fields, 2004). For instance, Zheng et al. (2018) showed a clear increase in CH$_4$ fluxes from mangroves that received high nutrient loading than from pristine ones based on a recent review of 24 mangroves sites (Fig. 3.2). However, some other studies reported lower CH$_4$ fluxes from coastal wetlands following chronic nutrient loading (Chmura et al., 2016; Moseman-Valtierra et al., 2016; Geoghegan et al., 2018; Pfeifer-Meister et al., 2018), which could be attributed to the stimulation of plant growth and denser rhizosphere that promoted soil oxygenation and CH$_4$ oxidation. Another possible cause for the decrease in CH$_4$ emission with higher nutrient load is the increase in the availability of alternative electron acceptors for more competitive metabolic pathways of organic matter degradation such as denitrification that prevents the onset of methanogenesis.

3.5 Biological controls of GHG emissions from coastal wetland soils

3.5.1 Benthic biofilm

The soil surface of coastal wetlands provides a suitable habitat for the development of microbial biofilm, which is mainly composed of diatoms, green algae, cyanobacteria, and flagellates that form the microphytobenthos (MPB), but archaea and other prokaryotes may also develop within the upper several millimeters of photic soils (Macintyre et al., 1996; Janousek, 2009). MPB biomass may be assessed by measuring the content of pigments at the soil surface, especially chlorophyll-a (chl-a) that highlights the oxygenic phototrophs (Kromkamp and Foster, 2006). Soil chl-a measurements provide evidence of the variations in biofilm development both in space and time. Temperature, light intensity, and soil moisture are some major factors controlling biofilm growth (Zedler, 1980; Joye and Lee, 2004; Leopold et al., 2015; Ha et al., 2018; Chen et al., 2019). Other parameters may also influence biofilm growth, including (i) nutrient availability and soil carbon content (Van Raalte et al., 1976; Kristensen et al., 1988), (ii) grain size and sediment texture (Balasubramaniam et al., 2017), (iii) grazing pressure (Daggers et al., 2020), (iv) hydrodynamics, notably resuspension and currents induced by tides and wind (Ha et al., 2020; Redzuan and Underwood, 2020), (v) intertidal position (Leopold et al., 2015), and (vi) the type of vegetated wetland (Pinckney and Zingmark, 1993). Because of the shading effect induced by vegetation, especially in dense mangrove forests, the development of biofilm, particularly microphytobenthic organisms, is generally low (Kristensen et al., 1988). However, some studied reported higher MPB biomass at the soil surface underneath the mangrove canopy than in exposed mud- or salt-flats, depending on the micro-climatic conditions at the sampling site that are related to the seasons and wetland zonations (Leopold et al., 2015; Chen et al., 2019; Kwon et al., 2020). Tolhurst et al. (2020) even showed that shading increased MPB biomass, even if the changes would not be directly related to the light intensity, but to the covariation in temperature and moisture (Liu et al., 2013; Leopold et al., 2015). In addition, the role of mangrove roots on sediment deposition and tidal current has been suggested to result in an accumulation of MPB under the canopy (Kwon et al., 2020).

MPB biomass and microenvironment in coastal wetlands are important in governing the relative magnitude of autotrophic and heterotrophic metabolisms of MPB. Because of the presence of photoautotrophic benthic organisms, the surface soil of coastal wetlands can become a net sink of CO_2. Negative or low CO_2 emissions have been reported in light conditions at the soil-air interface in salt-marshes and mangroves under different climates (Leopold et al., 2013, 2015; Chen et al., 2019; Kwon et al., 2020). Consequently, light CO_2 fluxes are lower than dark fluxes (Jacotot et al., 2019; Fig. 3.3).

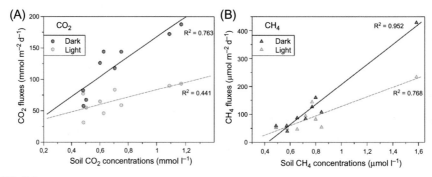

FIG. 3.3

GHG fluxes vs. soil GHG concentrations (mean values on 60 cm deep cores) in the dark and at light in a semiarid *Rhizophora* mangrove forest.

Modified from Jacotot, A., Marchand, C., Allenbach, M., 2019. Biofilm and temperature controls on greenhouse gas (CO_2 and CH_4) emissions from a Rhizophora mangrove soil (New Caledonia). Sci. Total Environ. 650, 1019–1028. https:/doi.org/10.1016/j.scitotenv.2018.09.093.

Differences in soil CO_2 emissions are especially clear during the winter or cold season (Leopold et al., 2015; Grellier et al., 2017, Chen et al., 2019), which could be attributed to (i) lower CO_2 production in the soil column during the cold season and subsequent offset of MPB photosynthesis (Chen et al., 2019) and (ii) higher MPB biomass and/or higher photosynthetic activity during the cold season (Liu et al., 2013; Leopold et al., 2015; Chen et al., 2019; Kwon et al., 2020). Even if high MPB biomass have been measured during the summer, the rate of CO_2 production may outweigh that of CO_2 consumption. Indeed, very high light intensity and temperature, high evaporation rate, and desiccation may lead to photo-inhibition of MPB activity and a decrease of CO_2 consumption and/or high MPB death and decomposition, while CO_2 production in the soil may increase because of the optimal micro-climatic conditions. Clear relationships between chl-a, magnitude of CO_2 fluxes in light conditions (Kristensen and Alongi, 2006; Leopold et al., 2015; Ouyang et al., 2017; Chen et al., 2019), and ^{13}C-depleted values of emitted CO_2 in light conditions (Jacotot et al., 2019) highlighted the importance of photoautotrophic productivity at the soil-air interface of coastal wetlands (Fig. 3.4).

Soil respiration in the dark has been shown to increase to a maximum at 25–27°C and then decline with a further increase in temperature, which could be attributed to the higher biofilm biomass (Lovelock, 2008). After removing the top 0.5 cm of soil prior to repeated measurement of CO_2 fluxes, soil respiration rate was found to be greatly improved, suggesting a significant impact of biofilm on CO_2 fluxes from soil (Lovelock, 2008). Other studies in temperate, subtropical, and tropical wetlands also highlighted a similar trend when removing the upper mm of soil (Leopold et al., 2013, 2015; Bulmer et al., 2017; Grellier et al., 2017; Hien et al., 2018), suggesting

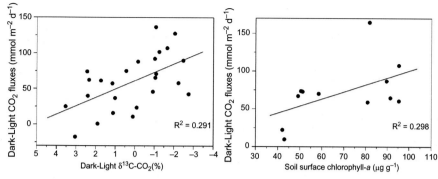

FIG. 3.4

Relationship between the difference in dark and light CO_2 fluxes (mmol m^{-2} day^{-1}) and: (A) the difference between dark and light $\delta^{13}C\text{-}CO_2$ (‰) and (B) chlorophyll-a concentrations at the soil surface ($\mu g\ g^{-1}$) in a *Rhizophora* mangrove forest.

From *Jacotot, A., Marchand, C., Allenbach, M., 2019. Biofilm and temperature controls on greenhouse gas (CO$_2$ and CH$_4$) emissions from a Rhizophora mangrove soil (New Caledonia). Sci. Total Environ. 650, 1019–1028. https:/doi.org/10.1016/j.scitotenv.2018.09.093.*

a major role of biofilm on CO_2 fluxes at the soil surface, even in dark conditions. Preshading experiments showed that lower CO_2 emission and/or CO_2 consumption were not related to the continuation of photosynthesis in dark conditions (supposed to last until the absence of coenzymes) (Bulmer et al., 2017). The role of ebb with high tidal range (Krauss and Whitbeck, 2012) and the role of diffusion and dissolution processes driven by soil pH and moisture were suggested for the low or negative CO_2 emissions observed in dark conditions. However, the CO_2 fluxes under dark conditions were systematically higher and positive after biofilm removal, emphasizing their biological control (Leopold et al., 2015). Microphytobenthos are known for exuding a part of its production as extracellular polysaccharides (Stal, 2003; de Brouwer et al., 2006; Oakes and Eyre, 2014). The cohesive organic matrix so formed, in which organisms and particles are embedded, allows biofilm to respond quickly to changing conditions in coastal wetlands (de Brouwer et al., 2006). Moreover, it acts as "glue," reducing erosion and increasing sediment stabilization in intertidal areas (Stal, 2003, 2010; de Brouwer et al., 2005, 2006). Taking advantage of these characteristics, notably its capacity for clogging the permeability of substrate, biofilm is used in ecological engineering to create a physical barrier to mass transport. Consequently, biofilm developing at the surface of intertidal soils is seen as a barrier limiting gas fluxes between the soil and the atmosphere (Leopold et al., 2013). This hypothesis has not been confirmed yet for CH_4, but preliminary results tend to support it. Actually, Jacotot et al. (2019) obtained higher CH_4 fluxes in dark conditions after biofilm removal when compared with those measured without biofilm removal. The biofilm at the sediment surface might have limited O$_2$ diffusion from the atmosphere to the soil and promoted anoxic conditions, thereby increasing CH$_4$ production in the soil (Chen et al., 2019).

Lower soil CO$_2$ fluxes were obtained at light than in dark in different wetlands around the world, with MBP assuming to play a role in impacting soil CO$_2$ fluxes at the soil-air interface of wetlands. We suggest that by understanding and quantifying the variation of light/dark CO$_2$ ratios with edaphic constraints at different latitudes, further studies could improve the consideration of MBP in the coastal wetland budget. Assuming a global mangrove area of 86,495 km^2, Alongi (2020c), based on previous estimations (Alongi, 2014), indicated that the gross primary production of benthic algae could account for more than 10% of mangrove ecosystems, of which half is probably respired. However, since benthic metabolisms are highly variable at diurnal, seasonal, and annual scales, incorporating biofilm into the carbon budget of coastal vegetated wetlands is challenging. Biofilm is crucial in the storage of blue carbon in coastal vegetated wetlands owing to its high photosynthetic efficiencies and ubiquitous distribution, respiration, and supply of labile carbon, putative physical role on GHG production and diffusion, and carbon retention in soil. Further studies should focus on quantifying and understanding the role of biofilm on the carbon cycle in coastal wetlands, notably CO$_2$ and CH$_4$ consumption and production.

3.5.2 Biogenic structures

Among the macrofaunal communities that colonize soils in the coastal wetlands, crabs are considered as major actors of the biogeochemical functioning of these systems and are thus called "ecosystem engineers" (Kristensen, 2008). Crab burrowing activity strongly influences carbon cycle, notably GHG fluxes at the soil-air interface of both saltmarshes and mangroves (Agusto et al., 2020). However, the relationship between burrow density and GHG emissions is highly variable according to the sampling site, seasons, and abundance and typology of burrows (Pülmanns et al., 2014; Ouyang et al., 2017). Montague (1982) indicated that burrow respiration accounted for 20%–90% of saltmarsh soil respiration, while Pülmanns et al. (2014) mentioned that CO$_2$ fluxes from soils with burrows were 15%–500% higher than plain soils. Kristensen et al. (2008b) showed that 36%–62% of CO$_2$ emissions were related to burrows, of which 58% were due to CO$_2$ diffusion from the surrounding soil.

Deep soil layers in coastal wetlands are mainly anoxic, especially at low tidal elevation (Marchand et al., 2005; Leopold et al., 2013). Macrofaunal burrowing activity modifies the soil column by changing the redox potential, increasing O$_2$ vertical penetration in soil, and extending artificially the surface exchange areas available for diffusive transport of CO$_2$ to the atmosphere (Kristensen et al., 2008a, b; Pülmanns et al., 2014; Martinetto et al., 2016; Ouyang et al., 2017; Guimond et al., 2020). Moreover, burrowing activity promotes iron respiration as compared to sulfate reduction in both saltmarshes (Kostka et al., 2002a,b; Gribsholt et al., 2003) and mangroves (Kristensen, 2007). Burrowing activity was also shown to modify the emissions, production, and oxidation rates of CH$_4$ (Kristensen et al., 2008b; Chen et al., 2021). Despite CH$_4$ is mainly originated from surrounding anaerobic wetland soils, crabs could increase CH_4 production inside burrows. On the one hand, crab presence leads to organic matter accumulation in burrows that favors the growth and reproduction of both methanogens and methanotrophs. Das et al. (2018) showed that the addition of organic matter had a stronger effect on methanogens than methanotrophs. On the other hand, by increasing the contact between the soil and flood water, crab burrows could increase the ammonia formation rate and subsequently inhibit the activity of methanotrophs (Chen et al., 2021). The influence of crab presence on CH_4 production and emissions also depends on the position of burrows along the intertidal gradient and flooding frequency. By increasing tidal flooding, sea-level rise as a result of global climate change could have positive feedbacks on global warming by favoring GHG production and emissions through crab activity.

Crab burrows can also indirectly influence the carbon cycle by increasing the hydraulic conductivity of soils. They affect porewater flows by creating preferential flow paths (Ridd, 1996; Susilo et al., 2005; Xin et al., 2009; Xiao et al., 2019). In mangrove ecosystems, Stieglitz et al. (2013) named the tidal pumping of water through animal burrows the "mangrove pump." At the regional scale, these authors estimated that 20% of the total annual river discharge at central Great Barrier Reef was circulated by the mangrove pump through animal burrows. Clearly, CO_2 produced by microbial respiration can dissolve in water-filled burrows and be exported through porewater seepage during the ebb tide (Call et al., 2015; Guimond et al., 2020), which is considered a major fate of dissolved inorganic carbon in coastal vegetated wetlands (e.g., Maher et al., 2013; Sadat-Noori et al., 2017).

GHG emissions at the soil-air interface also vary according to root architecture and functioning. It is obvious that roots participate in CO_2 fluxes through autotrophic respiration (Troxler et al., 2015). Meanwhile, root functioning related to wetland plant productivity can also impact both CO_2 and CH_4 production in the rhizospheric soil by releasing root exudates. These labile carbon inputs can directly influence CO_2 and CH_4 production in the soil, and subsequently GHG emissions at the air-soil interface, by fueling both aerobic and anaerobic heterotrophic respiration (Whiting and Chanton, 1993; Girkin et al., 2018; Al-Haj and Fulweiler, 2020; Yu et al., 2020). Moreover, these inputs could influence GHG production in wetland soils by modifying edaphic characteristics such as soil pH and redox conditions, as well as microbial community structure and activity (Liu et al., 2016; Girkin et al., 2018; Waldo et al., 2019). Very few is known about the feedback loops of root exudates on GHG fluxes in coastal vegetated wetlands, whereas temperature, light, CO_2, and vegetation composition are known to modify both the quantity and quality of root exudates, which can subsequently exert impact on rhizospheric soil, soil microbiome, and GHG production (Zhai et al., 2013; Girkin et al., 2018; Martin et al., 2018b; Sánchez-Carrillo et al., 2018; Mueller et al., 2020; Yu et al., 2020).

Pneumatophores and aerial roots of mangroves can strongly increase the magnitude of CO_2 fluxes at the soil-air interface (Kreuzwieser et al., 2003; Kristensen and Alongi, 2006; Troxler et al., 2015; Arai et al., 2016; Ouyang et al., 2017). Troxler et al. (2015) indicated that soil CO_2 fluxes were twice higher in the presence of pneumatophores, while Kristensen et al. (2008b) mentioned that pneumatophores were responsible for 29% of the total CO$_2$ emissions from soils during exposed conditions (i.e., low tide). Prop roots (Kreuzwieser et al., 2003), as well as pneumatophores especially in *Avicennia* stands (Purvaja et al., 2004; Lin et al., 2020, 2021), also act as a transport path for CH$_4$. Thus, pneumatophore density alone explained approximately 48% of the variation in CH$_4$ emissions in an *Avicennia* mangrove (Chen et al., 2021). The role of pneumatophores in CH$_4$ emissions is especially highlighted when considering that CH$_4$ emissions in *A. marina* mangroves were 50- to 100-fold higher than those of *Kandelia obovata* mangroves or mudflats (Chen et al., 2021). Kristensen et al. (2008b) showed that burrows and pneumatophores were responsible of 93%–100% of total CH$_4$ emissions from mangrove soils, respectively. While pneumatophores act as a way for CO$_2$ in deep soil to ascend, some studies suggested that (i) they would release as much or less CO$_2$ as prop roots for a given area (Kitaya et al., 2002; Troxler et al., 2015) and (ii) they have a higher photosynthetic rate during ebb and at low tide than prop roots (Kitaya et al., 2002). Indeed, these studies showed an increasing rate of photosynthesis in the following sequence: knee roots of *Bruguiera gymnorhiza* < prop roots of *Rhizophora stylosa* < pneumatophores of *A. marina* < pneumatophores of *Sonneratia alba*. Pneumatophores can assimilate CO$_2$ from the belowground parts of the plants, as well as promote the diffusion of O$_2$ in anoxic soils because of their permeable tissues. They can subsequently modify carbon cycling in the rhizospheric soil by increasing the oxidation of organic matter (Scholander et al., 1955; Andersen and Kristensen, 1988; Kitaya et al., 2002).

Lastly, other structures from mangrove and saltmarsh vegetation can mediate GHG emissions. The aerenchyma of wetland plants can act as a CH$_4$ transport pathway bypassing the potential sedimentary CH$_4$ oxidation processes (Bartlett and Harriss, 1993; Carmichael et al., 2014; Olsson et al., 2015). Moreover, recent studies suggest that tree stems could be a very important pathway for wetland CH$_4$ emissions. Pangala et al. (2017) showed that tree stems from living trees mediated approximately half of total CH$_4$ emissions in the Amazon floodplain wetland. Some studies also highlighted the role of dead trees as conduits for GHG emissions (Carmichael and Smith, 2016; Carmichael et al., 2018). Jeffrey et al. (2019) showed that dead mangrove tree stems that lack root-mediated oxygen transfer are a significant component of the net ecosystem CH$_4$ flux, emitting CH$_4$ at a rate eightfold higher than that of living mangrove tree stems (249.2 ± 41.0 vs 37.5 ± 5.8 μmol m^{-2} day^{-1}).

Biogenic structures significantly influence GHG production and emissions, but their impacts are more pronounced during the ebb and low tide because of the opening of lenticels on pneumatophores and burrows. Their impacts on CO$_2$ and CH$_4$ fluxes are expected to depend on the fauna species, the type of bioturbation activity, flora species, tidal range, and intertidal location. While these different structures can represent major drivers in the carbon cycling of coastal wetlands, studies partitioning their specific contribution are still lacking in both saltmarshes and mangroves globally.

3.6 **Synthesis and research directions**

Coastal wetlands, which belong to the blue carbon ecosystems, play a key role in climate change mitigation by fixing and storing carbon. Until recently, GHG fluxes in these ecosystems were less frequently quantified than C stocks. Since fluxes are more variable in space and time than stocks, the absence of accurate flux quantification impedes the determination of precise C budget for coastal wetlands. Over the past two decades, an increasing number of case studies have significantly increased our knowledge of GHG dynamics in coastal wetlands, especially GHG fluxes at the soil-air interface, owing to the advancement of techniques that support the in situ measurement of GHG fluxes. GHG concentrations can be determined on discrete samples using gas chromatograph at the laboratory, but they can also be directly measured in the field using flux chambers that are connected to infrared gas analyzers or cavity ring-down spectroscopes. These instruments can perform high-frequency measurements at 1 Hz or above, thereby enhancing the precision, reproducibility, and representativeness of flux measurements. However, some precautions must be taken in their use in coastal wetlands, because the tides will induce the migration of the saturation zone and thus reduce the time available to perform comparative measurements. In addition, in order to better assess GHG fluxes at the soil-air interface in coastal wetlands, we suggest using both opaque and transparent chambers that will allow the determination of the role of benthic biofilm on the GHG fluxes measured.

Recent studies using flux chambers confirmed that CO_2 emissions in coastal wetlands are low compared to those in terrestrial ecosystems. The mean soil CO_2 emission from mangroves is $257.1\,g\,C\,m^{-2}\,year^{-1}$, while that from terrestrial ecosystems is higher by more than twice. The low sediment CO_2 emissions from mangroves can be attributed to the anaerobic, waterlogged, and saline conditions that are prevalent in these ecosystems, which inhibit carbon mineralization and subsequently CO_2 production. The mean soil CH_4 emission from global mangroves is estimated to be $1.69\pm0.66\,g\,C\,m^{-2}\,year^{-1}$, which is higher than that from terrestrial forests, but one order of magnitude lower than that from freshwater wetlands and saltmarshes. In coastal wetlands, the sulfate ions from tidal inputs limit methanogenesis. In addition, CH_4 produced in the deep layers of the soil is subjected to strong anaerobic and aerobic oxidation as it diffuses through the soil column that can reduce its net emission to the atmosphere by up to 90%.

GHG emissions from coastal wetland soils depend on a series of biotic and abiotic parameters (Fig. 3.5). Temperature is a major driver of these emissions, acting as a catalyst for chemical reactions, accelerating organic matter decomposition and thus GHG production. Q_{10} values reported in coastal wetlands vary between 1.3 and 2.6 for CO_2 and between 2.5 and 5 for CH_4. Soil water content can also strongly affect soil GHG emissions by influencing microbial activity, chemical reaction rates, and gas diffusion rates throughout the soil column. The combined effects of these two parameters induce a seasonal and latitudinal variability of CO_2 and CH_4 fluxes at

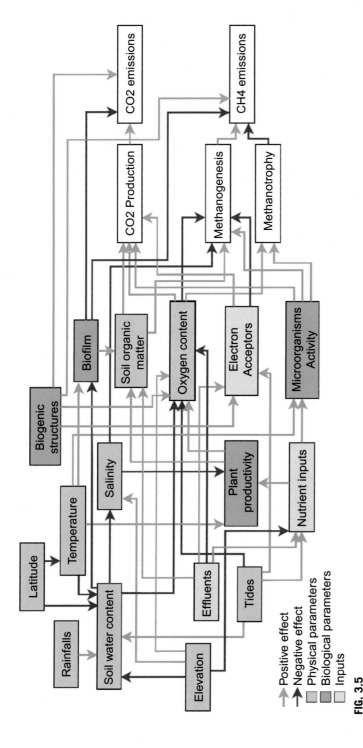

FIG. 3.5

Flow diagram showing the influence of biotic and abiotic parameters on CO_2 and CH_4 emissions at the soil-air interface in coastal wetlands. Important controlling drivers are indicated, and marked with *green* and *red* arrows if they have a positive and a negative effect, respectively, on processes.

Modified from Kristensen, E, Connolly, R., Ferreira, T.O., Marchand, C., Otero, X.L., Rivera-Monroy, V.H., 2017. Biogeochemical cycles; global approaches and perspectives. In: Mangrove Ecosystems: A Global Biogeographic Perspective Structure, Function and Services. Springer.

the soil-air interface in coastal wetlands, with the highest emissions being reported at low latitudes in tropical humid climate. The elevation in the intertidal zone of the wetland ecosystem also affects the GHG fluxes at the soil-air interface by controlling: (i) plant productivity and thus the soil organic carbon content and (ii) the tidal immersion and thus the soil water content. Nutrient inputs from shrimp ponds, sewage, and livestock effluents or fertilizers can alter the soil GHG emissions in coastal wetlands. An increase in nutrient loadings can enhance ecosystem productivity and thus organic matter accumulation in soils, higher organic matter lability, and higher availability of electron acceptors, leading to greater CO_2 emissions but lower CH_4 fluxes.

GHG emissions from coastal wetland soils are also subjected to biological controls. At light, the relationships between chl-a content, magnitude of CO_2 fluxes, and ^{13}C-CO_2 depletion demonstrate the role of microphytobenthos, which develop at the soil surface, in reducing CO_2 fluxes. On the one hand, the biofilm is composed notably by photoautotrophic organisms that can consume the CO_2 produced within the soil, while on the other hand, the biofilm can form a cohesive organic matrix composed of extracellular polysaccharides that can act as a physical barrier reducing GHG fluxes from the soil to the atmosphere. Conversely, the presence of crab burrows may enhance GHG emissions. Crab burrowing activity modifies the soil redox conditions, increasing the vertical penetration of oxygen in soil that results in higher CO_2 production. Pneumatophores, prop roots, and stem trees can also promote the upward transport of GHG from deeper soils, increasing the magnitude of CO_2 and CH_4 fluxes at the soil-air interface.

In spite of a better understanding of the role of physical and biological parameters on GHG production and emissions in coastal wetlands, we still lack a complete understanding of the mechanisms and origin of the GHG fluxes measured. We suggest developing more in-depth studies of GHG partitioning, using in situ isotopic analyzers, in order to quantify the respective contribution of organic carbon mineralization, root respiration, and microphytobenthic activity. Additionally, measuring GHG concentrations in different soil layers, comparing them with the fluxes, as well as identifying and characterizing the layers of CH_4 production and oxidation would be highly relevant. Lastly, studying GHG emissions in coastal wetlands in response to combined pressures, e.g., urbanization and climate changes, notably through greenhouse experiments, would help to assess the future ability of these systems to act as a carbon sink. For instance, the clearing of coastal wetlands results in strong organic carbon mineralization and enhanced GHG fluxes to the atmosphere. Eutrophication may have the same result, while sea level rise may reduce the emissions by increasing the soil water content.

Acknowledgments

Xiaoguang Ouyang is sponsored by an Impact Postdoctoral Fellowship from The Chinese University of Hong Kong. The help of two anonymous reviewers and of the associate editor (Derrick Y.F. Lai) to improve this chapter is gratefully acknowledged.

References

Abdul-Aziz, O.I., Ishtiaq, K.S., Tang, J., Moseman-Valtierra, S., Kroeger, K.D., Gonneea, M. E., Mora, J., Morkeski, K., 2018. Environmental controls, emergent scaling, and predictions of greenhouse gas (GHG) fluxes in coastal salt marshes. J. Geophys. Res. Biogeo. 123, 2234–2256. https://doi.org/10.1029/2018JG004556.

Agusto, L.E., Thibodeau, B., Tang, J., Wang, F., Canincci, S., 2020. Fiddling with the carbon budget: fiddler crab burrowing activity increases wetland's carbon flux. Earth Space Sci. https://doi.org/10.1002/essoar.10503504.1.

Al-Haj, A., Fulweiler, R.W., 2020. A synthesis of methane emissions from shallow vegetated coastal ecosystems. Glob. Chang. Biol. https://doi.org/10.1111/gcb.15046.

Allen, D.E., Dalal, R.C., Rennenberg, H., Meyer, R.L., Reeves, S., Schmidt, S., 2007. Spatial and temporal variation of nitrous oxide and methane flux between subtropical mangrove sediments and the atmosphere. Soil Biol. Biochem. 39, 622–631. https://doi.org/10.1016/j. soilbio.2006.09.013.

Allen, D., Dalal, R.C., Rennenberg, H., Schmidt, S., 2011. Seasonal variation in nitrous oxide and methane emissions from subtropical estuary and coastal mangrove sediments, Australia. Plant Biol. 13, 126–133. https://doi.org/10.1111/j.1438-8677.2010.00331.x.

Alongi, D.M., 2014. Carbon cycling and storage in mangrove forests. Ann. Rev. Mar. Sci. 6, 195–219.

Alongi, D.M., 2020a. Global significance of mangrove blue carbon in climate change mitigation. Science 2 (3), 67. https://doi.org/10.3390/sci2030067.

Alongi, D.M., 2020b. Carbon balance in salt marsh and mangrove ecosystems: a global synthesis. J. Mar. Sci. Eng. 8, 767.

Alongi, D.M., 2020c. Carbon cycling in the world's mangrove ecosystems revisited: significance of non-steady state diagenesis and subsurface linkages between the forest floor and the coastal ocean. Forests 2020 (11), 977. https://doi.org/10.3390/f11090977.

Alongi, D.M., Wattayakorn, G., Pfitzner, J., Tirendi, F., Zagorskis, I., Brunskill, G.J., Davidson, A., Clough, B.F., 2001. Organic carbon accumulation and metabolic pathways in sediments of mangrove forests in southern Thailand. Mar. Geol. 179, 85–103. https://doi.org/ 10.1016/S0025-3227(01)00195-5.

Andersen, F.O., Kristensen, E., 1988. Oxygen micro gradients in the rhizosphere of the mangrove Avicennia marina. Mar. Ecol. Prog. Ser. 44, 201–204.

Arai, H., Yoshioka, R., Hanazawa, S., Minh, V., Vo, T., Tran, K.T., Truong, P., Jha, C., Reddy, S., Dadhwal, V., Mano, M., Inubushi, K., 2016. Function of the methanogenic community in mangrove soils as influenced by the chemical properties of the hydrosphere. Soil Sci. Plant Nutr. 62, 150–163. https://doi.org/10.1080/00380768.2016.1165598.

Balasubramaniam, J., Prasath, D., Jayaraj, K.A., 2017. Microphytobenthic biomass, species composition and environmental gradients in the mangrove intertidal region of the Andaman Archipelago, India. Environ. Monit. Assess. 189, 231. https://doi.org/10.1007/ s10661-017-5936-0.

Baldocchi, D.D., 2020. How eddy covariance flux measurements have contributed to our understanding of Global Change Biology. Glob. Chang. Biol. 26, 242–260. https://doi. org/10.1111/gcb.14807.

Barr, J.G., Engel, V., Fuentes, J.D., Zieman, J.C., O'Halloran, T.L., Smith, T.J., Anderson, G. H., 2010. Controls on mangrove forest-atmosphere carbon dioxide exchanges in western Everglades National Park. J. Geophys. Res. Biogeo. 115, G02020. https://doi.org/10.1029/ 2009JG001186.

Bartlett, K.B., Harriss, R.C., 1993. Review and assessment of methane emissions from wetlands. Chemosphere 26 (1–4), 261–320.

Boetius, A., Ravenschlag, K., Schubert, C., et al., 2000. A marine microbial consortium apparently mediating anaerobic oxidation of methane. Nature 407, 623–626. https://doi.org/10.1038/35036572.

Bond-Lamberty, B., Thomson, A., 2010. A global database of soil respiration measurements. Biogeosciences 7, 1321–1344.

Bond-Lamberty, B., Christianson, D.S., Malhotra, A., Pennington, S.C., Sihi, D., AghaKouchak, A., Anjileli, H., Altaf Arain, M., Armesto, J.J., Ashraf, S., Ataka, M., Baldocchi, D., Andrew Black, T., Buchmann, N., Carbone, M.S., Chang, S.C., Crill, P., Curtis, P.S., Davidson, E.A., Desai, A.R., Drake, J.E., El-Madany, T.S., Gavazzi, M., Görres, C.M., Gough, C.M., Goulden, M., Gregg, J., Gutiérrez del Arroyo, O., He, J.S., Hirano, T., Hopple, A., Hughes, H., Järveoja, J., Jassal, R., Jian, J., Kan, H., Kaye, J., Kominami, Y., Liang, N., Lipson, D., Macdonald, C.A., Maseyk, K., Mathes, K., Mauritz, M., Mayes, M.A., McNulty, S., Miao, G., Migliavacca, M., Miller, S., Miniat, C.F., Nietz, J.G., Nilsson, M.B., Noormets, A., Norouzi, H., O'Connell, C.S., Osborne, B., Oyonarte, C., Pang, Z., Peichl, M., Pendall, E., Perez-Quezada, J.F., Phillips, C.L., Phillips, R.P., Raich, J.W., Renchon, A.A., Ruehr, N.K., Sánchez-Cañete, E.P., Saunders, M., Savage, K.E., Schrumpf, M., Scott, R.L., Seibt, U., Silver, W.L., Sun, W., Szutu, D., Takagi, K., Takagi, M., Teramoto, M., Tjoelker, M.G., Trumbore, S., Ueyama, M., Vargas, R., Varner, R.K., Verfaillie, J., Vogel, C., Wang, J., Winston, G., Wood, T.E., Wu, J., Wutzler, T., Zeng, J., Zha, T., Zhang, Q., Zou, J., 2020. COSORE: a community database for continuous soil respiration and other soil-atmosphere greenhouse gas flux data. Glob. Chang. Biol. 26, 7268–7283. https://doi.org/10.1111/gcb.15353.

Bonnett, S.A.F., Blackwell, M.S.A., Leah, R., Cook, V., O'Connor, M., Maltby, E., 2013. Temperature response of denitrification rate and greenhouse gas production in agricultural river marginal wetland soils. Geobiology 11, 252–267. https://doi.org/10.1111/gbi.12032.

Born, M., Dörr, H., Levin, I., 1990. Methane consumption in aerated soils of the temperate zone. Tellus B Chem. Phys. Meteorol. 42, 2–8. https://doi.org/10.3402/tellusb.v42i1.15186.

Bouillon, S., Borges, A.V., Castañeda-Moya, E., Diele, K., Dittmar, T., Duke, N.C., Kristensen, E., Lee, S.Y., Marchand, C., Middelburg, J.J., Rivera-Monroy, V.H., Smith III, T.J., Twilley, R.R., 2008. Mangrove production and carbon sinks: a revision of global budget estimates. Global Biogeochem. Cycles 22 (2), GB2013.

Bourgeois, C., Alfaro, A.C., Leopold, A., Andreoli, R., Bisson, E., Desnues, A., Duprey, J.-L., Marchand, C., 2019. Sedimentary and elemental dynamics as a function of the elevation profile in a semi-arid mangrove toposequence. Catena 173, 289–301.

Breithaupt, J.L., Smoak, J.M., Smith, T.J., Sanders, C.J., Hoare, A., 2012. Organic carbon burial rates in mangrove sediments: strengthening the global budget: mangrove organic carbon burial rates. Global Biogeochem. Cycles 26. https://doi.org/10.1029/2012GB004375.

Bulmer, R.H., Schwendenmann, L., Lohrer, A.M., Lundquist, C.J., 2017. Sediment carbon and nutrient fluxes from cleared and intact temperate mangrove ecosystems and adjacent sandflats. Sci. Total Environ. 599–600, 1874–1884. https://doi.org/10.1016/j.scitotenv.2017.05.139.

Cai, Y., Wang, X., Ding, W., Tian, L., Zhao, H., Lu, X., 2013. Potential short-term effects of yak and Tibetan sheep dung on greenhouse gas emissions in two alpine grassland soils under laboratory conditions. Biol. Fertil. Soils 49, 1215–1226. https://doi.org/10.1007/s00374-013-0821-7.

Call, M., Maher, D.T., Santos, I.R., Ruiz-Halpern, S., Mangion, P., Sanders, C.J., Erler, D. V., Oakes, J.M., Rosentreter, J., Murray, R., Eyre, B.D., 2015. Spatial and temporal variability of carbon dioxide and methane fluxes over semi-diurnal and spring–neap–spring timescales in a mangrove creek. Geochim. Cosmochim. Acta 150, 211–225. https://doi.org/10.1016/j.gca.2014.11.023.

Carmichael, M.J., Smith, W.K., 2016. Standing dead trees: a conduit for the atmospheric flux of greenhouse gases from wetlands? Wetlands 36 (6), 1183–1188. https://doi.org/10.1007/s13157-016-0845.

Carmichael, M.J., Bernhardt, E.S., Bräuer, S.L., Smith, W.K., 2014. The role of vegetation in methane flux to the atmosphere: should vegetation be included as a distinct category in the global methane budget? Biogeochemistry 119 (1–3), 1–24. https://doi.org/10.1007/s10533-014-9974-1.

Carmichael, M.J., Helton, A.M., White, J.C., Smith, W.K., 2018. Standing dead trees are a conduit for the atmospheric flux of CH$_4$ and CO$_2$ from wetlands. Wetlands 38 (1), 133–143. https://doi.org/10.1007/s13157-017-0963-8.

Chanda, A., Akhand, A., Manna, S., Dutta, S., Das, I., Hazra, S., Rao, K.H., Dadhwal, V.K., 2013. Measuring daytime CO$_2$ fluxes from the inter-tidal mangrove soils of Indian Sundarbans. Environ. Earth Sci. 72, 417–427. https://doi.org/10.1007/s12665-013-2962-2.

Chang, T.-C., Yang, S.-S., 2003. Methane emissions from wetlands in Thailand. Atmos. Environ. 37, 4551–4558.

Charles, H., Dukes, J.S., 2009. Effects of warming and altered precipitation on plant and nutrient dynamics of a New England salt marsh. Ecol. Appl. 19, 1758–1773. https://doi.org/10.1890/08-0172.1.

Chauhan, R., Datta, A., Ramanathan, A., Adhya, T.K., 2015. Factors influencing spatiotemporal variation of methane and nitrous oxide emission from a tropical mangrove of eastern coast of India. Atmos. Environ. 107, 95–106. https://doi.org/10.1016/j.atmosenv.2015.02.006.

Chen, G.C., Tam, N.F.Y., Ye, Y., 2010. Summer fluxes of atmospheric greenhouse gases N$_2$O, CH$_4$ and CO$_2$ from mangrove soil in South China. Sci. Total Environ. 408, 2761–2767. https://doi.org/10.1016/j.scitotenv.2010.03.007.

Chen, G.C., Tam, N.F.Y., Ye, Y., 2012. Spatial and seasonal variations of atmospheric N$_2$O and CO$_2$ fluxes from a subtropical mangrove swamp and their relationships with soil characteristics. Soil Biol. Biochem. 48, 175–181. https://doi.org/10.1016/j.soilbio.2012.01.029.

Chen, G., Yang, Y., Robinson, D., 2014. Allometric constraints on, and trade-offs in, belowground carbon allocation and their control of soil respiration across global forest ecosystems. Glob. Chang. Biol. 20 (5), 1674–1684.

Chen, S., Chmura, G.L., Wang, Y., Yu, D., Ou, D., Chen, B., Ye, Y., Chen, G., 2019. Benthic microalgae offset the sediment carbon dioxide emission in subtropical mangrove in cold seasons. Limnol. Oceanogr. 64, 1297–1308. https://doi.org/10.1002/lno.11116.

Chen, X., Wiesmeier, M., Sardans, J., Van Zwieten, L., Fang, Y., Gargallo-Garriga, A., Chen, Y., Chen, S., Zeng, C., Peñuelas, J., Wang, W., 2021. Effects of crabs on greenhouse gas emissions, soil nutrients, and stoichiometry in a subtropical estuarine wetland. Biol. Fertil. Soils 57, 131–144.

Chmura, G.L., Kellman, L., van Ardenne, L., Guntenspergen, G.R., 2016. Greenhouse gas fluxes from salt marshes exposed to chronic nutrient enrichment. PLoS One 11. https://doi.org/10.1371/journal.pone.0149937.

Cotovicz, L.C., Knoppers, B.A., Brandini, N., Poirier, D., Santos, S.J.C., Abril, G., 2016. Spatio-temporal variability of methane (CH_4) concentrations and diffusive fluxes from a tropical coastal embayment surrounded by a large urban area (Guanabara Bay, Rio de Janeiro, Brazil). Limnol. Oceanogr. 61, S238–S252. https://doi.org/10.1002/lno.10298.

Cui, X., Liang, J., Lu, W., Chen, H., Liu, F., Lin, G., Xu, F., Luo, Y., Lin, G., 2018. Stronger ecosystem carbon sequestration potential of mangrove wetlands with respect to terrestrial forests in subtropical China. Agric. For. Meteorol. 249, 71–80. https://doi.org/10.1016/j.agrformet.2017.11.019.

Daggers, T.D., van Oevelen, D., Herman, P.M.J., Boschker, H.T.S., van der Wal, D., 2020. Spatial variability in macrofaunal diet composition and grazing pressure on microphytobenthos in intertidal areas. Limnol. Oceanogr. https://doi.org/10.1002/lno.11554.

Dalal, R.C., Wang, W., Robertson, G.P., Parton, W.J., 2003. Nitrous oxide emission from Australian agricultural lands and mitigation options: a review. Soil Res. 41, 165. https://doi.org/10.1071/SR02064.

Das, S., Ganguly, D., Chakraborty, S., Mukherjee, A., Kumar De, T., 2018. Methane flux dynamics in relation to methanogenic and methanotrophic populations in the soil of Indian Sundarban mangroves. Mar. Ecol. 39. https://doi.org/10.1111/maec.12493, e12493.

David, F., Meziane, T., Tran-Thi, N.-T., Truong Van, V., Thanh-Nho, N., Taillardat, P., Marchand, C., 2018. Carbon biogeochemistry and CO_2 emissions in a human impacted and mangrove dominated tropical estuary (Can Gio, Vietnam). Biogeochemistry. https://doi.org/10.1007/s10533-018-0444-z.

Davidson, E.A., Janssens, I.A., 2006. Temperature sensitivity of soil carbon decomposition and feedbacks to climate change. Nature 440, 165–173. https://doi.org/10.1038/nature04514.

Davidson, E.A., Savage, K., Verchot, L.V., Navarro, R., 2002. Minimizing artifacts and biases in chamber-based measurements of soil respiration. Agric. For. Meteorol. 113, 21–37.

de Brouwer, J.F.C., Wolfstein, K., Ruddy, G.K., Jones, T.E.R., Stal, L.J., 2005. Biogenic stabilization of intertidal sediments: the importance of extracellular polymeric substances produced by benthic diatoms. Microb. Ecol. 49, 501–512. https://doi.org/10.1007/s00248-004-0020-z.

de Brouwer, J.F.C., Neu, T.R., Stal, L.J., 2006. On the function of secretion of extracellular polymeric substances by benthic diatoms and their role in intertidal mudflats: a review of recent insights and views. In: Kromkamp, J.C., De Brouwer, J.F.C., Blanchard, G.F., Forster, R.M., Créach, V. (Eds.), Functioning of Microphytobenthos in Estuaries : Proceedings of the Colloquium, Amsterdam, 21-23 August 2003. Royal Netherlands Academy of Arts and Sciences, Amsterdam, pp. 45–61, https://doi.org/10.1086/586980.

Denier van der Gon, H.A.C., Neue, H.U., 1995. Methane emission from a wetland rice field as affected by salinity. Plant and Soil 170, 307–313. https://doi.org/10.1007/bf00010483.

Drewitt, G.B., Black, T.A., Nesic, Z., Humphreys, E.R., Jork, E.M., Swanson, R., Ethier, G.J., Griffis, T., Morgenstern, K., 2002. Measuring forest floor CO_2 fluxes in a Douglas-fir forest. Agric. For. Meteorol. 110, 299–317. https://doi.org/10.1016/S0168-1923(01)00294-5.

Duarte, C.M., Middelburg, J.J., Caraco, N., 2005. Major role of marine vegetation on the oceanic carbon cycle. Biogeosciences 2, 1–8.

Dunfield, P., Knowles, R., Dumont, R., Moore, T., 1993. Methane production and consumption in temperate and subarctic peat soils: response to temperature and pH. Soil Biol. Biochem. 25, 321–326. https://doi.org/10.1016/0038-0717(93)90130-4.

Emery, H.E., Angell, J.H., Fulweiler, R.W., 2019. Salt marsh greenhouse gas fluxes and microbial communities are not sensitive to the first year of precipitation change. J. Geophys. Res. Biogeo. 124, 1071–1087. https://doi.org/10.1029/2018JG004788.

Fang, C., Moncrieff, J.B., 2001. The dependence of soil CO$_2$ efflux on temperature. Soil Biol. Biochem. 33, 155–165. https://doi.org/10.1016/S0038-0717(00)00125-5.

Fariña, J.M., He, Q., Silliman, B.R., Bertness, M.D., 2018. Biogeography of salt marsh plant zonation on the Pacific coast of South America. J. Biogeogr. 45, 238–247. https://doi.org/10.1111/jbi.13109.

Fields, S., 2004. Global nitrogen: cycling out of control. Environ. Health Perspect. 112, A556–A563.

Gebremichael, A.W., Osborne, B., Orr, P., 2017. Flooding-related increases in CO$_2$ and N$_2$O emissions from a temperate coastal grassland ecosystem. Biogeosciences 14, 2611–2626. https://doi.org/10.5194/bg-14-2611-2017.

Geoghegan, E.K., Caplan, J.S., Leech, F.N., Weber, P.E., Bauer, C.E., Mozdzer, T.J., 2018. Nitrogen enrichment alters carbon fluxes in a New England salt marsh. Ecosyst. Health Sustain. 4, 277–287. https://doi.org/10.1080/20964129.2018.1532772.

Girkin, N.T., Turner, B.L., Ostle, N., Craigon, J., Sjögersten, S., 2018. Root exudate analogues accelerate CO$_2$ and CH$_4$ production in tropical peat. Soil Biol. Biochem. 117, 48–55.

Gnanamoorthy, P., Selvam, V., Ramasubramanian, R., Nagarajan, R., Chakraborty, S., Deb Burman, P.K., Karipot, A., 2019. Diurnal and seasonal patterns of soil CO$_2$ efflux from the Pichavaram mangroves, India. Environ. Monit. Assess. 191, 258. https://doi.org/10.1007/s10661-019-7407-2.

Grellier, S., Janeau, J.-L., Dang Hoai, N., Nguyen Thi Kim, C., Le Thi Phuong, Q., Pham Thi Thu, T., Tran-Thi, N.-T., Marchand, C., 2017. Changes in soil characteristics and C dynamics after mangrove clearing (Vietnam). Sci. Total Environ. 593–594, 654–663. https://doi.org/10.1016/j.scitotenv.2017.03.204.

Gribsholt, B., Kostka, J., Kristensen, E., 2003. Impact of fiddler crabs and plant roots on sediment biogeochemistry in a Georgia saltmarsh. Mar. Ecol. Prog. Ser., 237–251. https://doi.org/10.3354/meps259237.

Guckland, A., Flessa, H., Prenzel, J., 2009. Controls of temporal and spatial variability of methane uptake in soils of a temperate deciduous forest with different abundance of European beech (Fagus sylvatica L.). Soil Biol. Biochem. 41 (8), 1659–1667.

Guimond, J.A., Seyfferth, A.L., Moffett, K.B., Michael, H.A., 2020. A physical-biogeochemical mechanism for negative feedback between marsh crabs and carbon storage. Environ. Res. Lett. 15. https://doi.org/10.1088/1748-9326/ab60e2, 034024.

Ha, H., Kim, H., Noh, J., Ha, H., 2018. Rainfall effects on the erodibility of sediment and microphytobenthos in the intertidal flat. Environ. Pollut. 242. https://doi.org/10.1016/j.envpol.2018.06.079.

Ha, H.J., Kim, H., Kwon, B.-O., Khim, J.S., Ha, H.K., 2020. Influence of tidal forcing on microphytobenthic resuspension dynamics and sediment fluxes in a disturbed coastal environment. Environ. Int. 139. https://doi.org/10.1016/j.envint.2020.105743, 105743.

Heinemeyer, A., Di Bene, C., Lloyd, A.R., Tortorella, D., Baxter, R., Huntley, B., Gelsomino, A., Ineson, P., 2011. Soil respiration: implications of the plant-soil continuum and respiration chamber collar-insertion depth on measurement and modelling of soil CO2 efflux rates in three ecosystems. Eur. J. Soil Sci. 62, 82–94. https://doi.org/10.1111/j.1365-2389.2010.01331.x.

Hien, H.T., Marchand, C., Aimé, J., Cuc, N.T.K., 2018. Seasonal variability of CO_2 emissions from sediments in planted mangroves (Northern Viet Nam). Estuar. Coast. Shelf Sci. 213, 28–39. https://doi.org/10.1016/j.ecss.2018.08.006.

Hirota, M., Senga, Y., Seike, Y., Nohara, S., Kunii, H., 2007. Fluxes of carbon dioxide, methane and nitrous oxide in two contrastive fringing zones of coastal lagoon, Lake Nakaumi, Japan. Chemosphere 68, 597–603. https://doi.org/10.1016/j.chemosphere.2007.01.002.

Hoehler, T.M., Alperin, M.J., 2014. Biogeochemistry: methane minimalism. Nature 507 (7493), 436–437. https://doi.org/10.1038/nature13215.

Hosono, T., Nouchi, I., 1997. The dependence of methane transport in rice plants on the root zone temperature. Plant and Soil 191, 233–240. https://doi.org/10.1023/A:1004203208686.

Hu, W., Jiang, Y., Chen, D., Lin, Y., Han, Q., Cui, Y., 2018. Impact of pore geometry and water saturation on gas effective diffusion coefficient in soil. Appl. Sci. 8, 2097. https://doi.org/10.3390/app8112097.

Hu, M., Sardans, J., Yang, X., Peñuelas, J., Tong, C., 2020. Patterns and environmental drivers of greenhouse gas fluxes in the coastal wetlands of China: a systematic review and synthesis. Environ. Res. 186. https://doi.org/10.1016/j.envres.2020.109576, 109576.

Huang, J., Luo, M., Liu, Y., Zhang, Y., Tan, J., 2019. Effects of tidal scenarios on the methane emission dynamics in the subtropical tidal marshes of the min river estuary in southeast China. Int. J. Environ. Res. Public Health 16, 2790. https://doi.org/10.3390/ijerph16152790.

Huertas, I.E., de la Paz, M., Perez, F.F., Navarro, G., Flecha, S., 2019. Methane emissions from the salt marshes of Doñana wetlands: spatio-temporal variability and controlling factors. Front. Ecol. Evol. 7, 32. https://doi.org/10.3389/fevo.2019.00032.

Hütsch, B.W., 1998. Tillage and land use effects on methane oxidation rates and their vertical profiles in soil. Biol. Fertil. Soils 27 (3), 284–292. https://doi.org/10.1007/s003740050435.

Inglett, K.S., Inglett, P.W., Reddy, K.R., Osborne, T.Z., 2012. Temperature sensitivity of greenhouse gas production in wetland soils of different vegetation. Biogeochemistry 108, 77–90. https://doi.org/10.1007/s10533-011-9573-3.

Jacotot, A., Marchand, C., Allenbach, M., 2019. Biofilm and temperature controls on greenhouse gas (CO_2 and CH_4) emissions from a Rhizophora mangrove soil (New Caledonia). Sci. Total Environ. 650, 1019–1028. https://doi.org/10.1016/j.scitotenv.2018.09.093.

Janousek, C.N., 2009. Taxonomic composition and diversity of microphytobenthos in southern California marine wetland habitats. Wetlands 29, 163–175. https://doi.org/10.1672/08-06.1.

Jeffrey, L.C., Reithmaier, G., Sippo, J.Z., Johnston, S.G., Tait, D.R., Harada, Y., Maher, D.T., 2019. Are methane emissions from mangrove stems a cryptic carbon loss pathway? Insights from a catastrophic forest mortality. New Phytol. 224 (1), 146–154.

Jensen, L.S., Mueller, T., Tate, K.R., Ross, D.J., Magid, J., Nielsen, N.E., 1996. Soil surface CO_2 flux as an index of soil respiration in situ: a comparison of two chamber methods. Soil Biol. Biochem. 28, 1297–1306. https://doi.org/10.1016/S0038-0717(96)00136-8.

Jian, J., Vargas, R., Anderson-teixeira, K., Stell, E., Herrmann, V., Horn, M., Kholod, N., Manzon, J., Marchesi, R., Paredes, D., 2020. A restructured and updated global soil respiration database (SRDB-V5). Earth Syst. Sci. Data, 1–19.

Joye, S., Lee, R., 2004. Benthic microbial mats: important sources of fixed nitrogen and carbon to the Twin Cays, Belize ecosystem. Atoll Res. Bull. 528. https://doi.org/10.5479/si.00775630.528.1.

Kabwe, L.K., Hendry, M.J., Wilson, G.W., Lawrence, J.R., 2002. Quantifying CO_2 fluxes from soil surfaces to the atmosphere. J. Hydrol. 260, 1–14. https://doi.org/10.1016/S0022-1694(01)00601-1.

Keeling, C.D., 1958. The concentration and isotopic abundances of atmospheric carbon dioxide in rural areas. Geochim. Cosmochim. Acta 13, 322–334.

Keeling, C.D., 1961. The concentration and isotopic abundances of carbon dioxide in rural and marine air. Geochim. Cosmochim. Acta 24, 277–298. https://doi.org/10.1016/0016-7037(61)90023-0.

Kirui, B., Huxham, M., Kairo, J.G., Mencuccini, M., Skov, M.W., 2009. Seasonal dynamics of soil carbon dioxide flux in a restored young mangrove plantation at Gazi Bay. Adv. Coast. Ecol, 122.

Kirwan, M.L., Guntenspergen, G.R., Langley, J.A., 2014. Temperature sensitivity of organic-matter decay in tidal marshes. Biogeosciences 11, 4801–4808. https://doi.org/10.5194/bg-11-4801-2014.

Kitaya, Y., Yabuki, K., Kiyota, M., Tani, A., Hirano, T., Aiga, I., 2002. Gas exchange and oxygen concentration in pneumatophores and prop roots of four mangrove species. Trees 16, 155–158. https://doi.org/10.1007/s00468-002-0167-5.

Kostka, J.E., Gribsholt, B., Petrie, E., Dalton, D., Skelton, H., Kristensen, E., 2002a. The rates and pathways of carbon oxidation in bioturbated saltmarsh sediments. Limnol. Oceanogr. 47, 230–240. https://doi.org/10.4319/lo.2002.47.1.0230.

Kostka, J.E., Roychoudhury, A., Van Cappellen, P., 2002b. Rates and controls of anaerobic microbial respiration across spatial and temporal gradients in saltmarsh sediments. Biogeochemistry 60, 49–76. https://doi.org/10.1023/A:1016525216426.

Krauss, K.W., Whitbeck, J.L., 2012. Soil greenhouse gas fluxes during wetland forest retreat along the lower Savannah River, Georgia (USA). Wetlands 32 (1), 73–81. https://doi.org/10.1007/s13157-011-0246-8.

Kreuzwieser, J., Buchholz, J., Rennenberg, H., 2003. Emission of methane and nitrous oxide by australian mangrove ecosystems. Plant Biol. 5, 423–431. https://doi.org/10.1055/s-2003-42712.

Kristensen, E., 2007. Carbon balance in mangrove sediments: the driving processes and their controls. In: Tateda, Y., et al. (Eds.), Greenhouse Gas and Carbon Balances in Mangrove Coastal Ecosystems. Gendai Tosho, Kanagawa, Japan, pp. 61–78 (257 pp).

Kristensen, E., 2008. Mangrove crabs as ecosystem engineers; with emphasis on sediment processes. J. Sea Res. 59, 30–43. https://doi.org/10.1016/j.seares.2007.05.004.

Kristensen, E., Alongi, D., 2006. Control by fiddler crabs (Uca vocans) and plant roots (Avicennia marina) on carbon, iron, and sulfur biogeochemistry in mangrove sediment. Limnol. Oceanogr. 51, 1557–1571. https://doi.org/10.4319/lo.2006.51.4.1557.

Kristensen, E., Andersen, F., Kofoed, L., 1988. Preliminary assessment of benthic community metabolism in a Southeast Asian mangrove swamp. Mar. Ecol. Prog. Ser. 48, 137–145. https://doi.org/10.3354/meps048137.

Kristensen, E., Bouillon, S., Dittmar, T., Marchand, C., 2008a. Organic carbon dynamics in mangrove ecosystems: a review. Aquat. Bot. 89, 201–219. https://doi.org/10.1016/j.aquabot.2007.12.005.

Kristensen, E., Flindt, M., Ulomi, S., Borges, A., Abril, G., Bouillon, S., 2008b. Emission of CO_2 and CH_4 to the atmosphere by sediments and open waters in two Tanzanian mangrove forests. Mar. Ecol. Prog. Ser. 370, 53–67. https://doi.org/10.3354/meps07642.

Kristensen, E., Connolly, R., Ferreira, T.O., Marchand, C., Otero, X.L., Rivera-Monroy, V.H., 2017. Biogeochemical cycles; global approaches and perspectives. In: Mangrove Ecosystems: A Global Biogeographic Perspective Structure, Function and Services. Springer.

Kromkamp, J.C., Foster, R.M., 2006. Developments in microphytobenthos primary productivity studies. In: Kromkamp, J.C., de Brouwer, J.F.C., Blanchard, G.F., Forster, R.M., Créach, V. (Eds.), Functioning of Microphytobenthos in Estuaries. Royal Netherlands Academy of Arts and Sciences.

Kwon, B.-O., Kim, H., Noh, J., Lee, S., Nam, J., Khim, J., 2020. Spatiotemporal variability in microphytobenthic primary production across bare intertidal flat, saltmarsh, and mangrove forest of Asia and Australia. Mar. Pollut. Bull. 151. https://doi.org/10.1016/j.marpolbul.2019.110707, 110707.

Lee, S.Y., 1995. Mangrove outwelling: a review. Hydrobiologia 295, 203–212.

Lee, J.-S., Kim, J.-W., Lee, S.H., Myeong, H.-H., Lee, J.-Y., Cho, J.S., 2016. Zonation and soil factors of salt marsh halophyte communities. J. Ecol. Environ. 40, 4. https://doi.org/10.1186/s41610-016-0010-3.

Le Mer, J., Roger, P., 2001. Production, oxidation, emission and consumption of methane by soils: a review. Eur. J. Soil Biol. 37, 25–50. https://doi.org/10.1016/S1164-5563(01)01067-6.

Leopold, A., Marchand, C., Deborde, J., Chaduteau, C., Allenbach, M., 2013. Influence of mangrove zonation on CO_2 fluxes at the sediment–air interface (New Caledonia). Geoderma 202–203, 62–70. https://doi.org/10.1016/j.geoderma.2013.03.008.

Leopold, A., Marchand, C., Deborde, J., Allenbach, M., 2015. Temporal variability of CO_2 fluxes at the sediment-air interface in mangroves (New Caledonia). Sci. Total Environ. 502, 617–626. https://doi.org/10.1016/j.scitotenv.2014.09.066.

Lin, C.-W., Kao, Y.-C., Chou, M.-C., Wu, H.-H., Ho, C.-W., Lin, H.-J., 2020. Methane emissions from subtropical and tropical mangrove ecosystems in Taïwan. Forests 11, 470. https://doi.org/10.3390/f11040470.

Lin, C.-W., Kao, Y.-C., Lin, W.-J., Ho, W.-W., Lin, H.-J., 2021. Effects of pneumatophore density on methane emissions in mangroves. Forests 12 (3), 314. https://doi.org/10.3390/f12030314.

Liu, W., Zhang, J., Tian, G., Xu, H., Yan, X., 2013. Temporal and vertical distribution of microphytobenthos biomass in mangrove sediments of Zhujiang (Pearl River) Estuary. Acta Oceanol. Sin. 32. https://doi.org/10.1007/s13131-013-0302-8.

Liu, B., Liu, X., Huo, S., Chen, X., Wu, L., Chen, M., Zhou, K., Li, Q., Peng, L., 2016. Properties of root exudates and rhizosphere sediment of *Bruguiera gymnorrhiza* (L.). J. Soil. Sediment. 17, 266–276.

Liu, Y., Liu, G., Xiong, Z., Liu, W., 2017. Response of greenhouse gas emissions from three types of wetland soils to simulated temperature change on the Qinghai-Tibetan Plateau. Atmos. Environ. 171, 17–24. https://doi.org/10.1016/j.atmosenv.2017.10.005.

Lloyd, J., Taylor, J.A., 1994. On the temperature dependence of soil respiration. Funct. Ecol., 315–323.

Lovelock, C., 2008. Soil respiration and belowground carbon allocation in mangrove forests. Ecosystems 11, 342–354. https://doi.org/10.1007/s10021-008-9125-4.

Lovelock, C.E., Feller, I.C., Reef, R., Ruess, R.W., 2014. Variable effects of nutrient enrichment on soil respiration in mangrove forests. Plant and Soil 379, 135–148. https://doi.org/10.1007/s11104-014-2036-6.

Luyssaert, S., Inglima, I., Jung, M., Richardson, A.D., Reichstein, M., et al., 2007. CO_2 balance of boreal, temperate, and tropical forests. Glob. Chang. Biol. 13, 2509–2537.

Lyimo, T.J., Pol, A., Op den Camp, H.J.M., 2002. Sulfate reduction and methanogenesis in sediments of Mtoni Mangrove Forest, Tanzania. AMBIO J. Hum. Environ. 31, 614–616. https://doi.org/10.1579/0044-7447-31.7.614.

Macintyre, H., Geider, R., Miller, D., 1996. Microphytobenthos: the ecological role of the "secret garden" of unvegetated, shallow-water marine habitats. I. Distribution, abundance and primary production. Estuaries Coast 19, 186–201. https://doi.org/10.2307/1352224.

Maher, D., Golsby-Smith, L., Gleeson, J., Eyre, B., 2013. Groundwater-derived dissolved inorganic and organic carbon exports from a mangrove tidal creek: the missing mangrove carbon sink? Limnol. Oceanogr. 58, 475–488. https://doi.org/10.4319/lo.2013.58.2.0475.

Malhi, Y., 2012. The productivity, metabolism and carbon cycle of tropical forest vegetation. J. Ecol. 100, 65–75.

Marchand, C., Disnar, J.R., Lallier-Vergès, E., Lottier, N., 2005. Early diagenesis of carbohydrates and lignin in mangrove sediments subject to variable redox conditions (French Guiana). Geochim. Cosmochim. Acta 69, 131–142. https://doi.org/10.1016/j.gca.2004.06.016.

Marchand, C., Allenbach, M., Lallier-Vergès, E., 2011. Relationships between heavy metals distribution and organic matter cycling in mangrove sediments (Conception Bay, New Caledonia). Geoderma 160, 444–456. https://doi.org/10.1016/j.geoderma.2010.10.015.

Marchand, C., Fernandez, J.-M., Moreton, B., Landi, L., Lallier-Vergès, E., Baltzer, F., 2012. The partitioning of transitional metals (Fe, Mn, Ni, Cr) in mangrove sediments downstream of a ferralitized ultramafic watershed (New Caledonia). Chem. Geol. 300–301, 70–80. https://doi.org/10.1016/j.chemgeo.2012.01.018.

Martin, R.M., Moseman-Valtierra, S., 2017. Different short-term responses of greenhouse gas fluxes from salt marsh mesocosms to simulated global change drivers. Hydrobiologia 802, 71–83. https://doi.org/10.1007/s10750-017-3240-1.

Martin, R.M., Wigand, C., Elmstrom, E., Lloret, J., Valiela, I., 2018a. Long-term nutrient addition increases respiration and nitrous oxide emissions in a New England salt marsh. Ecol. Evol. 8, 4958–4966. https://doi.org/10.1002/ece3.3955.

Martin, B.C., Gleeson, D., Statton, J., Siebers, A.R., Grierson, P., Hyan, M.H., Kendrick, G.A., 2018b. Low light availability alters root exudation and reduces putative beneficial microorganisms in seagrass roots. Front. Microbiol. https://doi.org/10.3389/fmicb.2017.02667.

Martinetto, P., Montemayor, D.I., Alberti, J., Costa, C.S.B., Iribarne, O., 2016. Crab bioturbation and herbivory may account for variability in carbon sequestration and stocks in south west Atlantic salt marshes. Front. Mar. Sci. 3, 122. https://doi.org/10.3389/fmars.2016.00122.

Mcleod, E., Chmura, G.L., Bouillon, S., Salm, R., Björk, M., Duarte, C.M., Lovelock, C.E., Schlesinger, W.H., Silliman, B.R., 2011. A blueprint for blue carbon: toward an improved understanding of the role of vegetated coastal habitats in sequestering CO_2. Front. Ecol. Environ. 9, 552–560. https://doi.org/10.1890/110004.

Montague, C.L., 1982. The influence of fiddler crab burrows and burrowing on metabolic processes in salt marsh sediments. In: Kennedy, V.S. (Ed.), Estuarine Comparisons. Academic Press, pp. 283–301, https://doi.org/10.1016/B978-0-12-404070-0.50023-5.

Morris, J.T., Bradley, P.M., 1999. Effects of nutrient loading on the carbon balance of coastal wetland sediments. Limnol. Oceanogr. 44, 699–702. https://doi.org/10.4319/lo.1999.44.3.0699.

Moseman-Valtierra, S., Abdul-Aziz, O.I., Tang, J., Ishtiaq, K.S., Morkeski, K., Mora, J., Quinn, R.K., Martin, R.M., Egan, K., Brannon, E.Q., Carey, J., Kroeger, K.D., 2016. Carbon dioxide fluxes reflect plant zonation and belowground biomass in a coastal marsh. Ecosphere 7. https://doi.org/10.1002/ecs2.1560, e01560.

Mueller, P., Mozdzer, T.J., Langley, J.A., Aoki, L.R., Noyce, G.L., Megonigal, J.P., 2020. Plant species determine tidal wetland methane response to sea level rise. Nat. Commun. 11, 5154.

Murray, R.H., Erler, D.V., Eyre, B.D., 2015. Nitrous oxide fluxes in estuarine environments: response to global change. Glob. Chang. Biol. 21, 3219–3245. https://doi.org/10.1111/gcb.12923.

Muzuka, A.N.N., Shunula, J.P., 2006. Stable isotope compositions of organic carbon and nitrogen of two mangrove stands along the Tanzanian coastal zone. Estuar. Coast. Shelf Sci. 66, 447–458.

Oakes, J.M., Eyre, B.D., 2014. Transformation and fate of microphytobenthos carbon in subtropical, intertidal sediments: potential for long-term carbon retention revealed by 13C-labeling. Biogeosciences 11, 1927–1940. https://doi.org/10.5194/bg-11-1927-2014.

O'Flaherty, V., Mahony, T., O'Kennedy, R., Colleran, E., 1998. Effect of pH on the growth kinetics and sulphide toxicity thresholds of a range of methanogenic, syntrophic and sulphate reducing bacteria. Process Biochem 33, 555–569.

Olsson, L., Ye, S., Yu, X., Wei, M., Krauss, K.W., Brix, H., 2015. Factors influencing CO_2 and CH_4 emissions from coastal wetlands in the Liaohe Delta, Northeast China. Biogeosciences 12, 4965–4977. https://doi.org/10.5194/bg-12-4965-2015.

Ouyang, X., Lee, S.Y., 2014. Updated estimates of carbon accumulation rates in coastal marsh sediments. Biogeosciences 11, 5057–5071. https://doi.org/10.5194/bg-11-5057-2014.

Ouyang, X., Lee, S.Y., 2020. Improved estimates on global carbon stock and carbon pools in tidal wetlands. Nat. Commun. 11, 317. https://doi.org/10.1038/s41467-019-14120-2.

Ouyang, X., Lee, S.Y., Connolly, R.M., 2017. Structural equation modelling reveals factors regulating surface sediment organic carbon content and CO2 efflux in a subtropical mangrove. Sci. Total Environ. 578, 513–522. https://doi.org/10.1016/j.scitotenv.2016.10.218.

Ouyang, X., Lee, S.Y., Connolly, R.M., 2018. Using isotope labeling to partition sources of CO_2 efflux in newly established mangrove seedlings. Limnol. Oceanogr. 63 (2), 731–740. https://doi.org/10.1002/lno.10663.

Pangala, S.R., Enrich-Prast, A., Basso, L.S., Peixoto, R.B., Bastviken, D., Hornibrook, E.R.C., Gatti, L.V., Marotta, H., Calazans, L.S.B., Sakuragui, C.M., Bastos, W.R., Malm, O., Gloor, E., Miller, J.B., Gauci, V., 2017. Large emissions from floodplain trees close the Amazon methane budget. Nature 552 (7684), 230.

Pataki, D.E., Ehleringer, J.R., Flanagan, L.B., Yakir, D., Bowling, D.R., Still, C.J., Buchmann, N., Kaplan, J.O., Berry, J.A., 2003. The application and interpretation of Keeling plots in terrestrial carbon cycle research. Global Biogeochem. Cycles. 17.

Pennings, S.C., Callaway, R.M., 1992. Salt marsh plant zonation: the relative importance of competition and physical factors. Ecology 73, 681–690. https://doi.org/10.2307/1940774.

Pennings, S.C., Grant, M.-B., Bertness, M.D., 2005. Plant zonation in low-latitude salt marshes: disentangling the roles of flooding, salinity and competition. J. Ecol. 93, 159–167. https://doi.org/10.1111/j.1365-2745.2004.00959.x.

Pfeifer-Meister, L., Gayton, L.G., Roy, B.A., Johnson, B.R., Bridgham, S.D., 2018. Greenhouse gas emissions limited by low nitrogen and carbon availability in natural, restored, and agricultural Oregon seasonal wetlands. PeerJ 6. https://doi.org/10.7717/peerj.5465.

Pinckney, J., Zingmark, R.G., 1993. Biomass and production of benthic microalgal communities in estuarine habitats. Estuaries 16, 887–897. https://doi.org/10.2307/1352447.

Poffenbarger, H.J., Needelman, B.A., Megonigal, J.P., 2011. Salinity influence on methane emissions from tidal marshes. Wetlands 31, 831–842. https://doi.org/10.1007/s13157-011-0197-0.

Priemé, A., Christensen, S., 1997. Seasonal and spatial variation of methane oxidation in a Danish spruce forest. Soil Biol. Biochem. 29 (8), 1165–1172. https://doi.org/10.1016/S0038-0717(97)00038-2.

Pülmanns, N., Diele, K., Mehlig, U., Nordhaus, I., 2014. Burrows of the semi-terrestrial crab Ucides cordatus enhance CO$_2$ release in a North Brazilian mangrove forest. PLoS One 9, e109532. https://doi.org/10.1371/journal.pone.0109532.

Purvaja, R., Ramesh, R., 2001. Natural and anthropogenic methane emission from coastal wetlands of south India. Environ. Manag. 27, 547–557. https://doi.org/10.1007/s002670010169.

Purvaja, R., Ramesh, R., Frenzel, P., 2004. Plant-mediated methane emission from an Indian mangrove. Glob. Chang. Biol. 10, 1825–1834. https://doi.org/10.1111/j.1365-2486.2004.00834.x.

Redzuan, N., Underwood, G., 2020. Movement of microphytobenthos and sediment between mudflats and salt marsh during spring tides. Front. Mar. Sci. 7, 496. https://doi.org/10.3389/fmars.2020.00496.

Ridd, P.V., 1996. Flow through animal burrows in mangrove creeks. Estuar. Coast. Shelf Sci. 43, 617–625. https://doi.org/10.1006/ecss.1996.0091.

Rochette, P., Gregorich, E.G., Desjardins, R.L., 1992. Comparison of static and dynamic closed chambers for measurement of soil respiration under field conditions. Can. J. Soil Sci. 72 (4), 605–609. https://doi.org/10.4141/cjss92-050.

Roehm, C.L., 2005. In: del Giorgio, P.A., Williams, P.J.L.B. (Eds.), Respiration in Wetland Ecosystems. Oxford University Press, Oxford, pp. 83–102.

Rosentreter, J.A., Maher, D.T., Erler, D.V., Murray, R.H., Eyre, B.D., 2018. Methane emissions partially offset "blue carbon" burial in mangroves. Sci. Adv. 4 (6), eaao4985.

Rosentreter, J.A., Al-Haj, A.N., Fulweiler, R.W., Williamson, P., 2021. Methane and nitrous oxide emissions complicate coastal blue carbon assessments. Global Biogeochem. Cycles 35. https://doi.org/10.1029/2020GB006858, e2020GB006858.

Saari, A., Smolander, A., Martikainen, P., 2004. Methane consumption in a frequently nitrogen-fertilized and limed spruce forest soil after clear-cutting. Soil Use Manag 20, 65–73. https://doi.org/10.1111/j.1475-2743.2004.tb00338.x.

Sadat-Noori, M., Santos, I.R., Tait, D.R., Reading, M.J., Sanders, C.J., 2017. High porewater exchange in a mangrove-dominated estuary revealed from short-lived radium isotopes. J. Hydrol. 553, 188–198. https://doi.org/10.1016/j.jhydrol.2017.07.058.

Sánchez-Carrillo, S., Álvarez-Cobelas, M., Angeler, D.G., Serrano-Grijalva, L., Sánchez-Andrés, R., Cirujano, S., Schmid, T., 2018. Elevated atmospheric CO$_2$ increases root exudation of carbon in wetlands: results from the first free-air CO$_2$ enrichment facility (FACE) in a Marshland. Ecosystems 21, 852–867. https://doi.org/10.1007/s10021-017-0189-x.

Sasmito, S.D., Taillardat, P., Clendenning, J.N., et al., 2019. Effect of land-use and land-cover change on mangrove blue carbon: a systematic review. Glob Change Biol 25, 4291–4302. https://doi.org/10.1111/gcb.14774.

Schaufler, G., Kitzler, B., Schindlbacher, A., Skiba, U., Sutton, M.A., Zechmeister-Boltenstern, S., 2010. Greenhouse gas emissions from European soils under different land use: effects of soil moisture and temperature. Eur. J. Soil Sci. 61, 683–696. https://doi.org/10.1111/j.1365-2389.2010.01277.x.

Scholander, P.F., van Dam, L., Scholander, S.I., 1955. Gas exchange in the roots of mangroves. Am. J. Bot. 42, 92–98. https://doi.org/10.1002/j.1537-2197.1955.tb11097.x.

Smith, K.A., Ball, T., Conen, F., Dobbie, K.E., Massheder, J., Rey, A., 2003. Exchange of greenhouse gases between soil and atmosphere: interactions of soil physical factors and biological processes. Eur. J. Soil Sci. 54, 779–791. https://doi.org/10.1046/j.1351-0754.2003.0567.x.

Snedaker, S.C., 1982. Mangrove species zonation: why? In: Sen, D.N., Rajpurohit, K.S. (Eds.), Contributions to the Ecology of Halophytes, Tasks for Vegetation Science. Springer, Netherlands, Dordrecht, pp. 111–125, https://doi.org/10.1007/978-94-009-8037-2_8.

Sotomayor, D., Corredor, J.E., Morell, J.M., 1994. Methane flux from mangrove sediments along the Southwestern coast of Puerto Rico. Estuaries 17, 140–147. https://doi.org/10.2307/1352563.

Stal, L.J., 2003. Microphytobenthos, their extracellular polymeric substances, and the morphogenesis of intertidal sediments. Geomicrobiol J. 20, 463–478. https://doi.org/10.1080/713851126.

Stal, L.J., 2010. Microphytobenthos as a biogeomorphological force in intertidal sediment stabilization. Ecol. Eng. 36, 236–245. https://doi.org/10.1016/j.ecoleng.2008.12.032.

Stark, J.M., Firestone, M.K., 1995. Mechanisms for soil moisture effects on activity of nitrifying bacteria. Appl. Environ. Microbiol. 61, 218–221.

Stieglitz, T.C., Clark, J.F., Hancock, G.J., 2013. The mangrove pump: the tidal flushing of animal burrows in a tropical mangrove forest determined from radionuclide budgets. Geochim. Cosmochim. Acta 102, 12–22. https://doi.org/10.1016/j.gca.2012.10.033.

Sun, Z., Jiang, H., Wang, L., Mou, X., Sun, W., 2013. Seasonal and spatial variations of methane emissions from coastal marshes in the northern Yellow River estuary, China. Plant and Soil 369, 317–333. https://doi.org/10.1007/s11104-012-1564-1.

Susilo, A., Ridd, P.V., Thomas, S., 2005. Comparison between tidally driven groundwater flow and flushing of animal burrows in tropical mangrove swamps. Wetl. Ecol. Manag. 13, 377–388. https://doi.org/10.1007/s11273-004-0164-0.

Sutton-Grier, A.E., Ariana, E., Keller, J.K., Koch, R., Gilmour, C., Megonigal, J.P., 2011. Electron donors and acceptors influence anaerobic soil organic matter mineralization in tidal marshes. Soil Biol. Biochem. 43, 1576–1583. https://doi.org/10.1016/j.soilbio.2011.04.008.

Taillardat, P., Friess, D.A., Lupascu, M., 2018a. Mangrove blue carbon strategies for climate change mitigation are most effective at the national scale. Biol. Lett. 14, 20180251. https://doi.org/10.1098/rsbl.2018.0251.

Taillardat, P., Ziegler, A.D., Friess, D., Widory, D., Truong, V.V., David, F., Nho, N.T., Marchand, C., 2018b. Surface and porewater export from a mangrove forest during contrasting tidal cycles and seasons: implications for tidal creek carbon dynamics. Geochim. Cosmochim. Acta 237, 32–48.

Tang, X., Pei, X., Lei, N., Luo, X., Liu, L., Shi, L., Chen, G., Liang, J., 2020. Global patterns of soil autotrophic respiration and its relation to climate, soil and vegetation characteristics. Geoderma 369, 114339.

Tolhurst, T.J., Chapman, M.G., Murphy, R.J., 2020. The effect of shading and nutrient addition on the microphytobenthos, macrofauna, and biogeochemical properties of intertidal flat sediments. Front. Mar. Sci. 7. https://doi.org/10.3389/fmars.2020.00419.

Tong, C., Zhang, L., Wang, W., Gauci, V., Marrs, R., Liu, B., Jia, R., Zeng, C., 2011. Contrasting nutrient stocks and litter decomposition in stands of native and invasive species in a sub-tropical estuarine marsh. Environ. Res. 111, 909–916. https://doi.org/10.1016/j.envres.2011.05.023.

Tong, C., Wang, W.-Q., Huang, J.-F., Gauci, V., Zhang, L.-H., Zeng, C.-S., 2012. Invasive alien plants increase CH4 emissions from a subtropical tidal estuarine wetland. Biogeochemistry 111, 677–693. https://doi.org/10.1007/s10533-012-9712-5.

Tong, C., Huang, J.F., Hu, Z.Q., Jin, Y.F., 2013. Diurnal variations of carbon dioxide, methane, and nitrous oxide vertical fluxes in a subtropical estuarine marsh on neap and spring tide days. Estuaries Coast 36, 633–642. https://doi.org/10.1007/s12237-013-9596-1.

Treat, C.C., Wollheim, W.M., Varner, R.K., Grandy, A.S., Talbot, J., Frolking, S., 2014. Temperature and peat type control CO2 and CH4 production in Alaskan permafrost peats. Glob. Chang. Biol. 20, 2674–2686. https://doi.org/10.1111/gcb.12572.

Troxler, T.G., Barr, J.G., Fuentes, J.D., Engel, V., Anderson, G., Sanchez, C., Lagomasino, D., Price, R., Davis, S.E., 2015. Component-specific dynamics of riverine mangrove CO2 efflux in the Florida coastal Everglades. Agric. For. Meteorol. 213, 273–282. https://doi.org/10.1016/j.agrformet.2014.12.012.

Van Raalte, C.D., Valiela, I., Teal, J.M., 1976. Production of epibenthic salt marsh algae: light and nutrient limitation1. Limnol. Oceanogr. 21, 862–872. https://doi.org/10.4319/lo.1976.21.6.0862.

Vinh, T.V., Allenbach, M., Aimée, J., Marchand, C., 2019. Seasonal variability of CO2 fluxes at different interfaces and vertical CO2 concentrations profiles in a Rhizophora mangrove stand (Can Gio, Viet Nam). Atmos. Environ. https://doi.org/10.1016/j.atmosenv.2018.12.049.

Vizza, C., West, W.E., Jones, S.E., Hart, J.A., Lamberti, G.A., 2017. Regulators of coastal wetland methane production and responses to simulated global change. Biogeosciences 14, 431–446. https://doi.org/10.5194/bg-14-431-2017.

Waldo, N.B., Hunt, B.K., Fadely, E.C., Moran, J.J., Neumann, R.B., 2019. Plant root exudates increase methane emissions through direct and indirect pathways. Biogeochemistry 145 (213), 234.

Wang, C., Lai, D.Y.F., Tong, C., Wang, W., Huang, J., Zeng, C., 2015. Variations in temperature sensitivity (Q10) of CH4 emission from a Subtropical Estuarine Marsh in Southeast China. PLoS One 10. https://doi.org/10.1371/journal.pone.0125227, e0125227.

Warner, D.L., Bond-Lamberty, B., Jian, J., Stell, E., Vargas, R., 2019. Spatial predictions and associated uncertainty of annual soil respiration at the global scale. Global Biogeochem. Cycles 33 (12), 1733–1745.

Watson, E., Oczkowski, A.J., Wigand, C., Hanson, A., Davey, E.W., Crosby, S., Johnson, R., Andrews, H., 2014. Nutrient enrichment and precipitation changes do not enhance resiliency of salt marshes to sea level rise in the Northeastern U.S. Clim. Change. https://doi.org/10.1007/s10584-014-1189-x.

Welti, N., Hayes, M., Lockington, D., 2017. Seasonal nitrous oxide and methane emissions across a subtropical estuarine salinity gradient. Biogeochemistry 132, 55–69. https://doi.org/10.1007/s10533-016-0287-4.

Whiting, G.J., Chanton, J.P., 1993. Primary production control of methane emission from wetlands. Nature 364, 794–795.

Wigand, C., Brennan, P., Stolt, M., Holt, M., Ryba, S., 2009. Soil respiration rates in coastal marshes subject to increasing watershed nitrogen loads in southern New England, USA. Wetlands 29, 952–963. https://doi.org/10.1672/08-147.1.

Xiao, K., Wilson, A.M., Li, H., Ryan, C., 2019. Crab burrows as preferential flow conduits for groundwater flow and transport in salt marshes: a modeling study. Adv. Water Resour. 132. https://doi.org/10.1016/j.advwatres.2019.103408, 103408.

Xin, P., Jin, G., Li, L., Barry, D.A., 2009. Effects of crab burrows on pore water flows in salt marshes. Adv. Water Resour. 32, 439–449. https://doi.org/10.1016/j.advwatres.2008.12.008.

Xu, C., Wong, V.N.L., Reef, R.E., 2020. Effect of inundation on greenhouse gas emissions from temperate coastal wetland soils with different vegetation types in southern Australia. Sci. Total Environ., 142949. https://doi.org/10.1016/j.scitotenv.2020.142949.

Yang, P., Wang, M.H., Lai, D.Y.F., Chun, K.P., Huang, J.F., Wan, S.A., Bastviken, D., Tong, C., 2019. Methane dynamics in an estuarine brackish Cyperus malaccensis marsh: production and porewater concentration in soils, and net emissions to the atmosphere over five years. Geoderma 337, 132–142. https://doi.org/10.1016/j.geoderma.2018.09.019.

Yu, L., Huang, Y., Zhang, W., Li, T., Sun, W., 2017. Methane uptake in global forest and grassland soils from 1981 to 2010. Sci. Total Environ. 607, 1163–1172.

Yu, X., Yang, X., Wu, Y., Peng, Y., Yang, T., Xiao, F., Zhong, Q., Xu, K., Shu, L.S., He, Q., Tian, Y., Yan, Q., Wang, C., Wu, B., He, Z., 2020. *Sonneratia apetala* introduction alters methane cycling microbial communities and increases methane emissions in mangrove ecosystems. Soil Biol. Biochem. 144, 107775.

Zedler, J.B., 1980. Algal mat productivity: comparisons in a salt marsh. Estuaries 3, 122–131. https://doi.org/10.2307/1351556.

Zhai, X., Piwpuan, N., Arias, C.A., Headley, T., Brix, H., 2013. Can root exudates from emergent wetland plants fuel denitrification in subsurface flow constructed wetland systems? Ecol. Eng. 61, 555–563.

Zheng, X., Guo, J., Song, W., Feng, J., Lin, G., 2018. Methane emission from mangrove wetland soils is marginal but can be stimulated significantly by anthropogenic activities. Forests 9, 738. https://doi.org/10.3390/f9120738.

Biosphere-atmosphere exchange of CO_2 and CH_4 in mangrove forests and salt marshes

Jiangong Liu[a], Karina V.R. Schäfer[b], and Derrick Y.F. Lai[a]

[a]*Department of Geography and Resource Management, The Chinese University of Hong Kong, Hong Kong Special Administrative Region, China,*
[b]*Earth and Environmental Science Department, Rutgers University Newark, Newark, NJ, United States*

4.1 Overview of the eddy covariance technique

4.1.1 Principles of the eddy covariance technique

Over the past three decades, the eddy covariance technique has become increasingly popular as a direct measurement of the biosphere-atmosphere exchange of CO_2, CH_4, water vapor (H_2O), and energy (Aubinet et al., 2012; Baldocchi, 2003, 2014, 2020). The theoretical framework of the eddy covariance technique is based upon Reynolds' rules of decomposition, in which the instantaneous value of a scalar such as vertical wind velocity (w) and mixing ratios (c) can be decomposed into the time-average and the time-fluctuating components as shown in Eqs. (4.1) and (4.2) (Reynolds, 1895). Physically, the average and fluctuating components refer to the mean flow and turbulence, respectively.

$$w = \overline{w} + w' \qquad (4.1)$$

$$c = \overline{c} + c' \qquad (4.2)$$

The overbars and primes indicate time averaging and fluctuations from the mean, respectively. Through applying Reynolds' averaging and decomposition, the mean flux density of a scalar (mass or energy) over a certain time span (typically 0.5–1 h in eddy covariance studies) is computed as the product of mean dry air density (ρ_a) and the covariance between w and c (Eq. 4.3):

$$F = \overline{\rho_a} \cdot \overline{w'c'} \qquad (4.3)$$

The equation assumes that turbulent flux dominates the vertical exchanges. In terms of CO_2 flux, $c = \rho_c/\rho_a$ where ρ_c is the CO_2 density and ρ_a is the air density. In the atmospheric science community, positive and negative F values conventionally denote a net gain and a net loss, respectively, of mass or energy by the atmosphere.

The eddy covariance concept for measuring the vertical fluxes of heat, mass, and momentum was proposed and developed in the mid-20th century by Montgomery (1948), Swinbank (1951), and Obukhov (1951). Field implementation of this method first occurred in the early 1970s through the measurement of net ecosystem exchange of CO_2 (i.e., NEE), using anemometers and infrared gas analyzers with a relatively slow sampling frequency of $\sim 2\,Hz$ (Desjardins, 1974; Desjardins and Lemon, 1974). Yet, Garratt (1975) argued that failure in capturing high-frequency turbulence signals could introduce a significant bias of approximately 40% in the measured fluxes. Constraints in instrumentation capacity had greatly limited the application of eddy covariance measurements until the development of commercially available, fast-response sonic anemometers and infrared gas analyzers in the 1990s. Wofsy et al. (1993) were among the first group of scientists publishing yearlong ecosystem-scale NEE based on eddy covariance measurements. Since then, various regional flux networks have emerged in North America (AmeriFlux), Europe (CarboEurope, ICOS), Asia (AsiaFlux), Africa (CarboAfrica), China (ChinaFlux), Korea (KoFlux), Canada (Fluxnet-Canada), Mexico (MexFlux), as well as Australia and New Zealand (OzFlux). Up to date, over 900 sites have been registered in FLUXNET, a global network of eddy covariance-based flux measurement sites (Baldocchi et al., 2001a; Chu et al., 2017). FLUXNET-CH_4, a new synthesis data set for ecosystem-scale CH_4 fluxes, was launched in 2017 with over 70 participant sites (Delwiche et al., 2021; Knox et al., 2019).

To the best of our knowledge, the eddy covariance measurement of ecosystem-scale carbon (C) fluxes in coastal wetlands (Fig. 4.1) was first carried out in the mangroves in Florida Everglades in the United States in 2004 (Barr et al., 2010). Since then, the eddy covariance flux towers have been installed in other coastal wetlands in the United States (Forbrich and Giblin, 2015; Kathilankal et al., 2008; Knox et al., 2018; Krauss et al., 2016; Miao et al., 2017; Moffett et al., 2010; Nahrawi et al., 2020; Schäfer et al., 2014, 2019; Starr et al., 2018), France (Polsenaere et al., 2012), India (Jha et al., 2014; Rodda et al., 2016; Gnanamoorthy et al., 2020), New Caledonia (Leopold et al., 2016), Australia (Negandhi et al., 2019; Safari et al., 2020), China (Chen et al., 2014; Cui et al., 2018; Guo et al., 2009; Han et al., 2015; Lee et al., 2015; Li et al., 2014; Liu and Lai, 2019; Yan et al., 2008; Zhong et al., 2016; Zhou et al., 2009), and Mexico (Alvarado-Barrientos et al., 2020).

4.1.2 Advantages of eddy covariance measurements

The eddy covariance method can cover a larger spectrum of time series ranging from seconds to decades and a larger footprint area than the closed chamber method in making flux measurements (Baldocchi et al., 2001b; Stoy et al., 2005). The high-frequency measurements of fluxes (i.e., half-hourly) and ancillary variables (usually 1 min) by eddy covariance enable data analysis over multiple temporal resolutions, which helps address multiscale processes in complex ecosystems like the coastal

FIG. 4.1

Eddy covariance instruments (e.g., CO_2/H_2O and CH_4 analyzers and sonic anemometers) in (A) a mangrove forest (MP-MPM in FLUXNET) and (B) a salt marsh (US-MRM in FLUXNET).

wetlands. Coastal wetlands are affected by periodic dynamics of spring-neap tides and water quality issues such as wastewater discharge. Therefore, the hydrologic regime in coastal wetlands can exhibit different spectral characteristics from that in freshwater wetlands. Wavelet analysis, a signal processing technique, has been used to decompose multiscale variations of C fluxes measured by eddy covariance and their biophysical controls in coastal wetlands. For example, Knox et al. (2018) employed a combination of wavelet analysis and information theory in examining the NEE in a salt marsh, and demonstrated that tides had substantial direct and indirect effects on NEE across diel and multiday scales. Li et al. (2018) studied the controls of ecosystem-scale CH_4 flux (F_{CH_4}) using partial wavelet coherence and found that the dominant link between F_{CH_4} and its drivers always varied at different time scales. On the other hand, the spatial coverage of the eddy covariance technique can span from a radius of a few hundred meters to several kilometers (Schmid, 1994), while the chamber technique can only cover a source area with a diameter of several meters. The spatial extent of the flux footprints is compatible with the pixel size of most remote-sensing products, enabling the application of eddy covariance data in the modeling studies based on remote sensing images (e.g., Jiang and Ryu, 2016). The spatially aggregated flux can help reduce the upscaling uncertainty in heterogeneous ecosystems (e.g., wetlands). Furthermore, eddy covariance measurements can simultaneously measure multiple scalars (e.g., CO_2, CH_4, and H_2O), which provide insights into numerous ecosystem characteristics such as water use efficiency and energy partitioning.

In contrast, the chamber method has several limitations. First, ecophysiologists often measure CO_2 exchange at the leaf surface with cuvettes, and then upscale it to the canopy or ecosystem level with considerable uncertainty (Sprintsin et al., 2012). Given that leaf-level fluxes are influenced by a suite of factors such as canopy structural complexity, light conditions, and leaf angles relative to the sun (Norman, 1979; Rayment et al., 2000), an accurate flux upscaling will require at least a good characterization of the canopy and leaf conditions. Second, the deployment of closed chambers can alter the microclimate at the biosphere-atmosphere interface, leading to artifacts of local pressure, heat, humidity, and gas concentration gradient (Davidson et al., 2002; Welles et al., 2001). Third, chamber measurements are labor-intensive and hence challenging to cover a wide range of temporal (e.g., diurnal, seasonal) scales over a large number of sites, which are often required for ensuring representativeness in the spatially heterogeneous and hydrologically dynamic coastal wetlands.

In a first attempt to estimate the global CH_4 emissions from mangrove forests, Rosentreter et al. (2018) upscaled the CH_4 fluxes across the sediment-atmosphere and water-atmosphere interfaces measured by closed chambers and suggested that CH_4 emissions could potentially offset 20% of the C burial rates in mangroves. However, there were some critical uncertainties associated with their chamber measurements, which might reduce the accuracy of the upscaled estimates. First, the authors assumed a 1:1 ratio of inundation and air-exposure periods for all the mangrove wetlands worldwide due to a lack of observational data on tidal heights. However, most mangrove species are experiencing shorter inundation periods than this hypothetical scenario. For example, *Kandelia obovata*, a mangrove species that is commonly found in Asia, typically thrives with an inundation frequency of 0.4%–40% (Yang et al., 2013). Second, our knowledge regarding the biogeochemical processes of CH_4 dynamics in coastal wetlands is still nascent (Bridgham et al., 2013). Emerging evidence shows that soil and tidal water are not the only CH_4 sources in mangroves. For example, a recent finding suggested that CH_4 emissions from mangrove tree stems accounted for about 26% of the total F_{CH_4} (Jeffrey et al., 2019). Additionally, some field campaigns of chamber measurements concentrated in a specific season (e.g., Chen et al., 2010). Thus, they failed to characterize the seasonal F_{CH_4} variation, which can bias the upscaled estimates on an annual scale.

4.1.3 Disadvantages of eddy covariance measurements

Some implicit assumptions are made during the application of eddy covariance technique, which are the key sources of uncertainties. For example, an eddy covariance site requires a horizontally homogeneous canopy on flat terrain to nullify the advective transport, which is physically impossible. For nonideal sites like a forest on a steep mountain slope, advective fluxes have to be properly evaluated, which is still technologically challenging to be well presented in C budgets partly due to the low sensor accuracy (Dellwik et al., 2010a,b; Etzold et al., 2010; Moderow et al., 2011). Furthermore, the application of eddy covariance measurements in heterogeneous coastal wetlands necessitates a careful evaluation of the flux footprint to verify if the flux originates from the area of interest. Flux footprints refer to the source area of the flux detected by eddy covariance instruments (Hsieh et al., 2000). Flux footprints can be assessed using footprint models with instrument height, canopy height, canopy roughness dynamics, and turbulent conditions as model inputs (Arriga et al., 2017; Rey-Sanchez et al., 2018). Footprint analysis has been a fundamental tool for interpreting eddy covariance results in wetlands because NEE and F_{CH_4} can be physically linked to the spatially heterogeneous sources and sinks of CO_2 and CH_4 (Morin, 2019). The sampling of eddy covariance measurements in space poses a challenge of whether the available fetches in each direction are sufficiently long. Most eddy covariance studies in mangrove wetlands have carefully examined the footprint and available fetches in all directions, and discarded the fluxes that have originated beyond the fetch boundaries (Barr et al., 2010; Leopold et al., 2016; Liu and Lai, 2019), which can help detect spurious fluxes (Baldocchi et al., 2012). Moreover, some coastal wetlands are logistically challenging for access and power supply. The humid environment in coastal wetlands also poses a challenge for the maintenance of open-path gas analyzers. In addition, obtaining a high-quality signal of F_{CH_4} is challenging in saline wetlands due to a low signal-to-noise ratio.

Other inherent sources of errors in eddy covariance measurements can potentially result from violations of theoretical assumptions, nonsteady-state conditions or instrument limitations. These include spikes in raw time series (Vickers and Mahrt, 1997), anemometer tilt on uneven terrain (Wilczak et al., 2001), signal time lags induced by physical separation among analyzers (Fan et al., 1990), high-frequency and low-frequency signal loss due to relatively low response analyzers and averaging schemes (Moncrieff et al., 1997, 2006), air density fluctuations induced by temperature and humidity changes (Chamberlain et al., 2017; Webb et al., 1980), and low turbulent conditions. Most of these potential errors can be appropriately corrected or addressed using the methodologies introduced in the references therein. For nighttime measurements, the low shear stress under nonturbulent conditions can lead to artificial results (Papale and Valentini, 2003). A friction velocity (u_*) threshold, determined by the relationship between nighttime fluxes (e.g., CO_2) and u^*, has been widely used as a criterion to eliminate the fluxes obtained during the atmospherically stable and low turbulence periods. The procedure is generally known as "u_* threshold filtering" or "u_* correction" (Papale et al., 2006). Yet, it is under debate whether such u_* filtering approach should be applied to F_{CH_4} (Gu et al., 2005). More details about the abovementioned error sources can be found in Aubinet et al. (2012).

For quantifying annual C fluxes, a set of algorithms have been developed to fill the gaps in 30-min NEE and F_{CH_4} (Kim et al., 2020; Wutzler et al., 2018). Eddy covariance NEE can be partitioned into two main components, namely gross ecosystem productivity (GPP) and ecosystem respiration (Re) (Reichstein et al., 2005). Flux communities developed two general approaches to achieve partitioning. For the estimation of Re, one approach relies on the relationship between nighttime NEE and temperature, while an alternative one is based on light-response curves (Falge et al., 2002). A refinement of the first approach has been developed to account for water table fluctuations in wetlands (Knox et al., 2018). Details about the uncertainties related to gap-filling and NEE partitioning algorithms can be found in Papale et al. (2006) and Keenan et al. (2019).

The abovementioned limitations exist commonly in most eddy covariance studies. However, their impacts could vary depending on the climatic zones, ecosystem types, local micrometeorology, the degree of landscape heterogeneity, instrument settings (e.g., open-path or closed-path), and data correction and processing methods. On the other hand, the conventional cuvette or chamber method has been widely used in the measurement of greenhouse gas (GHG) fluxes in wetlands, owing to its relatively high feasibility, simple principles, low cost of operation, and strong ability to examine ecophysiological processes over a small, well-defined area (e.g., the interfaces of air and soil surfaces and air and water bodies) within a specific time frame. In this regard, the cuvette or chamber technique facilitates the replicated measurement of GHG fluxes over small plots in the field and the examination of their biological, biogeochemical, and climatic controls.

4.2 Temporal variability of ecosystem-scale CO_2 fluxes and its biophysical controls in coastal wetlands

Coastal wetlands are of disproportionate importance to global C cycling. Tidal marshes, seagrass meadows, and mangroves account for about 4.8%, 1.5%, and 1.2% of the total wetland area of the world, representing a relatively small areal extent (Ramsar Convention on Wetlands, 2018). Yet, they sustain the highest C sequestration rate among all vegetated ecosystems owing to their high C assimilation rate and long-term preservation of organic C in sediments (Rogers et al., 2019). For example, the C sink strength of coastal wetlands is around two times that of inland wetlands, according to a synthesis of 43 wetlands (Lu et al., 2017). The C burial rates in salt marshes and mangroves are about 47 times higher than those in terrestrial forests (McLeod et al., 2011). The C storage in vegetated coastal wetlands, known as "blue carbon" (Herr et al., 2012; Howard et al., 2014), is about 2–4 times that of tropical forests (Donato et al., 2011; McLeod et al., 2011). Policymakers and scientific communities have increasingly recognized the significant role of blue carbon ecosystems (i.e., mangroves, tidal marshes and seagrass meadows) in mitigating climate change. Meanwhile, we are losing these vital C sinks at a rate of 1%–2% in total area per year primarily due to anthropogenic disturbances and climate change (Duke et al., 2007; Waycott et al., 2009). For example, deforestation had led to a loss of about 35% of the world's mangrove area between the 1980s and 1990s (Friess et al., 2019). Coastal wetlands are now being priorizied for conservation under several recent international initiatives. For example, the Global Mangrove Alliance plans to increase the global mangrove area by 20% by 2030 (Friess et al., 2020).

4.2.1 Magnitude and temporal variability of ecosystem-scale CO_2 fluxes in mangroves

Eddy covariance measurements of ecosystem-scale CO_2 fluxes have been conducted in several mangrove sites. The existing sites are located in Florida, United States (Barr et al., 2010), southern China (Chen et al., 2014; Cui et al., 2018; Liu and Lai, 2019), India (Rodda et al., 2016; Gnanamoorthy et al., 2020), Mexico (Alvarado-Barrientos et al., 2020), and New Caledonia (Leopold et al., 2016), spanning a latitudinal gradient from 20°N to 30°N (Table 4.1). To better represent the mangroves geospatially, more flux towers are needed in the regions with extended mangrove coverages like southeastern Asia and South America. Several limitations have hindered a wider application of eddy covariance measurements in mangroves, such as logistical challanges, harsh environments, heterogeneous landscapes, and tropical cyclones.

Clear diel patterns of NEE, GPP, and Re have been observed for most of the eddy covariance studies in mangroves (Barr et al., 2013a; Liu and Lai, 2019; Rodda et al., 2016). Similar to other terrestrial vegetated ecosystems, minimum NEE (i.e., maximum C sequestration rate) and maximum GPP occur at noon over a 24-h cycle. The minimum NEE can reach $-20\,\mu mol\,CO_2\,m^{-2}\,s^{-1}$, which is comparable to the tropical rainforests in the Amazon (Hayek et al., 2018). Since the majority of tidal mangrove wetlands are distributed across tropical and subtropical coastal regions where the high temperature and radiation can lead to substantial evapotranspiration, mangrove GPP could be reduced by the frequent cloudy conditions and convective thunderstorms, especially during the summertime (Barr et al., 2010).

For the mangrove wetlands where seasonal variations of NEE were reported, they displayed a stronger C sink during the dry than the wet seasons (Liu and Lai, 2019). Traditionally, terrestrial ecosystems are assumed to sequestrate more C during the wet season due to higher availability of energy, light and water. However, the opposite seasonal trend has been found in several evergreen forests (especially in the Amazon forests) (Bonal et al., 2008; Goulden et al., 2004; Saleska et al., 2003; Zhang et al., 2010). Two primary mechanisms would account for such seasonality in the Amazon forests. First, high soil moisture in the wet seasons can stimulate heterotrophic respiration, which then dominates the ecosystem C dynamics (Saleska et al., 2003). Second, plant green-up together with water supply in the dry seasons can maintain high photosynthesis (Huete et al., 2006). Although mangroves are evergreen species that can maintain yearlong productivity, vegetation indices derived from remote sensing images, e.g., the Enhanced Vegetation Index (EVI) and the Normalized Difference Vegetation Index (NDVI), showed greater values during the dry seasons (Pastor-Guzman et al., 2018). Vegetation indices are important indicators of plant activities that are closely linked to the GPP of mangrove wetlands (Barr et al., 2013a). Liu and Lai (2019) demonstrated that the increase in Re was much greater than the increase in GPP in the wet season in a subtropical mangrove wetland because of the combined effect of temperature, tidal salinity, and plant phenology, which subsequently resulted in the reduction of net C uptake.

Table 4.1 Eddy covariance studies of ecosystem-scale CO_2 fluxes in mangrove wetlands where annual CO_2 fluxes have been reported.

Site	Lat. (°)	Lon. (°)	Year	MAT (°C)	MAP (mm)	NEE ($gCm^{-2}\,year^{-1}$)	GPP ($gCm^{-2}\,year^{-1}$)	R_{eco} ($gCm^{-2}\,year^{-1}$)	Reference
Gaoqiao, Zhanjiang National Mangrove Nature Reserve	21.57	109.76	2010–12	22.9	1770	−722 [−738, −692]	1764 [1698, 1890]	1096 [1027, 1215]	Chen et al. (2014)
Yunxiao, Zhangjiangkou National Mangrove Nature Reserve	23.92	117.42	2009–12	21.1	1285	−684 [−857, −540]	1871 [1763, 1919]	1287 [1238, 1337]	Chen et al. (2014)
Leizhou	20.93	110.17	2015–17	24.5	1619	−1105	~2000	~895	Cui et al. (2018)
Mai Po Nature Reserve	22.50	114.03	2016–17	23.3	2399	−824 [−890, −758]	2784 [2741, 2827]	1960 [1937, 1983]	Liu and Lai (2019)
Everglades National Park, Florida	25.36	−81.08	2004–10	22.5	1410	−981 [−1176, −806]	~2290	~1667	Barr et al. (2012)
Sundarbans	21.82	88.62	2012–13	~26	~1800	−249	1271	1022	Rodda et al. (2016)
Pichavaram	11.33	79.92	2017–18	28.2	653	−183	1466	1283	Gnanamoorthy et al. (2020)
Coeur de Voh	−20.94	163.66	2014–15	23.5	800	−74	~975	~903	Leopold et al. (2016)
Puerto Morelos	20.85	−86.90	2017–18	26.2	1222	−709	2473	1764	Alvarado-Barrientos et al. (2020)

Annual average and range (in brackets) are presented. GPP, gross ecosystem productivity; MAT, mean annual temperature; MAP, mean annual precipitation; NEE, net ecosystem CO_2 exchange; R_{eco}, ecosystem respiration. Positive and negative signs of NEE denote the net uptake and release of CO_2 by the ecosystem, respectively.

Annual NEE in seven mangrove wetlands ranged between −1176 and − 74 g C m^{-2} year^{-1}, with a semiarid mangrove in New Caledonia and a subtropical mangrove in Florida being the weakest and strongest C sinks, respectively (Table 4.1). Based on the analysis of over 1700 site years of data, Baldocchi et al. (2018) reported an average annual NEE of − 153 g C m^{-2} year^{-1} from global ecosystems, which was lower than that in all the mangrove sites except the semiarid mangrove in New Caledonia. Furthermore, the mangroves in Florida and Hong Kong were among the strongest C sinks with a net ecosystem productivity (NEP, equivalent to -NEE) larger than 97% of the flux sites in the FLUXNET2015 data set. The global mean annual GPP among the FLUXNET2015 sites was 1294 g C m^{-2} year^{-1} (Baldocchi et al., 2018), while the annual GPP in the Amazon forests could generally exceed 3000 g C m^{-2} year^{-1} (Zeri et al., 2014). Annual GPP in the mangrove wetlands in Southern China and Florida was greater than that across all forest biomes except for tropical evergreen forests with an average GPP of around 3551 g C m^{-2} year^{-1} (Luyssaert et al., 2007) (Table 4.1). The ratio of Re to GPP across most of the forest sites was close to 1, yet it could drop to 0.76 for the temperate humid forests and Mediterranean warm evergreen forests (Luyssaert et al., 2007). The Re to GPP ratio was around 0.74 in a subtropical mangrove wetland, which suggested that the possible influence of frequent tidal inundation in inhibiting Re (Liu and Lai, 2019).

The interannual variability of ecosystem-scale CO$_2$ fluxes in mangrove wetlands is still uncertain due to a lack of long-term measurements (Ray et al., 2013). The Gaoqiao mangrove in southern China displayed only small interannual variations with a standard deviation of around 26 g C m^{-2} year^{-1} based on seven site years of data (Chen et al., 2014; Cui et al., 2018). On the other hand, the disturbance generated by a hurricane dominated the interannual variation in NEE for a mangrove site in the Florida Everglades (Barr et al., 2012). At this site, a relatively large interannual variation was observed with a standard deviation of around 103 g C m^{-2} year^{-1} based on five years of data. The interannual variability of NEE is small in mangroves when compared to that in other ecosystems with a standard deviation of 162 g C m^{-2} year^{-1} based on 59 time-series (Baldocchi et al. 2018), owing to the high C sequestration rate and small environmental variations of mangroves in the tropical and subtropical regions. This indicates that mangrove wetlands can act as a strong C sink over the long term.

4.2.2 Magnitude and temporal variability of ecosystem-scale CO$_2$ fluxes in tidal marshes

Currently, eddy covariance measurements in tidal marshes have been done in East Asia, the United States, Australia, and Europe. The magnitude and temporal variations of ecosystem-scale CO$_2$ fluxes have been examined for a variety of tidal marshes under different climatic zones, hydrological conditions, dominant plant species, management strategies, and anthropogenic disturbances (Table 4.2).

Table 4.2 Eddy covariance studies of ecosystem-scale CO_2 fluxes in tidal marshes where annual CO_2 fluxes have been reported.

Site	Lat.	Lon.	Year	MAT (°C)	mm	NEE (g C m^{-2} year^{-1})	GPP (g C m^{-2} year^{-1})	R_{eco} (g C m^{-2} year^{-1})	Reference
Salt marsh, Virginia Coastal Reserve Long Term Ecological Research	37.40	−75.83	2007[a]	NA	537[a]	−130[a]	NA	NA	Kathilankal et al. (2008)
Freshwater tidal wetland, Panjin Wetland Ecosystem Research Station	41.13	121.09	2005	8.6	631	−65	1300	1235	Zhou et al. (2009)
Supratidal marsh, The Yellow River Delta	37.76	118.99	2010–13	12.9	560	−211 [−247, −164]	810 [653, 1004]	598 [430, 757]	Han et al. (2015)
Restored salt marsh, The Secaucus High School Wetlands Enhancement Site	40.81	−74.05	2011–12	~15	~115	−213	979	766	Artigas et al. (2015)
Para grass, Guandu	25.12	121.47	2011–14	23.0	2405	−53	1786	1732	Lee et al. (2015)
Reed, Guandu	25.12	121.47	2013–14	23.0	2405	−376	1728	1351	Lee et al. (2015)
Salt marsh, Plum Island Sound estuary in northeastern Massachusetts	42.74	−70.83	2013–17	8.7	1269	−179 [−233, −104]	816 [737, 900]	580 [532, 643]	Forbrich et al. (2018)
Reclaimed marsh, Chongming Island in the Yangtze estuary	31.63	121.97	2012	15.3	1004	−558	1298	739	Zhong et al. (2016)
Brackish marsh, Louisiana	29.50	−90.44	2012	20.8	1580	171	264	434	Krauss et al. (2016)
Freshwater marsh, Louisiana	29.86	−90.29	2012–13	20.8	1580	−337	1230	893	Krauss et al. (2016)
Brackish Marsh, San Francisco Bay National Estuarine Research Reserve	38.20	−122.02	2016	15.1	326	−225	NA	NA	Knox et al. (2018)

Site	Latitude	Longitude	Years	MAT	MAP	NEE	R_{eco}	GPP	Reference
Restored brackish marsh with organic matter amended, Marsh Resource Mitigation Bank	40.48	−74.04	2011–13	13.6	635	485 [455, 531]	1312 [1255, 1349]	1798 [1725, 1880]	Schäfer et al. (2019)
Restored brackish marsh without organic matter amended, Secaucus High School	40.48	74.02	2011–12	13.6	635	6 [−548, 540]	1417	1423 [998, 1798]	Schäfer et al. (2019)
Natural brackish marsh, Hawk Property	40.46	74.05	2012–13	13.6	635	73 [−382, 474]	673	746 [433, 1039]	Schäfer et al. (2019)
Salt marsh, St. Jones Reserve, Delaware	39.09	75.44	2015–17	14.0	576	138 [13,201]	NA	NA	Vázquez-Lule and Vargas (2021)

Annual average and range (in brackets) are presented. GPP, gross ecosystem productivity; MAT, mean annual temperature; MAP, mean annual precipitation; NA, not available; NEE, net ecosystem CO_2 exchange; R_{eco}, ecosystem respiration.
[a]Only eddy covariance measurements during the growing season (from May to October) were available.

Unlike the tropical and subtropical mangrove wetlands that can maintain yearlong productivity, tidal marshes located in the temperate zones have distinct growing seasons. Overall, the minimum diel NEE ranged between -30 and $-10 \,\mu mol \,CO_2 \,m^{-2} \,s^{-1}$ during the growing season (Duman and Schäfer, 2018; Guo et al., 2009; Han et al., 2015; Knox et al., 2018; Lee et al., 2015; Polsenaere et al., 2012; Zhong et al., 2016), which was comparable to that reported for mangroves. Tidal marshes act as a minor C source during the nongrowing season. The magnitude of diel NEE was commonly lower than $5 \,\mu mol \,CO_2 \,m^{-2} \,s^{-1}$ due to low temperature and frequent tidal flooding (Guo et al., 2009; Schäfer et al., 2019). For the seasonal variability in NEE, all temperate tidal marsh sites exhibited similar patterns. The length of the period during which the marshes act as C sinks depends on the local climate, hydrology, and vegetation types, ranging between 5 (Schäfer et al., 2019) and 9 months (Zhong et al., 2016). The duration of C uptake periods would largely determine the final annual C flux.

Large variability in annual C fluxes has been shown in tidal marshes (Table 4.2). Generally, annual NEE in tidal marshes ranged between -300 and $100 \,g \,C \,m^{-2}$ $year^{-1}$. Zhong et al. (2016) showed that a reclaimed coastal marsh in Eastern China was a very strong C sink that sequestrated $\sim 558 \,g \,C \,m^{-2} \,year^{-1}$. The monthly C sequestration rate could reach up to $100 \,g \,C \,m^{-2} \,month^{-1}$. On the other hand, a brackish marsh in Louisiana (Krauss et al., 2016) and an urban restored marsh in New Jersey (Schäfer et al., 2019) were C sources on an annual scale. The soil in both sites was rich in organic matter, resulting in a high C loss rate. In addition, wetland degradation (Krauss et al., 2016), management history (e.g., fill materials) (Schäfer et al., 2019), and the input of wastewater from urban areas (Duman and Schäfer, 2018) also contributed to the high C emissions.

Five years of eddy covariance fluxes measured in a salt marsh in northeastern Massachusetts indicated a relatively small interannual variation (standard deviation was $32 \,g \,C \,m^{-2} \,year^{-1}$) compared to the mean annual NEE ($-179 \,g \,C \,m^{-2} \,year^{-1}$) (Forbrich et al., 2018). The interannual variation was primarily attributable to the number of most active days during a year and early season rainfall.

4.2.3 Biophysical controls of ecosystem-scale CO_2 fluxes in coastal wetlands

Since the widespread application of eddy covariance measurements, the responses of NEE, GPP, and Re to environmental and phenological factors have been extensively discussed and reviewed over different types of ecosystems (Baldocchi, 2020; Biederman et al., 2016; Law et al., 2002; Luyssaert et al., 2007; Wright et al., 2004). However, most of the syntheses have focused on terrestrial ecosystems only, which is partly due to the small body of literature regarding C exchange in coastal wetlands. For example, although the FLUXNET2015 data set contains 212 flux sites worldwide, coastal wetland sites have not been included in the project (Pastorello et al., 2017). Linear statistical models, e.g., Pearson correlation (Krauss et al., 2016), multiple linear regression (Zhong et al., 2016), and path analysis

(Lee et al., 2015; Liu and Lai, 2019), as well as nonparametric models, e.g., mutual information (Knox et al., 2018) and Granger causality (Schäfer et al., 2014), have been used to quantify the relative importance of different biophysical controls over C fluxes in coastal wetlands. In this session, the biophysical drivers of ecosystem-scale CO_2 fluxes in coastal wetlands will be reviewed.

4.2.3.1 Light

Light is the most important factor controlling the biophysical processes in the terrestrial biosphere (Ryu et al., 2011). It has also been identified as the dominant driver of NEE in both mangroves (Liu and Lai, 2019) and salt marshes (Lee et al., 2015). Photosynthetic rates increase linearly with the absorbed light under low light, and then saturate when light reaches the saturation point (Baldocchi and Harley, 1995; Gilmanov et al., 2010). Light compensation points and maximum photosynthetic rates vary with leaf area index (LAI), photosynthetic capacity, and soil moisture deficit in different ecosystems (Leuning et al., 1995; Xu and Baldocchi, 2004). Cui et al. (2018) compared the light response functions between mangroves and upland forests and found that mangrove wetlands have relatively lower compensation points but higher maximum photosynthetic rates than terrestrial forests.

Plants absorb both the direct short-wave radiation and the diffuse short-wave radiation redistributed by clouds, aerosols, and smoke. Diffuse radiation is more efficient in driving canopy photosynthesis than direct radiation because it can better penetrate canopies and thus activate nonsaturated shade leaves (Gu et al., 2002; Hemes et al., 2020; Oliphant and Stoy, 2018). The light environment is altered in urban areas, and two-thirds of megacities with a population of larger than 5 million are located in the subtropical and temperate coastal zones. Therefore, the increased loadings of aerosols in the atmosphere that originate from the urban and marine areas could possibly enhance the diffuse fraction of light and hence strengthen C assimilation in the urban coastal wetlands, especially salt marshes. For tropical and subtropical mangroves, cloud formation arising from high evapotranspiration also facilitates the transmission of diffuse light. Greater light use efficiency (LUE) and GPP were found under the influence of diffuse radiation as compared to direct light in subtropical mangroves (Barr et al., 2010, 2013a; Liu and Lai, 2019). Hemes et al. (2020) showed that wildfire-induced smoke significantly enhanced diffuse radiation but only diminished the total radiation slightly in a set of wetlands in California, with the enhanced diffuse fraction leading to a 1%–4% increase in GPP.

4.2.3.2 Temperature

Temperature can govern ecosystem-scale C fluxes by altering a suite of metabolic processes, such as plant photosynthesis, autotrophic respiration, and heterotrophic respiration, which have different temperature sensitivities (Baldocchi, 2020). Furthermore, temperature has a significant effect on plant phenology. For example, warming tends to increase the length of the growing seasons and thus enhance the cumulative C uptake (Cleland et al., 2007; Menzel et al., 2006). Given that temperature is closely related to other physical variables such as solar radiation, vapor pressure deficit (VPD), and u_*, it is challenging to disentangle the effect of temperature from other confounding factors on C fluxes.

For temperate tidal marshes, warming will likely enhance C assimilation because the activity of photosynthetic enzymes will increase with temperature until an optimal temperature of around 20°C to 30°C is reached (Bernacchi et al., 2001; Berry and Bjorkman, 1980). There has been increasing evidence that the optimal temperature is not static and exhibits acclimation to global warming (Smith and Dukes, 2017). For warm and hot regions where the mangrove wetlands are, warming is likely to affect C sequestration negatively, owing to enhanced stomatal closure and enzyme denaturation that will reduce the photosynthetic rate (Fu et al., 2018). In addition, Re increases exponentially with warming, which will further contribute to a reduction in C sequestration rates in the coastal wetlands.

The temperature sensitivity (Q_{10}) of Re in coastal wetlands generally exceeds two (Cui et al., 2018; Han et al., 2015; Zhou et al., 2009), which is relatively higher than that in the terrestrial forests (Curiel Yuste et al., 2004). Therefore, global warming will likely exert a greater impact on C loss for coastal wetlands than terrestrial forests. Temperature sensitivity of Re is found to increase in response to flooding in a supratidal marsh (Han et al., 2015). Therefore, the complex interaction between flooding regime (e.g., frequency, depth, etc.) and temperature would add an additional layer of uncertainty to the overall response of wetland NEE to increasing temperature under climate change.

4.2.3.3 Water availability and atmospheric moisture

Water availability is the key driver of GPP and Re in terrestrial ecosystems (Jung et al., 2017). Synthesis studies have demonstrated that 60%–70% of the variation in GPP can be explained by annual mean temperature and annual rainfall across terrestrial ecosystems (Law et al., 2002; Luyssaert et al., 2007). Soil moisture and VPD impact the biological processes associated with GPP and Re. Zhong et al. (2016) compared the light response of NEE in a reclaimed coastal marsh under different soil moisture conditions and found that inundation would cause stomatal closure and reduce the C fixation rate. Zhou et al. (2009) found a positive correlation between Re and soil moisture in a tidal freshwater wetland. Mangroves always exhibit a higher photosynthetic rate under low than high VPD (Barr et al., 2010; Leopold et al., 2016; Liu and Lai, 2019). However, water availability does not appear to be a limiting factor for plant productivity in most coastal wetlands, except in some semiarid sites, due to frequent tidal inundation (Leopold et al., 2016). Overall, soil moisture and VPD play a minor role in determining NEE when compared to other biophysical factors in coastal wetlands (Krauss et al., 2016; Lee et al., 2015; Liu and Lai, 2019; Zhong et al., 2016).

Rainfall has been suggested as a factor governing annual C exchange in mangroves. Leopold et al. (2016) and Gnanamoorthy et al. (2020) found a linear correlation between annual NEP and annual rainfall in three mangrove sites, but this relationship vanished for the sites in the humid regions. It is interesting to note that the opposite trend was found on a monthly scale (Leopold et al., 2016; Liu and Lai,

2019). Rainfall could cause a positive effect on mangrove GPP on an annual scale due to two possible reasons. First, heavy rainfall corresponds to high evapotranspiration and frequent cloud formation, resulting in more diffuse radiation that supports canopy photosynthesis. Second, heavy rainfall events and tidal flooding are supposed to flush out toxic metabolites and dilute the water salinity, which would be beneficial for vegetation growth (Han et al., 2015; Kathilankal et al., 2008). Forbrich et al. (2018) identified rainfall in the early growing season as the best predictor of the interannual variation in NEE for a salt marsh in Massachusetts, United States. Moreover, rainfall timing and intensity may determine the length of the growing seasons and thus influence the accumulative annual C uptake in tidal marshes (Noe and Zedler, 2001; Zhang et al., 2020). However, annual rainfall does not seem to be a limiting factor of annual NEP in tidal wetlands across the world (Table 4.2). Thus, timing more than the amount of rainfall are determinants of NEP in wetlands (Forbrich et al., 2018).

4.2.3.4 Water table depth

The effect of water table depth (WTD) on C exchange in coastal wetlands has been widely examined. Previous studies have found that high tide events can inhibit Re by creating anoxic conditions in soil and thus reducing soil respiration in both tidal marshes and mangroves (Guo et al., 2009; Kathilankal et al., 2008; Li et al., 2014; Moffett et al., 2010; Wei et al., 2020). The effect of WTD on ecosystem-scale photosynthesis is different between coastal wetlands with short and tall canopies (Liu and Lai, 2019). For low-elevation salt marshes with short canopies and mangroves dominated by short species like *Acanthus ilicifolius*, high tide events could impede C uptake by reducing the photosynthetically active leaf area (Guo et al., 2009; Kathilankal et al., 2008; Knox et al., 2018; Moffett et al., 2010). Nahrawi et al. (2020) showed that high tide events reduced the net C uptake in a Georgia salt marsh by up to 60%. For tall mangroves, high tide events could exert a positive or neutral effect on ecosystem-scale photosynthesis (Li et al., 2014; Liu and Lai, 2019). The increase in net CO_2 uptake during high tide events can be attributed to the reduced heterotrophic respiration. Furthermore, tidal activities can dynamically alter species composition in coastal wetlands, and influence the C exchange rate in the long term. For example, Wang et al. (2006) found that inundation could strengthen the competitive advantage of *Spartina alterniflora* over *Phragmites australis* in salt marshes through enhanced productivity.

While numerous studies have reported a significant difference in both C fluxes and their response to light and temperature between flooded and nonflooded conditions (Guo et al., 2009; Kathilankal et al., 2008; Knox et al., 2018; Wei et al., 2020), tidal height is typically a minor predictor of C fluxes in coastal wetlands based on the results of traditional parameteric analyses (Krauss et al., 2016; Liu and Lai, 2019; Schäfer et al., 2014). Nonparametric approaches such as information theory and Granger causality when combined with signal decomposition tools (e.g., Fourier transformation and wavelet decomposition) have shown great capability in revealing the potential linkages between C fluxes and WTD at different time scales (Hatala et al., 2012; Schäfer et al., 2014; Sturtevant et al., 2016). For example, Knox et al. (2018) used a combination of wavelet analysis and mutual information analysis and showed that the strong link between NEE and spring-neap cycles in a brackish tidal marsh could only be found at the multiday scale through asynchronous processes.

4.2.3.5 Salinity

Experiments conducted at both the leaf and whole-plant levels demonstrated a consistent negative impact of high salinity on the photosynthetic rate of coastal plants through a reduction in stomatal conductance, transpiration, and leaf surface expansion (López-Hoffman et al., 2007; Maricle et al., 2007; Nguyen et al., 2015; Redondo-Gómez et al., 2007). At the ecosystem scale, a salinity larger than 15 ppt has been shown to reduce the maximum photosynthetic rate in a reclaimed tidal marsh (Zhong et al., 2016) and the LUE of subtropical mangroves (Barr et al., 2013b; Cui et al., 2018). On the other hand, salinity has no significant effect on LUE in a salt marsh (Knox et al., 2018). Liu and Lai (2019) found that the ecosystem-scale LUE for subtropical estuarine *K. obovata* mangrove increased with salinity over the range of 1–17 ppt, which could be attributed to the theoretical growth efficiency curve of halophytes. The photosynthetic rate of halophytes is expected to increase with salinity under low levels until an optimal salinity is reached (Krauss and Ball, 2013). The positive link between LUE and salinity observed in this mangrove is consistent with the finding that the optimal range of salinity for the growth of *K. obovata* is around 5–15 ppt (Hwang and Chen, 2001; Li et al., 2008). This can be explained by the ion requirement and water uptake capacity of coastal vegetation (Wang et al., 2011), as tidal water can supply chloride (Cl^-) and sodium (Na^+) ions that are essential in regulating the activities of photosystem II and osmosis in leaves (Maggio et al., 2000; Roose et al., 2016).

4.2.3.6 Disturbances

Different types of disturbances (e.g., fire, storms, insect outbreaks, etc.) tend to decrease the productivity of ecosystems (Amiro et al., 2010). Coastal wetlands in the tropical and subtropical regions are subject to tropical cyclones, which may cause defoliation, tree mortality, and intensive rainfall. Barr et al. (2012) found that in the Florida mangroves, Hurricane Wilma had a profound negative impact on GPP due to the reduced canopy LAI and a positive impact on Re due to the decomposition of litterfall and the increased soil temperature. Chen et al. (2014) suggested that the impacts of typhoons on GPP and Re in the mangroves in southern China were determined by the timing and strength of typhoons and the rainfall amount. Insect outbreak is another form of disturbance that can reduce canopy LAI and thus GPP. Lu et al. (2019) showed that the productivity of mangroves can reduce markedly during bud moth larvae outbreaks. A projected increase in tropical storms and hurricane activities under climate change will exacerbate the disturbances encountered in the coastal ecosystems and thus potentially diminish GPP.

4.3 Temporal variability of ecosystem-scale CH$_4$ fluxes and its biophysical controls in coastal wetlands

Wetlands are the largest natural source of CH$_4$ globally (Kirschke et al., 2013), and are responsible for the interannual variation of recent atmospheric CH$_4$ concentrations (Fletcher and Schaefer, 2019). However, the quantification of ecosystem-scale CH$_4$ fluxes in wetlands is difficult due to their high variability in time and space. With the advantage of continuous and high-frequency sampling and great spatial representativeness, the eddy covariance technique has become a powerful tool for scientific communities to investigate temporal variability of F_{CH_4} and its biophysical drivers (Knox et al., 2019; Morin, 2019).

Most of the eddy covariance measurements of F_{CH_4} have focused on freshwater wetlands rather than coastal wetlands (Baldocchi, 2014; Knox et al., 2019). One of the primary reasons is that coastal wetlands have long been considered as minor CH$_4$ sources due to their highly saline environments. Under a high salinity, sulfate-reducing bacteria can outcompete methanogenic archaea for electron donors (e.g., hydrogen and acetate) because they yield a higher energy (Ozuolmez et al., 2015). In a synthesis of F_{CH_4} over global eddy covariance networks, Knox et al. (2019) reported a median emission rate of $0.8\,g\,CH_4\text{-}C\,m^{-2}\,year^{-1}$ from coastal wetlands, which is significantly lower than that from freshwater marshes ($43.2\,g\,CH_4\text{-}C\,m^{-2}\,year^{-1}$). However, the magnitude of F_{CH_4} in coastal wetlands is largely dependent on local hydrology. For example, a large amount of freshwater discharge and heavy rainfall events can turn a tidal estuarine wetland into a significant CH$_4$ source during the rainy season (Allen et al., 2011; Negandhi et al., 2019). Poffenbarger et al. (2011) identified a salinity level of larger than 18 ppt as a reliable threshold for salt marshes to achieve a net radiative cooling effect over a century. Liu et al. (2019) synthesized chamber-based F_{CH_4} measurements from estuarine coastal wetlands and found that hourly F_{CH_4} could reach up to $3500\,\mu mol\,m^{-2}\,s^{-1}$ in these low-salinity coastal ecosystems. Therefore, understanding the long-term local hydrologic regime is the key to predict F_{CH_4} accurately in coastal wetlands.

4.3.1 Magnitude and temporal variability of ecosystem-scale CH$_4$ fluxes in coastal wetlands

The temporal variability of F_{CH_4} is in general higher than that of NEE (Morin et al., 2014). The production of CH$_4$ is often decoupled from the physical processes driving the emission, and F_{CH_4} is variable depending on when the produced CH$_4$ will be emitted from the biosphere and subsequently detected by the eddy covariance equipment. This decoupling causes time lags between the drivers of CH$_4$ production and emission (Knox et al., 2019). CH$_4$ production as a biological process is mostly driven by temperature and substrate supply, while CH$_4$ emission is often driven by changes in atmospheric pressure, friction velocity (momentum), and water table (Chen and Slater, 2016). The timescale at which these drivers act on CH$_4$ production and emission also changes, with the biological drivers such as NEE, latent heat fluxes (LE), or sensible heat (H) acting more on diel time scales while atmospheric pressure and friction velocity having a greater influence on the multiday to seasonal time scales (Knox et al., 2019; Sturtevant et al., 2016). This decoupling on different time scales adds another layer of complexity in the modelling of CH_4 emission. Depending on the ecosystems, the diel variations of F_{CH_4} can be as large as the seasonal changes. Particularly, coastal ecosystems that are influenced by semidiurnal tides can exhibit episodic CH_4 emissions driven by matrix collapse of the marshes (Chen et al., 2017). Yamamoto et al. (2009) found that the spring-neap tidal cycle can significantly alter the diel pattern of chamber-based F_{CH_4} in the littoral zone of a brackish lake. Marshes can have surprisingly high concentrations of CH_4 belowground (Reid et al., 2013; Walter and Heimann, 2000). Thus, the diel pattern of F_{CH_4} in wetlands can be determined by a series of factors including temperature, air pressure, near-surface turbulence, plant-mediated transport, plant productivity, and WTD, which has been summarized by Koebsch et al. (2015) and Li et al. (2018).

Results of eddy covariance studies have shown that the seasonal variability of F_{CH_4} in wetlands is largely determined by temperature (Chu et al., 2014; Koebsch et al., 2015; Meijide et al., 2011; Wille et al., 2008), WTD (Dalmagro et al., 2019; Koebsch et al., 2015), salinity (Chamberlain et al., 2019; Holm et al., 2016; Krauss et al., 2016), and plant activities (Meijide et al., 2011). Long-term eddy covariance measurements have shown negligible F_{CH_4} in salt marshes during the nongrowing season due to high salinity and low temperature (Holm et al., 2016; Krauss et al., 2016; Li et al., 2018). On the other hand, F_{CH_4} can increase markedly during the growing season when riverine discharges and rainfall events reduce the local salinity, the air temperature increases and plants are photosynthetically active (Holm et al., 2016; Krauss et al., 2016; Li et al., 2018).

Annual F_{CH_4} in the tidal marshes varied with local salinity (Table 4.3). For tidal brackish and salt marshes with saltwater intrusion, annual F_{CH_4} ranged between 0 and 18 g CH_4-C m^{-2} year^{-1} (Holm et al., 2016; Krauss et al., 2016; Li et al., 2018; Negandhi et al., 2019). The salinity level and F_{CH_4} during the growing season can largely determine the cumulative CH_4 emission on an annual scale. If local salinity remains high throughout the year, the salt marsh can be expected to be a minor CH_4 source (Negandhi et al., 2019). On the other hand, tidal freshwater marshes can act as a significant CH_4 source with an annual F_{CH_4} of up to 60 g CH_4-C m^{-2} year^{-1} (Holm et al., 2016), which is larger than that in most of the FLUXNET-CH_4 flux sites (Knox et al., 2019). A similar magnitude of annual F_{CH_4} with 50 g CH_4-C m^{-2} year^{-1} was reported from a freshwater marsh in the coastal area of northwestern Ohio, United States (Chu et al., 2014).

Eddy covariance measurements of CH_4 fluxes are available for some tropical mangroves in India and subtropical mangroves in China. Jha et al. (2014) reported that the Sundarban mangrove in India could potentially act as a significant CH_4 source based on short-term measurements of F_{CH_4} for over two months. Daily average F_{CH_4} in mangroves can reach 150 mg CH_4-C m^{-2} day^{-1}, which is

Table 4.3 Eddy covariance studies of ecosystem-scale CH_4 fluxes in coastal wetlands where annual CH_4 fluxes have been reported.

Site	Lat.	Lon.	Year	MAT (°C)	MAP (mm)	Salinity range (ppt)	F_{CH_4} (g C m^{-2} year^{-1})	Reference
Tidal freshwater marsh, Louisiana, United States	29.86	−90.29	2012–13	20.8	1580	0.1–0.4	47.1–62.3	Holm et al. (2016) and Krauss et al. (2016)
Tidal brackish marsh, Louisiana, United States	29.50	−90.45	2012	20.8	1580	2–17	11.1–13.8	Holm et al. (2016) and Krauss et al. (2016)
Tidal salt marsh, Chongming Island, China	31.52	121.96	2011–12	17.1	868–1532	0–25	17.6	Li et al. (2018)
Tidal brackish marsh, Tomago wetland, Australia	−32.95	151.83	2015–16	NA	NA	7–20	0.1–1.6	Negandhi et al. (2019)
Salt marsh, St. Jones Reserve, Delaware	39.09	75.44	2015–17	14.0	576	8.5–18.2	8.5–15.2	Vázquez-Lule and Vargas (2021)
Mai Po Nature Reserve	22.50	114.03	2016–18	23.3	2399	0.7–18.3	11.4–11.8	Liu et al. (2020)

Annual average and range (in brackets) are presented. F_{CH_4}, net ecosystem CH_4 exchange; MAT, mean annual temperature; MAP, mean annual precipitation; NA, not available.

comparable to that in freshwater marshes that are known to be strong CH_4 sources (Knox et al., 2019). However, the magnitude of F_{CH_4} obtained from eddy covariance measurements is four orders of magnitude greater than that obtained from a bottom-up estimate ($0.06\,\text{mg}\,CH_4\text{-C}\,\text{m}^{-2}\,\text{day}^{-1}$) from the same region (Dutta et al., 2015), highlighting the urgent need to reconcile the differences in CH_4 emission estimates obtained from the bottom-up and top-down approaches (Kirschke et al., 2013). The discrepancies in these results may partly stem from CH_4 emission from the mangrove canopy itself that is not captured using a bottom-up approach (Megonigal et al., 2020). This tropical mangrove exhibited no clear diel pattern of F_{CH_4}, but soil temperature surprisingly exhibited a significant and negative effect on F_{CH_4}.

4.3.2 Biophysical controls of ecosystem-scale CH_4 fluxes in coastal wetlands

Morin (2019) has reviewed the biophysical factors that control wetland F_{CH_4}, including temperature, air pressure, light, humidity, WTD, turbulence, CO_2 fluxes, footprint, and LE. Atmospheric pressure has been identified as a significant driver of F_{CH_4} for 78% of the eddy covariance studies (Morin, 2019). A drop in atmospheric pressure can reduce CH_4 solubility and increase the volume of CH_4 in the gas phase, which subsequently induce CH_4 release via the release of gas bubbles (i.e., ebullition) (Sachs et al., 2010; Tokida et al., 2005; Waddington et al., 2009). There are other confounding factors affecting wetland F_{CH_4}. For example, temperature, light, and humidity can jointly impact leaf stomatal conductance and photosynthesis, which can in turn alter plant productivity and evapotranspiration. LE has been shown to be an important driver of F_{CH_4}, indicating either a linkage to photosynthesis, i.e., substrate supply, or stomatal conductance and thus transport mechanism for F_{CH_4}. The relative importance of these biophysical factors in governing F_{CH_4} is often site-specific.

The interactions between wetland F_{CH_4} and its biophysical drivers involve a set of nonlinear and asynchronous processes (Ford et al., 2012; Sturtevant et al., 2016). Moreover, the dominant interaction may vary across different time scales (Koebsch et al., 2015; Li et al., 2018). Therefore, it is challenging for traditional statistical approaches to fully characterize the biophysical controls of F_{CH_4}. Alternative approaches such as mutual information, Granger causality, and wavelet decomposition have emerged as promising statistical methods for wetland F_{CH_4} studies. Here we presented and discussed some potential regulating factors of F_{CH_4} in coastal wetlands.

4.3.2.1 Temperature
Temperature can directly affect the microbial activities associated with methanogenesis and methanotrophy. It has long been considered as one of the key drivers of F_{CH_4} and its effects on F_{CH_4} has been well incorporated into CH_4 biogeochemical models (Bridgham et al., 2013; Lai, 2009; Turetsky et al., 2014; Whiting and Chanton, 1993). Wetland F_{CH_4} can increase either linearly or exponentially with temperature

(Morin, 2019). Yvon-Durocher et al. (2014) indicated that the response of F_{CH_4} to temperature is consistent across microbial to ecosystem scales.

It is noteworthy that wetland F_{CH_4} can exhibit different responses to the three temperature indicators, i.e., air temperature, soil temperature, and water temperature. Air temperature is more directly related to photosynthetic processes and autotrophic respiration than microbial metabolisms in the wetland sediment. Soil temperature, especially at greater depths, has a strong influence on methanogenesis in coastal wetlands (Alongi, 2009) because the methanogenic archaea will be outcompeted by sulfate-reducing bacteria for substrates and thus shifted to the deeper soil layers based on the thermodynamic theory (Froelich et al., 1979). Microbial methanotrophy is believed to be more strongly correlated with surface soil temperature and water temperature (Osudar et al., 2015; Rey-Sanchez et al., 2018). In wetlands, soil temperature and water temperature are likely to lag behind air temperature by several hours for diel cycles (House et al., 2015) and by several days for daily values (Sun et al., 2018) due to the different heat capacities of air, soil, and water. Therefore, simply correlating temperature with F_{CH_4} over the same time frame could lead to a biased or erroneous conclusion about the dominant driver of wetland F_{CH_4} given the complexity of asynchronous processes involved in CH$_4$ dynamics.

As temperature also affects the physical transport processes such as gas diffusion and gas dissolution in water, it adds a secondary layer of complexity to the biophysical processes by modulating the transport mechanism of CH$_4$ from the deep soils or sediments to the atmosphere.

4.3.2.2 Salinity

Salinity has long been known as a critical driver of CH$_4$ emissions in wetland ecosystems. Bartlett et al. (1987) found that a sediment salinity of larger than 13 ppt could significantly inhibit CH$_4$ emissions in the salt marshes. Nevertheless, the salinity threshold above which F_{CH_4} would remarkably reduce is also related to the availability of substrates (Araujo et al., 2018). There has been growing evidence that methanogens and sulfate-reducing bacteria can coexist in the organic-rich sediment when there is abundant supply of common substrates (e.g., acetate) (Chuang et al., 2017; Holmer and Kristensen, 1994; Ozuolmez et al., 2015). Therefore, modeling the CH$_4$ emission from a specific coastal wetland may require a full understanding of the spatial variability of salinity and organic C contents in the sediment.

Poffenbarger et al. (2011) compiled the data of salinity and chamber-based CH$_4$ flux from 31 salt marshes, and found that the coastal wetlands with an average salinity of above 18 ppt would have very low CH$_4$ emissions and act as a net GHG sink over a 100-year time scale. Besides average salinity, the minimum salinity is another good indicator of F_{CH_4} in a specific wetland site based on both chamber and eddy covariance measurements. Coastal wetlands with a minimum salinity value of larger than 7 ppt are consistent minor CH$_4$ sources with an annual F_{CH_4} of smaller than $2\,\mathrm{g\,CH_4}\text{-}\mathrm{C\,m}^{-2}\,\mathrm{year}^{-1}$ (Bartlett et al., 1985; Holmquist et al., 2018; Negandhi et al., 2019; Sun et al., 2013a,b) (Table 4.3).

Wetlands around the world are now experiencing frequent extremes of drought and flooding in addition to rising sea levels and saltwater intrusion (USGCRP, 2018). Drought, flooding and sea-level rise can lead to a change in water salinity and influence ecosystem productivity and F_{CH_4} not only in coastal wetlands but also inland wetlands. For example, Chamberlain et al. (2019) applied information theory to eddy covariance fluxes in a restored freshwater wetland and found that a moderate drought-induced increase in salinity can reduce ecosystem productivity and F_{CH_4} by 64% and 10%, respectively. However, since salinity is often not measured in most inland wetlands, the ecological impact of salinization on global wetlands remains poorly understood (Herbert et al., 2015).

4.3.2.3 Plant activities

Plant activities can regulate F_{CH_4} in wetlands through root exudation and transpiration. The roots of wetland plants can release labile organic C to the soil, which provides additional substrates for methanogenesis (Whiting and Chanton, 1993). The root exudation rate of plants is highly dependent on the photosynthetic rate (Zhai et al., 2013). A positive correlation between GPP and F_{CH_4} has been documented in wetlands (Hatala et al., 2012; Sturtevant et al., 2016), pointing to the potential role of substrate supply in supporting CH_4 emissions. Li et al. (2018) indicated that GPP dominated the seasonal variation of F_{CH_4} in a coastal salt marsh. Yet, the effect of plant productivity on F_{CH_4} in coastal wetlands has not been well characterized. Mitra et al. (2020) found that plant activities have a more profound impact on F_{CH_4} than environmental factors across different time scales in a coastal freshwater swamp. On the other hand, the influence of substrate supply from plants on methanogenesis can be weakened by the competition with sulfate-reducing bacteria and lateral C transport. Likewise, plant roots in wetlands can release oxygen into the anoxic soil down the concentration gradient (Watanabe et al., 1997), thereby facilitating methanotrophy and inhibiting net CH_4 emissions. Girkin et al. (2020) found that this CH_4 oxidation pathway could account for up to 92% of the CH_4 flux reduction in tropical peatlands. However, the importance of this process in driving F_{CH_4} in coastal wetlands is still uncertain.

Porous aerenchyma tissues can connect the below-ground parts of wetland plants and the atmosphere, providing a conduit for an effective transport of CH_4 produced in the wetland sediment (Bridgham et al., 2013; Cicerone and Shetter, 1981; Dacey, 1981; Dacey and Klug, 1979). On the other hand, roots can actively take up the CH_4-dissolved in pore water, which will then be transported to the above-ground tissues through the transpiration stream. This potentially accounts for the high correlation observed between F_{CH_4} and LE among eddy covariance studies (Morin, 2019). Jeffrey et al. (2019) have identified the mangrove stem as a potential CH_4 source, with transpiration being one of the underlying transport mechanisms. Light and VPD are closely related to photosynthesis and transpiration, and therefore they may indirectly determine wetland F_{CH_4} (Morin, 2019; Villa et al., 2020).

4.3.2.4 Water table depth

The water table position is another key control of wetland F_{CH_4} by influencing the water-logged space in the soil that is thermodynamically favorable for methanogenesis (Moore and Roulet, 1993). A significant and positive relationship between WTD and annual F_{CH_4} has been reported in various chamber-based syntheses (Olefeldt et al., 2013; Treat et al., 2018; Turetsky et al., 2014). However, these chamber-based studies rely on the flux and WTD data collected from short-term field campaigns and thus do not fully reflect the seasonal or interannual variations. Moreover, a rapid drop in WTD can lead to large episodic bursts of CH_4 bubbles because of the pressure release in pore water (Bubier and Moore, 1994; Windsor et al., 1992), which are difficult to be captured by either chamber-based or eddy covariance measurements. Thus, the relative contribution of ebullition events to the overall CH_4 release is still poorly known. Based on long-term and continuous eddy covariance measurements, Knox et al. (2019) have suggested that WTD is a good predictor of annual F_{CH_4} only for the sites that have distinct inundation and air-exposed periods.

The interaction between WTD and F_{CH_4} is more complicated in the coastal wetlands than their inland counterparts. Holm et al. (2016) and Krauss et al. (2016) conducted eddy covariance measurements in both tidal freshwater and brackish wetlands and found that F_{CH_4} increased with WTD, which was consistent with the findings in a brackish lake (Yamamoto et al., 2009). On the other hand, high WTD can negatively impact F_{CH_4} in coastal wetlands by forming a physical barrier for CH_4 transport (Huang et al., 2019; Li et al., 2018; Tong et al., 2013). Dalmagro et al. (2019) suggested that WTD would only indirectly link to F_{CH_4} by altering the soil oxidation-reduction potentials. Indeed, WTD can be a poor indicator of soil oxidation-reduction potentials for tidal wetlands that are frequently inundated (Knox et al., 2019; Morin, 2019). More direct predictors of F_{CH_4}, such as soil oxidation-reduction potential and dissolved oxygen concentrations in soil or water, should be measured simultaneously with F_{CH_4} with a high temporal resolution in order to characterize the drivers of F_{CH_4} more accurately.

4.4 Synthesis

Overall, a similar set of biophysical drivers will govern the variations of NEE and F_{CH_4} in coastal wetlands, but their effects and relative importance can vary depending on the time scales and specific site conditions. Although the processes involved in NEE are fairly well characterized and can be modeled reasonably well, the governing factors and underlying mechanisms of ecosystem-scale F_{CH_4} in wetlands are still not fully understood. The machine learning technique has shown promise in gap-filling the F_{CH_4} dataset from eddy covariance measurements, with the advantage of being able to extract variable importance from the regression tree (Kim et al., 2020). While temperature seems to play an overarching role for CH_4 dynamics, solar radiation is the primary driver of photosynthesis and net CO_2 exchange in coastal wetlands. Continuous and simultaneous measurements of salinity and C fluxes can help fill the knowledge gap regarding the optimal salinity for the eco-physiological activities of mangroves, and can provide insights for solving the longstanding puzzle of whether mangrove trees are obligate or facultative halophytes (Krauss and Ball, 2013; Wang et al., 2011). Our still nascent understanding of the major drivers of ecosystem-scale F_{CH_4} suggests a need for more in-depth analysis of a larger dataset that is currently under development (Knox et al., 2019). Preliminary results suggest that WTD is an indirect indicator of F_{CH_4} through its link with redox potential and microbial community composition in both inland and coastal wetlands. The challenges ahead in enhancing our knowledge of CH_4 dynamics in coastal wetlands include quantifying the rate and relative contribution of ebullition to total F_{CH_4}, coupling the F_{CH_4} data obtained from eddy covariance measurements to the mechanistic processes of CH_4 production and oxidation, and incorporating the biogeochemical knowledge such as the time lags of F_{CH_4} to biotic controls, the proportion of photosynthate allocated to methanogenesis, the physical drivers of CH_4 emissions, etc. into wetland CH_4 models.

References

Allen, D., Dalal, R.C., Rennenberg, H., Schmidt, S., 2011. Seasonal variation in nitrous oxide and methane emissions from subtropical estuary and coastal mangrove sediments, Australia. Plant Biol. 13, 126–133. https://doi.org/10.1111/j.1438-8677.2010.00331.x.

Alongi, D.M., 2009. The Energetics of Mangrove Forests. Springer Press, London, https://doi.org/10.1007/978-1-4020-4271-3.

Alvarado-Barrientos, M.S., López-Adame, H., Lazcano-Hernández, H.E., Arellano-Verdejo, J., Hernández-Arana, H.A., 2020. Ecosystem-atmosphere exchange of CO_2, water and energy in a basin mangrove of the northeastern coast of the Yucatan Peninsula. J. Geophys. Res. Biogeo. 125. https://doi.org/10.1029/2020JG005811, e2020JG005811.

Amiro, B.D., Barr, A.G., Barr, J.G., Black, T.A., Bracho, R., Brown, M., Chen, J., Clark, K.L., Davis, K.J., Desai, A.R., Dore, S., Engel, V., Fuentes, J.D., Goldstein, A.H., Goulden, M.L., Kolb, T.E., Lavigne, M.B., Law, B.E., Margolis, H.A., Martin, T., McCaughey, J.H., Misson, L., Montes-Helu, M., Noormets, A., Randerson, J.T., Starr, G., Xiao, J., 2010. Ecosystem carbon dioxide fluxes after disturbance in forests of North America. J. Geophys. Res. Biogeo. 115. https://doi.org/10.1029/2010JG001390.

Araujo, J., Pratihary, A., Naik, R., Naik, H., Naqvi, S.W.A., 2018. Benthic fluxes of methane along the salinity gradient of a tropical monsoonal estuary: implications for CH_4 supersaturation and emission. Mar. Chem. 202, 73–85. https://doi.org/10.1016/j.marchem.2018.03.008.

Arriga, N., Rannik, Ü., Aubinet, M., Carrara, A., Vesala, T., Papale, D., 2017. Experimental validation of footprint models for eddy covariance CO_2 flux measurements above grassland by means of natural and artificial tracers. Agric. For. Meteorol. 242, 75–84. https://doi.org/10.1016/j.agrformet.2017.04.006.

Artigas, F., Shin, J.Y., Hobble, C., Marti-Donati, A., Schäfer, K.V.R., Pechmann, I., 2015. Long term carbon storage potential and CO_2 sink strength of a restored salt marsh in New Jersey. Agric. For. Meteorol. 200, 313–321. https://doi.org/10.1016/j.agrformet.2014.09.012.

Aubinet, M., Vesala, T., Papale, D., 2012. Eddy Covariance: A Practical Guide to Measurement and Data Analysis. Springer Science & Business Media.

Baldocchi, D.D., 2003. Assessing the eddy covariance technique for evaluating carbon dioxide exchange rates of ecosystems: past, present and future. Glob. Chang. Biol. 9, 479–492. https://doi.org/10.1046/j.1365-2486.2003.00629.x.

Baldocchi, D., 2014. Measuring fluxes of trace gases and energy between ecosystems and the atmosphere—the state and future of the eddy covariance method. Glob. Chang. Biol. 20, 3600–3609. https://doi.org/10.1111/gcb.12649.

Baldocchi, D.D., 2020. How eddy covariance flux measurements have contributed to our understanding of Global Change Biology. Glob. Chang. Biol. 26, 242–260. https://doi.org/10.1111/gcb.14807.

Baldocchi, D.D., Harley, P.C., 1995. Scaling carbon dioxide and water vapour exchange from leaf to canopy in a deciduous forest. II. Model testing and application. Plant Cell Environ. 18, 1157–1173. https://doi.org/10.1111/j.1365-3040.1995.tb00626.x.

Baldocchi, D., Falge, E., Gu, L., Olson, R., Hollinger, D., Running, S., Anthoni, P., Bernhofer, C., Davis, K., Evans, R., Fuentes, J., Goldstein, A., Katul, G., Law, B., Lee, X., Malhi, Y., Meyers, T., Munger, W., Oechel, W., Paw, U.K.T., Pilegaard, K., Schmid, H.P., Valentini, R., Verma, S., Vesala, T., Wilson, K., Wofsy, S., 2001a. FLUXNET: a new tool to study the temporal and spatial variability of ecosystem-scale carbon dioxide, water vapor, and energy flux densities. Bull. Am. Meteorol. Soc. 82, 2415–2434. https://doi.org/10.1175/1520-0477(2001)082<2415:FANTTS>2.3.CO;2.

Baldocchi, D., Falge, E., Wilson, K., 2001b. A spectral analysis of biosphere-atmosphere trace gas flux densities and meteorological variables across hour to multi-year time scales. Agric. For. Meteorol. 107, 1–27. https://doi.org/10.1016/S0168-1923(00)00228-8.

Baldocchi, D., Detto, M., Sonnentag, O., Verfaillie, J., Teh, Y.A., Silver, W., Kelly, N.M., 2012. The challenges of measuring methane fluxes and concentrations over a peatland pasture. Agric. For. Meteorol. 153, 177–187. https://doi.org/10.1016/j.agrformet.2011.04.013.

Baldocchi, D., Chu, H., Reichstein, M., 2018. Inter-annual variability of net and gross ecosystem carbon fluxes: a review. Agric. For. Meteorol. 249, 520–533. https://doi.org/10.1016/j.agrformet.2017.05.015.

Barr, J.G., Engel, V., Fuentes, J.D., Zieman, J.C., O'Halloran, T.L., Smith, T.J., Anderson, G.H., 2010. Controls on mangrove forest-atmosphere carbon dioxide exchanges in western Everglades National Park. J. Geophys. Res. Biogeo. 115, G02020. https://doi.org/10.1029/2009jg001186.

Barr, J.G., Engel, V., Smith, T.J., Fuentes, J.D., 2012. Hurricane disturbance and recovery of energy balance, CO_2 fluxes and canopy structure in a mangrove forest of the Florida Everglades. Agric. For. Meteorol. 153, 54–66. https://doi.org/10.1016/j.agrformet.2011.07.022.

Barr, J.G., Engel, V., Fuentes, J.D., Fuller, D.O., Kwon, H., 2013a. Modeling light use efficiency in a subtropical mangrove forest equipped with CO_2 eddy covariance. Biogeosciences 10, 2145–2158. https://doi.org/10.5194/bg-10-2145-2013.

Barr, J.G., Fuentes, J.D., Delonge, M.S., O'Halloran, T.L., Barr, D., Zieman, J.C., 2013b. Summertime influences of tidal energy advection on the surface energy balance in a mangrove forest. Biogeosciences 10, 501–511. https://doi.org/10.5194/bg-10-501-2013.

Bartlett, K.B., Harriss, R.C., Sebacher, D.I., 1985. Methane flux from coastal salt marshes. J. Geophys. Res. 90, 5710–5720. https://doi.org/10.1029/JD090iD03p05710.

Bartlett, K.B., Bartlett, D.S., Harriss, R.C., Sebacher, D.I., 1987. Methane emissions along a salt marsh salinity gradient. Biogeochemistry 4, 183–202. https://doi.org/10.1007/BF02187365.

Bernacchi, C.J., Singsaas, E.L., Pimentel, C., Portis, A.R., Long, S.P., 2001. Improved temperature response functions for models of Rubisco-limited photosynthesis. Plant Cell Environ. 24, 253–259. https://doi.org/10.1046/j.1365-3040.2001.00668.x.

Berry, J., Bjorkman, O., 1980. Photosynthetic response and adaptation to temperature in higher plants. Annu. Rev. Plant Physiol. 31, 491–543. https://doi.org/10.1146/annurev.pp.31.060180.002423.

Biederman, J.A., Scott, R.L., Goulden, M.L., Vargas, R., Litvak, M.E., Kolb, T.E., Yepez, E.A., Oechel, W.C., Blanken, P.D., Bell, T.W., Garatuza-Payan, J., Maurer, G.E., Dore, S., Burns, S.P., 2016. Terrestrial carbon balance in a drier world: the effects of water availability in southwestern North America. Glob. Chang. Biol. 22, 1867–1879. https://doi.org/10.1111/gcb.13222.

Bonal, D., Bosc, A., Ponton, S., Goret, J.Y., Burban, B.T., Gross, P., Bonnefond, J.M., Elbers, J., Longdoz, B., Epron, D., Guehl, J.M., Granier, A., 2008. Impact of severe dry season on net ecosystem exchange in the Neotropical rainforest of French Guiana. Glob. Chang. Biol. 14, 1917–1933. https://doi.org/10.1111/j.1365-2486.2008.01610.x.

Bridgham, S.D., Cadillo-Quiroz, H., Keller, J.K., Zhuang, Q., 2013. Methane emissions from wetlands: biogeochemical, microbial, and modeling perspectives from local to global scales. Glob. Chang. Biol. 19, 1325–1346. https://doi.org/10.1111/gcb.12131.

Bubier, J.L., Moore, T.R., 1994. An ecological perspective on methane emissions from northern wetlands. Trends Ecol. Evol. 9, 460–464. https://doi.org/10.1016/0169-5347(94)90309-3.

Chamberlain, S.D., Verfaillie, J., Eichelmann, E., Hemes, K.S., Baldocchi, D.D., 2017. Evaluation of density corrections to methane fluxes measured by open-path eddy covariance over contrasting landscapes. Bound.-Lay. Meteorol. 165, 197–210. https://doi.org/10.1007/s10546-017-0275-9.

Chamberlain, S.D., Hemes, K.S., Eichelmann, E., Szutu, D.J., Verfaillie, J.G., Baldocchi, D.D., 2019. Effect of drought-induced salinization on wetland methane emissions, gross ecosystem productivity, and their interactions. Ecosystems. https://doi.org/10.1007/s10021-019-00430-5.

Chen, X., Slater, L., 2016. Methane emission through ebullition from an estuarine mudflat: 1. A conceptual model to explain tidal forcing based on effective stress changes. Water Resour. Res. 52, 4469–4485. https://doi.org/10.1002/2015WR018058.

Chen, G.C., Tam, N.F.Y., Ye, Y., 2010. Summer fluxes of atmospheric greenhouse gases N$_2$O, CH$_4$ and CO$_2$ from mangrove soil in South China. Sci. Total Environ. 408, 2761–2767. https://doi.org/10.1016/j.scitotenv.2010.03.007.

Chen, H., Lu, W., Yan, G., Yang, S., Lin, G., 2014. Typhoons exert significant but differential impacts on net ecosystem carbon exchange of subtropical mangrove forests in China. Biogeosciences 11, 5323–5333. https://doi.org/10.5194/bg-11-5323-2014.

Chen, X., Schäfer, K.V.R., Slater, L., 2017. Methane emission through ebullition from an estuarine mudflat: 2. Field observations and modeling of occurrence probability. Water Resour. Res. 53, 6439–6453. https://doi.org/10.1002/2016WR019720.

Chu, H., Chen, J., Gottgens, J.F., Ouyang, Z., John, R., Czajkowski, K., Becker, R., 2014. Net ecosystem methane and carbon dioxide exchanges in a Lake Erie coastal marsh and a nearby cropland. J. Geophys. Res. Biogeo. 119, 722–740. https://doi.org/10.1002/2013JG002520.

Chu, H., Baldocchi, D.D., John, R., Wolf, S., Reichstein, M., 2017. Fluxes all of the time? A primer on the temporal representativeness of FLUXNET. J. Geophys. Res. Biogeo. 122, 289–307. https://doi.org/10.1002/2016JG003576.

Chuang, P.C., Young, M.B., Dale, A.W., Miller, L.G., Herrera-Silveira, J.A., Paytan, A., 2017. Methane fluxes from tropical coastal lagoons surrounded by mangroves, Yucatán, Mexico. J. Geophys. Res. Biogeo. 122, 1156–1174. https://doi.org/10.1002/2017JG003761.

Cicerone, R.J., Shetter, J.D., 1981. Sources of atmospheric methane: measurements in rice paddies and a discussion. J. Geophys. Res. 86, 7203–7209. https://doi.org/10.1029/jc086ic08p07203.

Cleland, E.E., Chuine, I., Menzel, A., Mooney, H.A., Schwartz, M.D., 2007. Shifting plant phenology in response to global change. Trends Ecol. Evol. https://doi.org/10.1016/j.tree.2007.04.003.

Cui, X., Liang, J., Lu, W., Chen, H., Liu, F., Lin, G., Xu, F., Luo, Y., Lin, G., 2018. Stronger ecosystem carbon sequestration potential of mangrove wetlands with respect to terrestrial forests in subtropical China. Agric. For. Meteorol. 249, 71–80. https://doi.org/10.1016/j.agrformet.2017.11.019.

Curiel Yuste, J., Janssens, I.A., Carrara, A., Ceulemans, R., 2004. Annual Q10 of soil respiration reflects plant phenological patterns as well as temperature sensitivity. Glob. Chang. Biol. 10, 161–169. https://doi.org/10.1111/j.1529-8817.2003.00727.x.

Dacey, J.W.H., 1981. Pressurized ventilation in the yellow waterlily. Ecology 62, 1137–1147. https://doi.org/10.2307/1937277.

Dacey, J.W.H., Klug, M.J., 1979. Methane efflux from lake sediments through water lilies. Science 203, 1253–1255. https://doi.org/10.1126/science.203.4386.1253.

Dalmagro, H.J., Zanella de Arruda, P.H., Vourlitis, G.L., Lathuillière, M.J., De S Nogueira, J., Couto, E.G., Johnson, M.S., 2019. Radiative forcing of methane fluxes offsets net carbon dioxide uptake for a tropical flooded forest. Glob. Chang. Biol. 25, 1967–1981. https://doi.org/10.1111/gcb.14615.

Davidson, E.A., Savage, K., Verchot, L.V., Navarro, R., 2002. Minimizing artifacts and biases in chamber-based measurements of soil respiration. Agric. For. Meteorol. 113, 21–37. https://doi.org/10.1016/S0168-1923(02)00100-4.

Dellwik, E., Mann, J., Bingöl, F., 2010a. Flow tilt angles near forest edges—Part 2: lidar anemometry. Biogeosciences 7, 1759–1768. https://doi.org/10.5194/bg-7-1759-2010.

Dellwik, E., Mann, J., Larsen, K.S., 2010b. Flow tilt angles near forest edges—Part 1: Sonic anemometry. Biogeosciences 7, 1745–1757. https://doi.org/10.5194/bg-7-1745-2010.

Delwiche, K.B., Knox, S.H., Malhotra, A., Fluet-Chouinard, E., McNicol, G., Feron, S., Ouyang, Z., Papale, D., Trotta, C., Canfora, E., Cheah, Y.-W., Christianson, D., Alberto, M.C.R., Alekseychik, P., Aurela, M., Baldocchi, D., Bansal, S., Billesbach, D.P., Bohrer, G., Bracho, R., Buchmann, N., Campbell, D.I., Celis, G., Chen, J., Chen, W., Chu, H., Dalmagro, H.J., Dengel, S., Desai, A.R., Detto, M., Dolman, H., Eichelmann, E., Euskirchen, E., Famulari, D., Fuchs, K., Goeckede, M., Gogo, S., Gondwe, M.J., Goodrich, J.P., Gottschalk, P., Graham, S.L., Heimann, M., Helbig, M., Helfter, C., Hemes, K.S., Hirano, T., Hollinger, D., Hörtnagl, L., Iwata, H., Jacotot, A., Jurasinski, G., Kang, M., Kasak, K., King, J., Klatt, J., Koebsch, F., Krauss, K.W., Lai, D.Y.F., Lohila, A., Mammarella, I., Belelli Marchesini, L., Manca, G., Matthes, J.H., Maximov, T., Merbold, L., Mitra, B., Morin, T.H., Nemitz, E., Nilsson, M.B., Niu, S., Oechel, W.C., Oikawa, P.Y., Ono, K., Peichl, M., Peltola, O., Reba, M.L., Richardson, A.D., Riley, W., Runkle, B.R.K., Ryu, Y., Sachs, T., Sakabe, A., Sanchez, C.R., Schuur, E.A., Schäfer, K.V.R., Sonnentag,

O., Sparks, J.P., Stuart-Haëntjens, E., Sturtevant, C., Sullivan, R.C., Szutu, D.J., Thom, J. E., Torn, M.S., Tuittila, E.-S., Turner, J., Ueyama, M., Valach, A.C., Vargas, R., Varlagin, A., Vazquez-Lule, A., Verfaillie, J.G., Vesala, T., Vourlitis, G.L., Ward, E.J., Wille, C., Wohlfahrt, G., Wong, G.X., Zhang, Z., Zona, D., Windham-Myers, L., Poulter, B., Jackson, R.B., 2021. FLUXNET-CH$_4$: a global, multi-ecosystem dataset and analysis of methane seasonality from freshwater wetlands. Earth Syst. Sci. Data 13, 3607–3689. https://doi.org/10.5194/essd-13-3607-2021.

Desjardins, R.L., 1974. A technique to measure CO$_2$ exchange under field conditions. Int. J. Biometeorol. 18, 76–83. https://doi.org/10.1007/BF01450667.

Desjardins, R.L., Lemon, E.R., 1974. Limitations of an eddy-correlation technique for the determination of the carbon dioxide and sensible heat fluxes. Bound.-Lay. Meteorol. 5, 475–488. https://doi.org/10.1007/BF00123493.

Donato, D.C., Kauffman, J.B., Murdiyarso, D., Kurnianto, S., Stidham, M., Kanninen, M., 2011. Mangroves among the most carbon-rich forests in the tropics. Nat. Geosci. 4, 293–297. https://doi.org/10.1038/ngeo1123.

Duke, N.C., Meynecke, J.-O., Dittmann, S., Ellison, A.M., Anger, K., Berger, U., Cannicci, S., Diele, K., Ewel, K.C., Field, C.D., Koedam, N., Lee, S.Y., Marchand, C., Nordhaus, I., Dahdouh-Guebas, F., 2007. A world without mangroves? Science 317, 41b–42b. https://doi.org/10.1126/science.317.5834.41b.

Duman, T., Schäfer, K.V.R., 2018. Partitioning net ecosystem carbon exchange of native and invasive plant communities by vegetation cover in an urban tidal wetland in the New Jersey Meadowlands (USA). Ecol. Eng. 114, 16–24. https://doi.org/10.1016/j.ecoleng.2017.08.031.

Dutta, M.K., Ray, R., Mukherjee, R., Jana, T.K., Mukhopadhyay, S.K., 2015. Atmospheric fluxes and photo-oxidation of methane in the mangrove environment of the Sundarbans, NE coast of India; a case study from Lothian Island. Agric. For. Meteorol. 213, 33–41. https://doi.org/10.1016/j.agrformet.2015.06.010.

Etzold, S., Buchmann, N., Eugster, W., 2010. Contribution of advection to the carbon budget measured by eddy covariance at a steep mountain slope forest in Switzerland. Biogeosciences 7, 2461–2475. https://doi.org/10.5194/bg-7-2461-2010.

Falge, E., Baldocchi, D., Tenhunen, J., Aubinet, M., Bakwin, P., Berbigier, P., Bernhofer, C., Burba, G., Clement, R., Davis, K.J., Elbers, J.A., Goldstein, A.H., Grelle, A., Granier, A., Gumundsson, J., Hollinger, D., Kowalski, A.S., Katul, G., Law, B.E., Malhi, Y., Meyers, T., Monson, R.K., Munger, J.W., Oechel, W., Kyaw Tha Paw, U., Pilegaard, K., Rannik, Ü., Rebmann, C., Suyker, A., Valentini, R., Wilson, K., Wofsy, S., 2002. Seasonality of ecosystem respiration and gross primary production as derived from FLUXNET measurements. Agric. For. Meteorol. 113, 53–74. https://doi.org/10.1016/S0168-1923(02)00102-8.

Fan, S.M., Wofsy, S.C., Bakwin, P.S., Jacob, D.J., Fitzjarrald, D.R., 1990. Atmosphere-biosphere exchange of CO$_2$ and O$_3$ in the central Amazon forest. J. Geophys. Res. 95, 16851–16864. https://doi.org/10.1029/jd095id10p16851.

Fletcher, S.E.M., Schaefer, H., 2019. Rising methane: a new climate challenge: the amount of the greenhouse gas methane in Earth's atmosphere is rising rapidly. Science 364, 932–933. https://doi.org/10.1126/science.aax1828.

Forbrich, I., Giblin, A.E., 2015. Marsh-atmosphere CO$_2$ exchange in a New England salt marsh. J. Geophys. Res. G: Biogeosci. 120, 1825–1838. https://doi.org/10.1002/2015JG003044.

Forbrich, I., Giblin, A.E., Hopkinson, C.S., 2018. Constraining marsh carbon budgets using long-term C burial and contemporary atmospheric CO_2 fluxes. J. Geophys. Res. Biogeo. 123, 867–878. https://doi.org/10.1002/2017JG004336.

Ford, H., Garbutt, A., Jones, L., Jones, D.L., 2012. Methane, carbon dioxide and nitrous oxide fluxes from a temperate salt marsh: grazing management does not alter Global Warming Potential. Estuar. Coast. Shelf Sci. 113, 182–191. https://doi.org/10.1016/j.ecss.2012.08.002.

Friess, D.A., Rogers, K., Lovelock, C.E., Krauss, K.W., Hamilton, S.E., Lee, S.Y., Lucas, R., Primavera, J., Rajkaran, A., Shi, S., 2019. The state of the world's mangrove forests: past, present, and future. Annu. Rev. Env. Resour. 44, 89–115. https://doi.org/10.1146/annurev-environ-101718-033302.

Friess, D.A., Yando, E.S., Abuchahla, G.M.O., Adams, J.B., Cannicci, S., Canty, S.W.J., Cavanaugh, K.C., Connolly, R.M., Cormier, N., Dahdouh-Guebas, F., Diele, K., Feller, I.C., Fratini, S., Jennerjahn, T.C., Lee, S.Y., Ogurcak, D.E., Ouyang, X., Rogers, K., Rowntree, J.K., Sharma, S., Sloey, T.M., Wee, A.K.S., 2020. Mangroves give cause for conservation optimism, for now. Curr. Biol. 30, R153–R154. https://doi.org/10.1016/j.cub.2019.12.054.

Froelich, P.N., Klinkhammer, G.P., Bender, M.L., Luedtke, N.A., Heath, G.R., Cullen, D., Dauphin, P., Hammond, D., Hartman, B., Maynard, V., 1979. Early oxidation of organic matter in pelagic sediments of the eastern equatorial Atlantic: suboxic diagenesis. Geochim. Cosmochim. Acta 43, 1075–1090. https://doi.org/10.1016/0016-7037(79)90095-4.

Fu, Z., Gerken, T., Bromley, G., Araújo, A., Bonal, D., Burban, B., Ficklin, D., Fuentes, J.D., Goulden, M., Hirano, T., Kosugi, Y., Liddell, M., Nicolini, G., Niu, S., Roupsard, O., Stefani, P., Mi, C., Tofte, Z., Xiao, J., Valentini, R., Wolf, S., Stoy, P.C., 2018. The surface-atmosphere exchange of carbon dioxide in tropical rainforests: sensitivity to environmental drivers and flux measurement methodology. Agric. For. Meteorol. 263, 292–307. https://doi.org/10.1016/j.agrformet.2018.09.001.

Garratt, J.R., 1975. Limitations of the Eddy-correlation technique for the determination of turbulent fluxes near the surface. Bound.-Lay. Meteorol. 8, 255–259. https://doi.org/10.1007/BF02153552.

Gilmanov, T.G., Aires, L., Barcza, Z., Baron, V.S., Belelli, L., Beringer, J., Billesbach, D., Bonal, D., Bradford, J., Ceschia, E., Cook, D., Corradi, C., Frank, A., Gianelle, D., Gimeno, C., Gruenwald, T., Guo, H., Hanan, N., Haszpra, L., Heilman, J., Jacobs, A., Jones, M.B., Johnson, D.A., Kiely, G., Li, S., Magliulo, V., Moors, E., Nagy, Z., Nasyrov, M., Owensby, C., Pinter, K., Pio, C., Reichstein, M., Sanz, M.J., Scott, R., Soussana, J.F., Stoy, P.C., Svejcar, T., Tuba, Z., Zhou, G., 2010. Productivity, respiration, and light-response parameters of world grassland and agroecosystems derived from flux-tower measurements. Rangel. Ecol. Manage. 63, 16–39. https://doi.org/10.2111/REM-D-09-00072.1.

Girkin, N., Vane, C.H., Turner, B., Ostle, N., Sjogersten, S., 2020. Root oxygen mitigates methane fluxes in tropical peatlands. Environ. Res. Lett. https://doi.org/10.1088/1748-9326/ab8495.

Gnanamoorthy, P., Selvam, V., Deb Burman, P.K., Chakraborty, S., Karipot, A., Nagarajan, R., Ramasubramanian, R., Song, Q., Zhang, Y., Grace, J., 2020. Seasonal variations of net ecosystem (CO_2) exchange in the Indian tropical mangrove forest of Pichavaram. Estuar. Coast. Shelf Sci. 243. https://doi.org/10.1016/j.ecss.2020.106828, 106828.

Goulden, M.L., Miller, S.D., Da Rocha, H.R., Menton, M.C., De Freitas, H.C., E Silva Figueira, A.M., Dias De Sousa, C.A., 2004. Diel and seasonal patterns of tropical forest CO_2 exchange. Ecol. Appl. 14, 42–54. https://doi.org/10.1890/02-6008.

Gu, L., Baldocchi, D., Verma, S.B., Black, T.A., Vesala, T., Falge, E.M., Dowty, P.R., 2002. Advantages of diffuse radiation for terrestrial ecosystem productivity. J. Geophys. Res. Atmos. 107, 4050. https://doi.org/10.1029/2001jd001242.

Gu, L., Falge, E.M., Boden, T., Baldocchi, D.D., Black, T.A., Saleska, S.R., Suni, T., Verma, S.B., Vesala, T., Wofsy, S.C., Xu, L., 2005. Objective threshold determination for nighttime eddy flux filtering. Agric. For. Meteorol. 128, 179–197. https://doi.org/10.1016/j.agrformet.2004.11.006.

Guo, H., Noormets, A., Zhao, B., Chen, J., Sun, G., Gu, Y., Li, B., Chen, J., 2009. Tidal effects on net ecosystem exchange of carbon in an estuarine wetland. Agric. For. Meteorol. 149, 1820–1828. https://doi.org/10.1016/j.agrformet.2009.06.010.

Han, G., Chu, X., Xing, Q., Li, D., Yu, J., Luo, Y., Wang, G., Mao, P., Rafique, R., 2015. Effects of episodic flooding on the net ecosystem CO_2 exchange of a supratidal wetland in the Yellow River Delta. J. Geophys. Res. Biogeo. 120, 1506–1520. https://doi.org/10.1002/2015JG002923.

Hatala, J.A., Detto, M., Baldocchi, D.D., 2012. Gross ecosystem photosynthesis causes a diurnal pattern in methane emission from rice. Geophys. Res. Lett. 39, 1–5. https://doi.org/10.1029/2012GL051303.

Hayek, M.N., Longo, M., Wu, J., Smith, M.N., Restrepo-Coupe, N., Tapajós, R., Da Silva, R., Fitzjarrald, D.R., Camargo, P.B., Hutyra, L.R., Alves, L.F., Daube, B., William Munger, J., Wiedemann, K.T., Saleska, S.R., Wofsy, S.C., 2018. Carbon exchange in an Amazon forest: from hours to years. Biogeosciences 15, 4833–4848. https://doi.org/10.5194/bg-15-4833-2018.

Hemes, K.S., Verfaillie, J., Baldocchi, D.D., 2020. Wildfire-smoke aerosols lead to increased light use efficiency among agricultural and restored wetland land uses in California's central valley. J. Geophys. Res. Biogeo. 125. https://doi.org/10.1029/2019JG005380.

Herbert, E.R., Boon, P., Burgin, A.J., Neubauer, S.C., Franklin, R.B., Ardon, M., Hopfensperger, K.N., Lamers, L.P.M., Gell, P., Langley, J.A., 2015. A global perspective on wetland salinization: ecological consequences of a growing threat to freshwater wetlands. Ecosphere 6, 1–43. https://doi.org/10.1890/ES14-00534.1.

Herr, D., Pidgeon, E., Laffoley, D., 2012. Blue Carbon Policy Framework: Based on the Discussion of the International Blue Carbon Policy Working Group. IUCN.

Holm, G.O., Perez, B.C., McWhorter, D.E., Krauss, K.W., Johnson, D.J., Raynie, R.C., Killebrew, C.J., 2016. Ecosystem level methane fluxes from tidal freshwater and brackish marshes of the Mississippi River Delta: implications for coastal wetland carbon projects. Wetlands 36, 401–413. https://doi.org/10.1007/s13157-016-0746-7.

Holmer, M., Kristensen, E., 1994. Coexistence of sulfate reduction and methane production in an organic-rich sediment. Mar. Ecol. Prog. Ser. 107, 177–184. https://doi.org/10.3354/meps107177.

Holmquist, J.R., Windham-Myers, L., Bernal, B., Byrd, K.B., Crooks, S., Gonneea, M.E., Herold, N., Knox, S.H., Kroeger, K.D., McCombs, J., Megonigal, J.P., Lu, M., Morris, J.T., Sutton-Grier, A.E., Troxler, T.G., Weller, D.E., 2018. Uncertainty in United States coastal wetland greenhouse gas inventorying. Environ. Res. Lett. 13. https://doi.org/10.1088/1748-9326/aae157.

House, A.R., Sorensen, J.P.R., Gooddy, D.C., Newell, A.J., Marchant, B., Mountford, J.O., Scarlett, P., Williams, P.J., Old, G.H., 2015. Exfiltrations discrètes en zone humide révélées par un modèle thermique tridimensionnel et par des indicateurs botaniques (Boxford, UK). Hydrgeol. J. 23, 775–787. https://doi.org/10.1007/s10040-015-1242-5.

Howard, J., Hoyt, S., Isensee, K., Pidgeon, E., Telszewski, M., 2014. Coastal blue carbon: methods for assessing carbon stocks and emissions factors in mangroves, tidal salt marshes, and seagrass meadows. In: Conserv. Int. Intergov. Oceanogr. Comm. UNESCO, Int. Union Conserv. Nature, Arlington, VA, pp. 1–180.

Hsieh, C.I., Katul, G., Chi, T.W., 2000. An approximate analytical model for footprint estimation of scalar fluxes in thermally stratified atmospheric flows. Adv. Water Resour. 23, 765–772. https://doi.org/10.1016/S0309-1708(99)00042-1.

Huang, J., Luo, M., Liu, Y., Zhang, Y., Tan, J., 2019. Effects of tidal scenarios on the methane emission dynamics in the subtropical tidal marshes of the min river estuary in southeast China. Int. J. Environ. Res. Public Health 16. https://doi.org/10.3390/ijerph16152790.

Huete, A.R., Didan, K., Shimabukuro, Y.E., Ratana, P., Saleska, S.R., Hutyra, L.R., Yang, W., Nemani, R.R., Myneni, R., 2006. Amazon rainforests green-up with sunlight in dry season. Geophys. Res. Lett. 33, 2–5. https://doi.org/10.1029/2005GL025583.

Hwang, Y.H., Chen, S.C., 2001. Effects of ammonium, phosphate, and salinity on growth, gas exchange characteristics, and ionic contents of seedlings of mangrove *Kandelia candel* (L.) Druce. Bot. Bull. Acad. Sin. 42, 131–139. https://doi.org/10.7016/BBAS.200104.0131.

Jeffrey, L.C., Reithmaier, G., Sippo, J.Z., Johnston, S.G., Tait, D.R., Harada, Y., Maher, D.T., 2019. Are methane emissions from mangrove stems a cryptic carbon loss pathway? Insights from a catastrophic forest mortality. New Phytol. 224, 146–154. https://doi.org/10.1111/nph.15995.

Jha, C.S., Rodda, S.R., Thumaty, K.C., Raha, A.K., Dadhwal, V.K., 2014. Eddy covariance based methane flux in Sundarbans mangroves, India. J. Earth Syst. Sci. 123, 1089–1096. https://doi.org/10.1007/s12040-014-0451-y.

Jiang, C., Ryu, Y., 2016. Multi-scale evaluation of global gross primary productivity and evapotranspiration products derived from Breathing Earth System Simulator (BESS). Remote Sens. Environ. 186, 528–547. https://doi.org/10.1016/j.rse.2016.08.030.

Jung, M., Reichstein, M., Schwalm, C.R., Huntingford, C., Sitch, S., Ahlström, A., Arneth, A., Camps-Valls, G., Ciais, P., Friedlingstein, P., Gans, F., Ichii, K., Jain, A.K., Kato, E., Papale, D., Poulter, B., Raduly, B., Rödenbeck, C., Tramontana, G., Viovy, N., Wang, Y.P., Weber, U., Zaehle, S., Zeng, N., 2017. Compensatory water effects link yearly global land CO_2 sink changes to temperature. Nature 541, 516–520. https://doi.org/10.1038/nature20780.

Kathilankal, J.C., Mozdzer, T.J., Fuentes, J.D., D'Odorico, P., McGlathery, K.J., Zieman, J.C., 2008. Tidal influences on carbon assimilation by a salt marsh. Environ. Res. Lett. 3. https://doi.org/10.1088/1748-9326/3/4/044010.

Keenan, T.F., Migliavacca, M., Papale, D., Baldocchi, D., Reichstein, M., Torn, M., Wutzler, T., 2019. Widespread inhibition of daytime ecosystem respiration. Nat. Ecol. Evol. 3, 407–415. https://doi.org/10.1038/s41559-019-0809-2.

Kim, Y., Johnson, M.S., Knox, S.H., Black, T.A., Dalmagro, H.J., Kang, M., Kim, J., Baldocchi, D., 2020. Gap-filling approaches for eddy covariance methane fluxes: a comparison of three machine learning algorithms and a traditional method with principal component analysis. Glob. Chang. Biol. 26, 1499–1518. https://doi.org/10.1111/gcb.14845.

Kirschke, S., Bousquet, P., Ciais, P., Saunois, M., Canadell, J.G., Dlugokencky, E.J., Berga-maschi, P., Bergmann, D., Blake, D.R., Bruhwiler, L., Cameron-Smith, P., Castaldi, S., Chevallier, F., Feng, L., Fraser, A., Heimann, M., Hodson, E.L., Houweling, S., Josse, B., Fraser, P.J., Krummel, P.B., Lamarque, J.F., Langenfelds, R.L., Le Quéré, C., Naik, V., O'doherty, S., Palmer, P.I., Pison, I., Plummer, D., Poulter, B., Prinn, R.G., Rigby, M., Ringeval, B., Santini, M., Schmidt, M., Shindell, D.T., Simpson, I.J., Spahni, R., Steele, L.P., Strode, S.A., Sudo, K., Szopa, S., Van Der Werf, G.R., Voulgarakis, A., Van Weele, M., Weiss, R.F., Williams, J.E., Zeng, G., 2013. Three decades of global methane sources and sinks. Nat. Geosci. https://doi.org/10.1038/ngeo1955.

Knox, S.H., Windham-Myers, L., Anderson, F., Sturtevant, C., Bergamaschi, B., 2018. Direct and indirect effects of tides on ecosystem-scale CO_2 exchange in a brackish tidal marsh in Northern California. J. Geophys. Res. Biogeo. 123, 787–806. https://doi.org/10.1002/2017JG004048.

Knox, S.H., Jackson, R.B., Poulter, B., McNicol, G., Fluet-Chouinard, E., Zhang, Z., Hugelius, G., Bousquet, P., Canadell, J.G., Saunois, M., Papale, D., Chu, H., Keenan, T.F., Baldocchi, D., Torn, M.S., Mammarella, I., Trotta, C., Aurela, M., Bohrer, G., Campbell, D.I., Cescatti, A., Chamberlain, S., Chen, J., Chen, W., Dengel, S., Desai, A.R., Euskirchen, E., Friborg, T., Gasbarra, D., Goded, I., Goeckede, M., Heimann, M., Helbig, M., Hirano, T., Hollinger, D.Y., Iwata, H., Kang, M., Klatt, J., Krauss, K.W., Kutzbach, L., Lohila, A., Mitra, B., Morin, T.H., Nilsson, M.B., Niu, S., Noormets, A., Oechel, W.C., Peichl, M., Peltola, O., Reba, M.L., Richardson, A.D., Runkle, B.R.K., Ryu, Y., Sachs, T., Schäfer, K.V.R., Schmid, H.P., Shurpali, N., Sonnentag, O., Tang, A.C.I., Ueyama, M., Vargas, R., Vesala, T., Ward, E.J., Windham-Myers, L., Wohlfahrt, G., Zona, D., 2019. FLUXNET-CH4 synthesis activity objectives, observations, and future directions. Bull. Am. Meteorol. Soc. 100, 2607–2632. https://doi.org/10.1175/BAMS-D-18-0268.1.

Koebsch, F., Jurasinski, G., Koch, M., Hofmann, J., Glatzel, S., 2015. Controls for multi-scale temporal variation in ecosystem methane exchange during the growing season of a permanently inundated fen. Agric. For. Meteorol. 204, 94–105. https://doi.org/10.1016/j.agrformet.2015.02.002.

Krauss, K.W., Ball, M.C., 2013. On the halophytic nature of mangroves. Trees 27, 7–11. https://doi.org/10.1007/s00468-012-0767-7.

Krauss, K.W., Holm, G.O., Perez, B.C., McWhorter, D.E., Cormier, N., Moss, R.F., Johnson, D.J., Neubauer, S.C., Raynie, R.C., 2016. Component greenhouse gas fluxes and radiative balance from two deltaic marshes in Louisiana: pairing chamber techniques and eddy covariance. J. Geophys. Res. Biogeo. 121, 1503–1521. https://doi.org/10.1002/2015JG003224.

Lai, D.Y.F., 2009. Methane dynamics in northern peatlands: a review. Pedosphere 19, 409–421. https://doi.org/10.1016/S1002-0160(09)00003-4.

Law, B.E., Falge, E., Gu, L., Baldocchi, D.D., Bakwin, P., Berbigier, P., Davis, K., Dolman, A.J., Falk, M., Fuentes, J.D., Goldstein, A., Granier, A., Grelle, A., Hollinger, D., Janssens, I.A., Jarvis, P., Jensen, N.O., Katul, G., Mahli, Y., Matteucci, G., Meyers, T., Monson, R., Munger, W., Oechel, W., Olson, R., Pilegaard, K., Kyaw Tha Paw, U., Thorgeirsson, H., Valentini, R., Verma, S., Vesala, T., Wilson, K., Wofsy, S., 2002. Environmental controls over carbon dioxide and water vapor exchange of terrestrial vegetation. Agric. For. Meteorol. 113, 97–120. https://doi.org/10.1016/S0168-1923(02)00104-1.

Lee, S.C., Fan, C.J., Wu, Z.Y., Juang, J.Y., 2015. Investigating effect of environmental controls on dynamics of CO_2 budget in a subtropical estuarial marsh wetland ecosystem. Environ. Res. Lett. 10. https://doi.org/10.1088/1748-9326/10/2/025005, 025005.

Leopold, A., Marchand, C., Renchon, A., Deborde, J., Quiniou, T., Allenbach, M., 2016. Net ecosystem CO_2 exchange in the "Coeur de Voh" mangrove, New Caledonia: effects of water stress on mangrove productivity in a semi-arid climate. Agric. For. Meteorol. 223, 217–232. https://doi.org/10.1016/j.agrformet.2016.04.006.

Leuning, R., Kelliher, F.M., De Pury, D.G.G., Schulze, E.D., 1995. Leaf nitrogen, photosynthesis, conductance and transpiration: scaling from leaves to canopies. Plant Cell Environ. 18, 1183–1200. https://doi.org/10.1111/j.1365-3040.1995.tb00628.x.

Li, N., Chen, S., Zhou, X., Li, C., Shao, J., Wang, R., Fritz, E., Hüttermann, A., Polle, A., 2008. Effect of NaCl on photosynthesis, salt accumulation and ion compartmentation in two mangrove species, *Kandelia candel* and *Bruguiera gymnorhiza*. Aquat. Bot. 88, 303–310. https://doi.org/10.1016/j.aquabot.2007.12.003.

Li, Q., Lu, W., Chen, H., Luo, Y., Lin, G., 2014. Differential responses of net ecosystem exchange of carbon dioxide to light and temperature between spring and neap tides in subtropical mangrove forests. Sci. World J. 2014. https://doi.org/10.1155/2014/943697.

Li, H., Dai, S., Ouyang, Z., Xie, X., Guo, H., Gu, C., Xiao, X., Ge, Z., Peng, C., Zhao, B., 2018. Multi-scale temporal variation of methane flux and its controls in a subtropical tidal salt marsh in eastern China. Biogeochemistry 137, 163–179. https://doi.org/10.1007/s10533-017-0413-y.

Liu, J., Lai, D.Y.F., 2019. Subtropical mangrove wetland is a stronger carbon dioxide sink in the dry than wet seasons. Agric. For. Meteorol. 278. https://doi.org/10.1016/j.agrformet.2019.107644, 107644.

Liu, L., Wang, D., Chen, S., Yu, Z., Xu, Y., Li, Y., Ge, Z., Chen, Z., 2019. Methane emissions from estuarine coastal wetlands: implications for global change effect. Soil Sci. Soc. Am. J. 83, 1368–1377. https://doi.org/10.2136/sssaj2018.12.047.

Liu, J., Zhou, Y., Valach, A., Shortt, R., Kasak, K., Rey-Sanchez, C., Hemes, K.S., Baldocchi, D., Lai, D.Y.F., 2020. Methane emissions reduce the radiative cooling effect of a subtropical estuarine mangrove wetland by half. Glob. Chang. Biol. 26, 4998–5016. https://doi.org/10.1111/gcb.15247.

López-Hoffman, L., Anten, N.P.R., Martínez-Ramos, M., Ackerly, D.D., 2007. Salinity and light interactively affect neotropical mangrove seedlings at the leaf and whole plant levels. Oecologia 150, 545–556. https://doi.org/10.1007/s00442-006-0563-4.

Lu, W., Xiao, J., Liu, F., Zhang, Y., Liu, C., Lin, G., 2017. Contrasting ecosystem CO_2 fluxes of inland and coastal wetlands: a meta-analysis of eddy covariance data. Glob. Chang. Biol. 23, 1180–1198. https://doi.org/10.1111/gcb.13424.

Lu, W., Xiao, J., Cui, X., Xu, F., Lin, G., Lin, G., 2019. Insect outbreaks have transient effects on carbon fluxes and vegetative growth but longer-term impacts on reproductive growth in a mangrove forest. Agric. For. Meteorol. 279. https://doi.org/10.1016/j.agrformet.2019.107747, 107747.

Luyssaert, S., Inglima, I., Jung, M., Richardson, A.D., Reichstein, M., Papale, D., Piao, S.L., Schulze, E.D., Wingate, L., Matteucci, G., Aragao, L., Aubinet, M., Beer, C., Bernhofer, C., Black, K.G., Bonal, D., Bonnefond, J.M., Chambers, J., Ciais, P., Cook, B., Davis, K.J., Dolman, A.J., Gielen, B., Goulden, M., Grace, J., Granier, A., Grelle, A., Griffis, T., Grünwald, T., Guidolotti, G., Hanson, P.J., Harding, R., Hollinger, D.Y., Hutyra, L.R., Kolari, P., Kruijt, B., Kutsch, W., Lagergren, F., Laurila, T., Law, B.E., Le Maire, G., Lindroth, A., Loustau, D., Malhi, Y., Mateus, J., Migliavacca, M., Misson, L., Montagnani, L., Moncrieff, J., Moors, E., Munger, J.W., Nikinmaa, E., Ollinger, S.V., Pita, G., Rebmann, C., Roupsard, O., Saigusa, N., Sanz, M.J., Seufert, G., Sierra, C., Smith, M.L., Tang, J., Valentini, R., Vesala, T., Janssens, I.A., 2007. CO_2 balance of boreal, temperate, and tropical forests derived from a global database. Glob. Chang. Biol. 13, 2509–2537. https://doi.org/10.1111/j.1365-2486.2007.01439.x.

Maggio, A., Reddy, M.P., Joly, R.J., 2000. Leaf gas exchange and solute accumulation in the halophyte Salvadora persica grown at moderate salinity. Environ. Exp. Bot. 44, 31–38. https://doi.org/10.1016/S0098-8472(00)00051-4.

Maricle, B.R., Cobos, D.R., Campbell, C.S., 2007. Biophysical and morphological leaf adaptations to drought and salinity in salt marsh grasses. Environ. Exp. Bot. 60, 458–467. https://doi.org/10.1016/j.envexpbot.2007.01.001.

McLeod, E., Chmura, G.L., Bouillon, S., Salm, R., Björk, M., Duarte, C.M., Lovelock, C.E., Schlesinger, W.H., Silliman, B.R., 2011. A blueprint for blue carbon: toward an improved understanding of the role of vegetated coastal habitats in sequestering CO_2. Front. Ecol. Environ. 9, 552–560. https://doi.org/10.1890/110004.

Megonigal, J.P., Brewer, P.E., Knee, K.L., 2020. Radon as a natural tracer of gas transport through trees. New Phytol. 225, 1470–1475. https://doi.org/10.1111/nph.16292.

Meijide, A., Manca, G., Goded, I., Magliulo, V., Di Tommasi, P., Seufert, G., Cescatti, A., 2011. Seasonal trends and environmental controls of methane emissions in a rice paddy field in Northern Italy. Biogeosciences 8, 3809–3821. https://doi.org/10.5194/bg-8-3809-2011.

Menzel, A., Sparks, T.H., Estrella, N., Koch, E., Aaasa, A., Ahas, R., Alm-Kübler, K., Bissolli, P., Braslavská, O., Briede, A., Chmielewski, F.M., Crepinsek, Z., Curnel, Y., Dahl, Å., Defila, C., Donnelly, A., Filella, Y., Jatczak, K., Måge, F., Mestre, A., Nordli, Ø., Peñuelas, J., Pirinen, P., Remišová, V., Scheifinger, H., Striz, M., Susnik, A., Van Vliet, A.J.H., Wielgolaski, F.E., Zach, S., Zust, A., 2006. European phenological response to climate change matches the warming pattern. Glob. Chang. Biol. 12, 1969–1976. https://doi.org/10.1111/j.1365-2486.2006.01193.x.

Miao, G., Noormets, A., Domec, J.C., Fuentes, M., Trettin, C.C., Sun, G., McNulty, S.G., King, J.S., 2017. Hydrology and microtopography control carbon dynamics in wetlands: implications in partitioning ecosystem respiration in a coastal plain forested wetland. Agric. For. Meteorol. 247, 343–355. https://doi.org/10.1016/j.agrformet.2017.08.022.

Mitra, B., Minick, K., Miao, G., Domec, J.C., Prajapati, P., McNulty, S.G., Sun, G., King, J.S., Noormets, A., 2020. Spectral evidence for substrate availability rather than environmental control of methane emissions from a coastal forested wetland. Agric. For. Meteorol. 291. https://doi.org/10.1016/j.agrformet.2020.108062, 108062.

Moderow, U., Feigenwinter, C., Bernhofer, C., 2011. Non-turbulent fluxes of carbon dioxide and sensible heat-a comparison of three forested sites. Agric. For. Meteorol. 151, 692–708. https://doi.org/10.1016/j.agrformet.2011.01.014.

Moffett, K.B., Wolf, A., Berry, J.A., Gorelick, S.M., 2010. Salt marsh-atmosphere exchange of energy, water vapor, and carbon dioxide: effects of tidal flooding and biophysical controls. Water Resour. Res. 46, 1–18. https://doi.org/10.1029/2009WR009041.

Moncrieff, J.B., Massheder, J.M., De Bruin, H., Elbers, J., Friborg, T., Heusinkveld, B., Kabat, P., Scott, S., Soegaard, H., Verhoef, A., 1997. A system to measure surface fluxes of momentum, sensible heat, water vapour and carbon dioxide. J. Hydrol. 188–189, 589–611. https://doi.org/10.1016/S0022-1694(96)03194-0.

Moncrieff, J., Clement, R., Finnigan, J., Meyers, T., 2006. Averaging, detrending, and filtering of eddy covariance time series. In: Lee, X., Massman, W., Law, B. (Eds.), Handbook of Micrometeorology. Springer, Netherlands, Dordrecht, pp. 7–31, https://doi.org/10.1007/1-4020-2265-4_2.

Montgomery, R.B., 1948. Vertical eddy flux of heat in the atmosphere. J. Meteorol. 5, 265–274. https://doi.org/10.1175/1520-0469(1948)005<0265:vefohi>2.0.co;2.

Moore, T.R., Roulet, N.T., 1993. Methane flux: water table relations in northern wetlands. Geophys. Res. Lett. 20, 587–590. https://doi.org/10.1029/93GL00208.

Morin, T.H., 2019. Advances in the eddy covariance approach to CH_4 monitoring over two and a half decades. J. Geophys. Res. Biogeo. 124, 453–460. https://doi.org/10.1029/2018JG004796.

Morin, T.H., Bohrer, G., Naor-Azrieli, L., Mesi, S., Kenny, W.T., Mitsch, W.J., Schäfer, K.V.R., 2014. The seasonal and diurnal dynamics of methane flux at a created urban wetland. Ecol. Eng. 72, 74–83. https://doi.org/10.1016/j.ecoleng.2014.02.002.

Nahrawi, H., Leclerc, M.Y., Pennings, S., Zhang, G., Singh, N., Pahari, R., 2020. Impact of tidal inundation on the net ecosystem exchange in daytime conditions in a salt marsh. Agric. For. Meteorol. 294, 108133. https://doi.org/10.1016/j.agrformet.2020.108133.

Negandhi, K., Edwards, G., Kelleway, J.J., Howard, D., Safari, D., Saintilan, N., 2019. Blue carbon potential of coastal wetland restoration varies with inundation and rainfall. Sci. Rep. 9, 1–9. https://doi.org/10.1038/s41598-019-40763-8.

Nguyen, H.T., Stanton, D.E., Schmitz, N., Farquhar, G.D., Ball, M.C., 2015. Growth responses of the mangrove Avicennia marina to salinity: development and function of shoot hydraulic systems require saline conditions. Ann. Bot. 115, 397–407. https://doi.org/10.1093/aob/mcu257.

Noe, G.B., Zedler, J.B., 2001. Variable rainfall limits the germination of upper intertidal marsh plants in southern California. Estuaries. https://doi.org/10.2307/1352810.

Norman, J.M.M., 1979. Modeling the complete crop canopy. In: Barfield, B.J., Gerber, J.F. (Eds.), Modification of the Aerial Environment of Plants. American Society of Agricultural Engineering, St. Joseph, MI, pp. 249–277.

Obukhov, A.M., 1951. Characteristics of the micro-structure of the wind in the surface layer of the atmosphere. Izv. AN SSSR, ser. Geofiz. 3, 49–68.

Olefeldt, D., Turetsky, M.R., Crill, P.M., Mcguire, A.D., 2013. Environmental and physical controls on northern terrestrial methane emissions across permafrost zones. Glob. Chang. Biol. 19, 589–603. https://doi.org/10.1111/gcb.12071.

Oliphant, A.J., Stoy, P.C., 2018. An evaluation of semiempirical models for partitioning photosynthetically active radiation into diffuse and direct beam components. J. Geophys. Res. Biogeo. 123, 889–901. https://doi.org/10.1002/2017JG004370.

Osudar, R., Matoušů, A., Alawi, M., Wagner, D., Bussmann, I., 2015. Environmental factors affecting methane distribution and bacterial methane oxidation in the German Bight (North Sea). Estuar. Coast. Shelf Sci. 160, 10–21. https://doi.org/10.1016/j.ecss.2015.03.028.

Ozuolmez, D., Na, H., Lever, M.A., Kjeldsen, K.U., Jørgensen, B.B., Plugge, C.M., 2015. Methanogenic archaea and sulfate reducing bacteria co-cultured on acetate: teamwork or coexistence? Front. Microbiol. 6. https://doi.org/10.3389/fmicb.2015.00492.

Papale, D., Valentini, R., 2003. A new assessment of European forests carbon exchanges by eddy fluxes and artificial neural network spatialization. Glob. Chang. Biol. 9, 525–535. https://doi.org/10.1046/j.1365-2486.2003.00609.x.

Papale, D., Reichstein, M., Aubinet, M., Canfora, E., Bernhofer, C., Kutsch, W., Longdoz, B., Rambal, S., Valentini, R., Vesala, T., Yakir, D., 2006. Towards a standardized processing of Net Ecosystem Exchange measured with eddy covariance technique: algorithms and uncertainty estimation. Biogeosciences 3, 571–583. https://doi.org/10.5194/bg-3-571-2006.

Pastorello, G., Papale, D., Chu, H., Trotta, C., Agarwal, D., Canfora, E., Baldocchi, D., Torn, M., 2017. A new data set to keep a sharper eye on land-air exchanges. Eos (Washington DC), 98. https://doi.org/10.1029/2017eo071597.

Pastor-Guzman, J., Dash, J., Atkinson, P.M., 2018. Remote sensing of mangrove forest phenology and its environmental drivers. Remote Sens. Environ. 205, 71–84. https://doi.org/10.1016/j.rse.2017.11.009.

Poffenbarger, H.J., Needelman, B.A., Megonigal, J.P., 2011. Salinity influence on methane emissions from tidal marshes. Wetlands 31, 831–842. https://doi.org/10.1007/s13157-011-0197-0.

Polsenaere, P., Lamaud, E., Lafon, V., Bonnefond, J.M., Bretel, P., Delille, B., Deborde, J., Loustau, D., Abril, G., 2012. Spatial and temporal CO_2 exchanges measured by Eddy Covariance over a temperate intertidal flat and their relationships to net ecosystem production. Biogeosciences 9, 249–268. https://doi.org/10.5194/bg-9-249-2012.

Ramsar Convention on Wetlands, 2018. Global wetland outlook: state of the world's wetlands and their services to people. Ramsar Conv. Wetl. 2018, 88.

Ray, R., Chowdhury, C., Majumder, N., Dutta, M.K., Mukhopadhyay, S.K., Jana, T.K., 2013. Improved model calculation of atmospheric CO_2 increment in affecting carbon stock of tropical mangrove forest. Tellus B Chem. Phys. Meteorol. 65, 1–11. https://doi.org/10.3402/tellusb.v65i0.18981.

Rayment, M.B., Loustau, D., Jarvis, P.G., 2000. Measuring and modeling conductances of black spruce at three organizational scales: shoot, branch and canopy. Tree Physiol. https://doi.org/10.1093/treephys/20.11.713.

Redondo-Gómez, S., Mateos-Naranjo, E., Davy, A.J., Fernández-Muñoz, F., Castellanos, E.M., Luque, T., Figueroa, M.E., 2007. Growth and photosynthetic responses to salinity of the salt-marsh shrub Atriplex portulacoides. Ann. Bot. 100, 555–563. https://doi.org/10.1093/aob/mcm119.

Reichstein, M., Falge, E., Baldocchi, D., Papale, D., Aubinet, M., Berbigier, P., Bernhofer, C., Buchmann, N., Gilmanov, T., Granier, A., Grünwald, T., Havránková, K., Ilvesniemi, H., Janous, D., Knohl, A., Laurila, T., Lohila, A., Loustau, D., Matteucci, G., Meyers, T., Miglietta, F., Ourcival, J.M., Pumpanen, J., Rambal, S., Rotenberg, E., Sanz, M., Tenhunen, J., Seufert, G., Vaccari, F., Vesala, T., Yakir, D., Valentini, R., 2005. On the separation of net ecosystem exchange into assimilation and ecosystem respiration: review and improved algorithm. Glob. Chang. Biol. 11, 1424–1439. https://doi.org/10.1111/j.1365-2486.2005.001002.x.

Reid, M.C., Tripathee, R., Schäfer, K.V.R., Jaffé, P.R., 2013. Tidal marsh methane dynamics: difference in seasonal lags in emissions driven by storage in vegetated versus unvegetated sediments. J. Geophys. Res. Biogeo. 118, 1802–1813. https://doi.org/10.1002/2013JG002438.

Reynolds, O., 1895. On the dynamical theory of incompressible viscous fluids and the determination of the criterion. Philos. Trans. R. Soc. Lond. A A174, 935–982. https://doi.org/10.1098/rsta.1895.0004.

Rey-Sanchez, A.C., Morin, T.H., Stefanik, K.C., Wrighton, K., Bohrer, G., 2018. Determining total emissions and environmental drivers of methane flux in a Lake Erie estuarine marsh. Ecol. Eng. 114, 7–15. https://doi.org/10.1016/j.ecoleng.2017.06.042.

Rodda, S.R., Thumaty, K.C., Jha, C.S., Dadhwal, V.K., 2016. Seasonal variations of carbon dioxide, water vapor and energy fluxes in tropical Indian mangroves. Forests 7, 1–18. https://doi.org/10.3390/f7020035.

Rogers, K., Kelleway, J.J., Saintilan, N., Megonigal, J.P., Adams, J.B., Holmquist, J.R., Lu, M., Schile-Beers, L., Zawadzki, A., Mazumder, D., Woodroffe, C.D., 2019. Wetland carbon storage controlled by millennial-scale variation in relative sea-level rise. Nature 567, 91–95. https://doi.org/10.1038/s41586-019-0951-7.

Roose, J.L., Frankel, L.K., Mummadisetti, M.P., Bricker, T.M., 2016. The extrinsic proteins of photosystem II: update. Planta 243, 889–908. https://doi.org/10.1007/s00425-015-2462-6.

Rosentreter, J.A., Maher, D.T., Erler, D.V., Murray, R.H., Eyre, B.D., 2018. Methane emissions partially offset "blue carbon" burial in mangroves. Sci. Adv. 4. https://doi.org/10.1126/sciadv.aao4985.

Ryu, Y., Baldocchi, D.D., Kobayashi, H., Van Ingen, C., Li, J., Black, T.A., Beringer, J., Van Gorsel, E., Knohl, A., Law, B.E., Roupsard, O., 2011. Integration of MODIS land and atmosphere products with a coupled-process model to estimate gross primary productivity and evapotranspiration from 1 km to global scales. Global Biogeochem. Cycles 25, 1–24. https://doi.org/10.1029/2011GB004053.

Sachs, T., Giebels, M., Boike, J., Kutzbach, L., 2010. Environmental controls on CH_4 emission from polygonal tundra on the microsite scale in the Lena river delta, Siberia. Glob. Chang. Biol. 16, 3096–3110. https://doi.org/10.1111/j.1365-2486.2010.02232.x.

Safari, D., Edwards, G., Gyabaah, F., 2020. Diurnal and seasonal variation of CO_2 and CH_4 fluxes in tomago wetland. Int. J. Sci. 9, 41–51. https://doi.org/10.18483/ijsci.2229.

Saleska, S.R., Miller, S.D., Matross, D.M., Goulden, M.L., Wofsy, S.C., Da Rocha, H.R., De Camargo, P.B., Crill, P., Daube, B.C., De Freitas, H.C., Hutyra, L., Keller, M., Kirchhoff, V., Menton, M., Munger, J.W., Pyle, E.H., Rice, A.H., Silva, H., 2003. Carbon in amazon forests: unexpected seasonal fluxes and disturbance-induced losses. Science 302, 1554–1557. https://doi.org/10.1126/science.1091165.

Schäfer, K.V.R., Tripathee, R., Artigas, F., Morin, T.H., Bohrer, G., 2014. Carbon dioxide fluxes of an urban tidal marsh in the Hudson-Raritan estuary. J. Geophys. Res. G: Biogeosci. 119, 2065–2081. https://doi.org/10.1002/2014JG002703.

Schäfer, K.V.R., Duman, T., Tomasicchio, K., Tripathee, R., Sturtevant, C., 2019. Carbon dioxide fluxes of temperate urban wetlands with different restoration history. Agric. For. Meteorol. 275, 223–232. https://doi.org/10.1016/j.agrformet.2019.05.026.

Schmid, H.P., 1994. Source areas for scalars and scalar fluxes. Bound.-Lay. Meteorol. 67, 293–318. https://doi.org/10.1007/BF00713146.

Smith, N.G., Dukes, J.S., 2017. Short-term acclimation to warmer temperatures accelerates leaf carbon exchange processes across plant types. Glob. Chang. Biol. 23, 4840–4853. https://doi.org/10.1111/gcb.13735.

Sprintsin, M., Chen, J.M., Desai, A., Gough, C.M., 2012. Evaluation of leaf-to-canopy upscaling methodologies against carbon flux data in North America. J. Geophys. Res. Biogeo. 117. https://doi.org/10.1029/2010JG001407.

Starr, G., Jarnigan, J.R., Staudhammer, C.L., Cherry, J.A., 2018. Variation in ecosystem carbon dynamics of saltwater marshes in the northern Gulf of Mexico. Wetl. Ecol. Manag. 26, 581–596. https://doi.org/10.1007/s11273-018-9593-z.

Stoy, P.C., Katul, G.G., Siqueira, M.B.S., Juang, J.Y., McCarthy, H.R., Kim, H.S., Oishi, A.C., Oren, R., 2005. Variability in net ecosystem exchange from hourly to inter-annual time scales at adjacent pine and hardwood forests: a wavelet analysis. Tree Physiol. 25, 887–902. https://doi.org/10.1093/treephys/25.7.887.

Sturtevant, C., Ruddell, B.L., Knox, S.H., Verfaillie, J., Matthes, J.H., Oikawa, P.Y., Baldocchi, D., 2016. Identifying scale-emergent, nonlinear, asynchronous processes of wetland methane exchange. J. Geophys. Res. Biogeo. 121, 188–204. https://doi.org/10.1002/2015JG003054.

Sun, Z., Jiang, H., Wang, L., Mou, X., Sun, W., 2013a. Seasonal and spatial variations of methane emissions from coastal marshes in the northern Yellow River estuary, China. Plant and Soil 369, 317–333. https://doi.org/10.1007/s11104-012-1564-1.

Sun, Z., Mou, X., Tian, H., Song, H., Jiang, H., Zhao, J., Sun, W., Sun, W., 2013b. Phosphorus biological cycle in the different Suaeda salsa marshes of the Yellow River estuary, China. Environ. Earth Sci. 69, 2595–2608. https://doi.org/10.1007/s12665-012-2081-5.

Sun, L., Song, C., Lafleur, P.M., Miao, Y., Wang, X., Gong, C., Qiao, T., Yu, X., Tan, W., 2018. Wetland-atmosphere methane exchange in Northeast China: a comparison of permafrost peatland and freshwater wetlands. Agric. For. Meteorol. 249, 239–249. https://doi.org/10.1016/j.agrformet.2017.11.009.

Swinbank, W.C., 1951. The measurement of vertical transfer of heat and water vapor by eddies in the lower atmosphere. J. Meteorol. 8, 135–145. https://doi.org/10.1175/1520-0469(1951)008<0135:tmovto>2.0.co;2.

Tokida, T., Miyazaki, T., Mizoguchi, M., 2005. Ebullition of methane from peat with falling atmospheric pressure. Geophys. Res. Lett. 32 (13). https://doi.org/10.1029/2005GL022949, L13823.

Tong, C., Huang, J.F., Hu, Z.Q., Jin, Y.F., 2013. Diurnal variations of carbon dioxide, methane, and nitrous oxide vertical fluxes in a subtropical estuarine marsh on neap and spring tide days. Estuaries Coast 36, 633–642. https://doi.org/10.1007/s12237-013-9596-1.

Treat, C.C., Bloom, A.A., Marushchak, M.E., 2018. Nongrowing season methane emissions–a significant component of annual emissions across northern ecosystems. Glob. Chang. Biol. 24, 3331–3343. https://doi.org/10.1111/gcb.14137.

Turetsky, M.R., Kotowska, A., Bubier, J., Dise, N.B., Crill, P., Hornibrook, E.R.C., Minkkinen, K., Moore, T.R., Myers-Smith, I.H., Nykänen, H., Olefeldt, D., Rinne, J., Saarnio, S., Shurpali, N., Tuittila, E.S., Waddington, J.M., White, J.R., Wickland, K.P., Wilmking, M., 2014. A synthesis of methane emissions from 71 northern, temperate, and subtropical wetlands. Glob. Chang. Biol. 20, 2183–2197. https://doi.org/10.1111/gcb.12580.

USGCRP, 2018. Impacts, Risks, and Adaptation in the United States: Fourth National Climate Assessment. U.S. Global Change Research Program, Washington, DC, https://doi.org/10.7930/NCA4.2018.

Vázquez-Lule, A., Vargas, R., 2021. Biophysical drivers of net ecosystem and methane exchange across phenological phases in a tidal salt marsh. Agric. For. Meteorol. 300. https://doi.org/10.1016/j.agrformet.2020.108309.

Vickers, D., Mahrt, L., 1997. Quality control and flux sampling problems for tower and aircraft data. J. Atmos. Oceanic Tech. 14, 512–526. https://doi.org/10.1175/1520-0426(1997)014<0512:QCAFSP>2.0.CO;2.

Villa, J.A., Ju, Y., Stephen, T., Rey-Sanchez, C., Wrighton, K.C., Bohrer, G., 2020. Plant-mediated methane transport in emergent and floating-leaved species of a temperate freshwater mineral-soil wetland. Limnol. Oceanogr. 65, 1635–1650. https://doi.org/10.1002/lno.11467.

Waddington, J.M., Harrison, K., Kellner, E., Baird, A.J., 2009. Effect of atmospheric pressure and temperature on entrapped gas content in peat. Hydrol. Processes 23, 2970–2980. https://doi.org/10.1002/hyp.7412.

Walter, B.P., Heimann, M., 2000. A process-based, climate-sensitive model to derive methane emissions from natural wetlands: application to five wetland sites, sensitivity to model parameters, and climate. Global Biogeochem. Cycles 14, 745–765. https://doi.org/10.1029/1999GB001204.

Wang, Q., Wang, C.H., Zhao, B., Ma, Z.J., Luo, Y.Q., Chen, J.K., Li, B., 2006. Effects of growing conditions on the growth of and interactions between salt marsh plants: implications for invasibility of habitats. Biol. Invasions 8, 1547–1560. https://doi.org/10.1007/s10530-005-5846-x.

Wang, W., Yan, Z., You, S., Zhang, Y., Chen, L., Lin, G., 2011. Mangroves: obligate or facultative halophytes? A review. Trees 25, 953–963. https://doi.org/10.1007/s00468-011-0570-x.

Watanabe, I., Hashimoto, T., Shimoyama, A., 1997. Methane-oxidizing activities and methanotrophic populations associated with wetland rice plants. Biol. Fertil. Soils 24, 261–265. https://doi.org/10.1007/s003740050241.

Waycott, M., Duarte, C.M., Carruthers, T.J.B., Orth, R.J., Dennison, W.C., Olyarnik, S., Calladine, A., Fourqurean, J.W., Heck, K.L., Hughes, A.R., Kendrick, G.A., Kenworthy, W.J., Short, F.T., Williams, S.L., 2009. Accelerating loss of seagrasses across the globe threatens coastal ecosystems. Proc. Natl. Acad. Sci. U. S. A. 106, 12377–12381. https://doi.org/10.1073/pnas.0905620106.

Webb, E.K., Pearman, G.I., Leuning, R., 1980. Correction of flux measurements for density effects due to heat and water vapour transfer. Q. J. Roy. Meteorol. Soc. 106, 85–100. https://doi.org/10.1002/qj.49710644707.

Wei, S., Han, G., Jia, X., Song, W., Chu, X., He, W., Xia, J., Wu, H., 2020. Tidal effects on ecosystem CO_2 exchange at multiple timescales in a salt marsh in the Yellow River Delta. Estuar. Coast. Shelf Sci. 238. https://doi.org/10.1016/j.ecss.2020.106727, 106727.

Welles, J.M., Demetriades-Shah, T.H., McDermitt, D.K., 2001. Considerations for measuring ground CO_2 effluxes with chambers. Chem. Geol. 177, 3–13. https://doi.org/10.1016/S0009-2541(00)00388-0.

Whiting, G.J., Chanton, J.P., 1993. Primary production control of methane emission from wetlands. Nature 364, 794–795. https://doi.org/10.1038/364794a0.

Wilczak, J.M., Oncley, S.P., Stage, S.A., 2001. Sonic anemometer tilt correction algorithms. Bound.-Lay. Meteorol. 99, 127–150. https://doi.org/10.1023/A:1018966204465.

Wille, C., Kutzbach, L., Sachs, T., Wagner, D., Pfeiffer, E.M., 2008. Methane emission from Siberian arctic polygonal tundra: eddy covariance measurements and modeling. Glob. Chang. Biol. 14, 1395–1408. https://doi.org/10.1111/j.1365-2486.2008.01586.x.

Windsor, J., Moore, T.R., Roulet, N.T., 1992. Episodic fluxes of methane from subarctic fens. Can. J. Soil Sci. 72, 441–452. https://doi.org/10.4141/cjss92-037.

Wofsy, S.C., Goulden, M.L., Munger, J.W., Fan, S.M., Bakwin, P.S., Daube, B.C., Bassow, S.L., Bazzaz, F.A., 1993. Net exchange of CO_2 in a mid-latitude forest. Science 260, 1314–1317. https://doi.org/10.1126/science.260.5112.1314.

Wright, I.J., Reich, P.B., Westoby, M., Ackerly, D.D., Baruch, Z., Bongers, F., Cavender-Bares, J., Chapin, T., Cornellssen, J.H.C., Diemer, M., Flexas, J., Garnier, E., Groom, P.K., Gulias, J., Hikosaka, K., Lamont, B.B., Lee, T., Lee, W., Lusk, C., Midgley, J.J., Navas, M.L., Niinemets, Ü., Oleksyn, J., Osada, H., Poorter, H., Pool, P., Prior, L., Pyankov, V.I., Roumet, C., Thomas, S.C., Tjoelker, M.G., Veneklaas, E.J., Villar, R., 2004. The worldwide leaf economics spectrum. Nature 428, 821–827. https://doi.org/10.1038/nature02403.

Wutzler, T., Lucas-Moffat, A., Migliavacca, M., Knauer, J., Sickel, K., Šigut, L., Menzer, O., Reichstein, M., 2018. Basic and extensible post-processing of eddy covariance flux data with REddyProc. Biogeosci. Discuss. 1–39. https://doi.org/10.5194/bg-2018-56.

Xu, L., Baldocchi, D.D., 2004. Seasonal variation in carbon dioxide exchange over a Mediterranean annual grassland in California. Agric. For. Meteorol. 123, 79–96. https://doi.org/10.1016/j.agrformet.2003.10.004.

Yamamoto, A., Hirota, M., Suzuki, S., Oe, Y., Zhang, P., Mariko, S., 2009. Effects of tidal fluctuations on CO_2 and CH_4 fluxes in the littoral zone of a brackish-water lake. Limnology 10, 229–237. https://doi.org/10.1007/s10201-009-0284-6.

Yan, Y., Zhao, B., Chen, J., Guo, H., Gu, Y., Wu, Q., Li, B., 2008. Closing the carbon budget of estuarine wetlands with tower-based measurements and MODIS time series. Glob. Chang. Biol. 14, 1690–1702. https://doi.org/10.1111/j.1365-2486.2008.01589.x.

Yang, S.C., Shih, S.S., Hwang, G.W., Adams, J.B., Lee, H.Y., Chen, C.P., 2013. The salinity gradient influences on the inundation tolerance thresholds of mangrove forests. Ecol. Eng. 51, 59–65. https://doi.org/10.1016/j.ecoleng.2012.12.049.

Yvon-Durocher, G., Allen, A.P., Bastviken, D., Conrad, R., Gudasz, C., St-Pierre, A., Thanh-Duc, N., Del Giorgio, P.A., 2014. Methane fluxes show consistent temperature dependence across microbial to ecosystem scales. Nature 507, 488–491. https://doi.org/10.1038/nature13164.

Zeri, M., Sá, L.D.A., Manzi, A.O., Araú, A.C., Aguiar, R.G., Von Randow, C., Sampaio, G., Cardoso, F.L., Nobre, C.A., 2014. Variability of carbon and water fluxes following climate extremes over a tropical forest in southwestern amazonia. PLoS One 9. https://doi.org/10.1371/journal.pone.0088130.

Zhai, X., Piwpuan, N., Arias, C.A., Headley, T., Brix, H., 2013. Can root exudates from emergent wetland plants fuel denitrification in subsurface flow constructed wetland systems? Ecol. Eng. 61, 555–563. https://doi.org/10.1016/j.ecoleng.2013.02.014.

Zhang, Y., Tan, Z., Song, Q., Yu, G., Sun, X., 2010. Respiration controls the unexpected seasonal pattern of carbon flux in an Asian tropical rain forest. Atmos. Environ. 44, 3886–3893. https://doi.org/10.1016/j.atmosenv.2010.07.027.

Zhang, Y., Parazoo, N.C., Williams, A.P., Zhou, S., Gentine, P., 2020. Large and projected strengthening moisture limitation on end-of-season photosynthesis. Proc. Natl. Acad. Sci., 1–7. https://doi.org/10.1073/pnas.1914436117.

Zhong, Q., Wang, K., Lai, Q., Zhang, C., Zheng, L., Wang, J., 2016. Carbon dioxide fluxes and their environmental control in a reclaimed coastal wetland in the Yangtze estuary. Estuaries Coast 39, 344–362. https://doi.org/10.1007/s12237-015-9997-4.

Zhou, L., Zhou, G., Jia, Q., 2009. Annual cycle of CO_2 exchange over a reed (*Phragmites australis*) wetland in Northeast China. Aquat. Bot. 91, 91–98. https://doi.org/10.1016/j.aquabot.2009.03.002.

Macrofaunal consumption as a mineralization pathway

Shing Yip Lee[a] and Cheuk Yan Lee[b]

[a]*Institute of Environment, Energy and Sustainability, Simon F.S. Li Marine Science Laboratory, School of Life Sciences, The Chinese University of Hong Kong, Hong Kong Special Administrative Region, China,*
[b]*Simon F. S. Li Marine Science Laboratory, School of Life Sciences, The Chinese University of Hong Kong, Hong Kong Special Administrative Region, China*

5.1 Introduction

Heterotrophic consumption and subsequent respiration of reduced carbon available in coastal wetlands is a major pathway of mineralization. Early studies on coastal wetland carbon dynamics focused on carbon sequestration by measuring primary productivity of vascular plants (salt marsh, mangrove, and seagrass species) (Odum, 1968), establishing the paradigm that coastal wetlands are globally among the most productive habitats. This organic production was then thought to be largely exported by tides to support offshore consumer communities (Teal, 1962; Odum, 1980) after gradual transformation by microbes (Tenore, 1977). This process of microbial enrichment, particularly of nitrogen, could be lengthy for vascular plant detritus (e.g., >16 weeks in *Spartina alterniflora*, Tenore et al., 1984). Despite the early emphasis on export, there was little direct evidence of both quantified actual export or assimilation of coastal wetland C by ex situ consumers. Lee (1995) suggested that most early studies actually indicated only modest levels of tidal export from mangrove forests. Duarte and Cebrian (1996) and Cebrián (2002) were some of the first studies to suggest at ecosystem level that the fate of primary production by different autotrophs (i.e., seagrass, mangrove and salt marsh plants, micro and macroalgae) differed significantly according to their origin, with heterotrophic metabolism being highest among the microphytobenthos (MPB) and the phytoplankton, whereas burial and storage were highest for vascular plant production. Earlier studies on particular wetland types had, however, documented widely different rates of organic matter accumulation. For example, mangrove forests with restricted tidal exchange were found to accumulate significantly higher proportions of their litter production (Twilley, et al., 1986; Lee, 1989b, 1990b). Low tidal export from salt marshes dominated by grasses is actually expected, as unlike mangroves, they do not abscise senescent leaves, which are usually only decomposed in situ as "standing dead" litter (Currin et al., 1995).

Work in the 1980s discovered that in situ consumption by macrofauna, particularly the diverse sesarmid crabs, could explain the apparent rapid loss of litter and thus low litter standing biomass, in tropical mangrove forests, previously attributed to tidal export (Robertson, 1986; Lee, 1989a; Robertson and Daniel, 1989) (Table 5.1). This consumption could account for up to >75% of the annual leaf litter production. This trophic link was later found to be significant in both the Indo-west-Pacific but also the less speciose Atlantic coast, where initial processing of mangrove C is mediated by ucidid crabs (Nordhaus et al., 2006). This link is also significant in that some crabs are behaviorally (Giddins et al., 1986; Harada and Lee, 2016) and/or physiologically (Bui and Lee, 2015a,b) adapted to capturing freshly senescent mangrove litter for consumption. This novel pathway drives further trophic links (Lee, 1997; Werry and Lee, 2005) and has helped revise the paradigm on C flow in tropical mangrove ecosystems (Alongi, 2009b).

Table 5.1 Consumption of mangrove litter production by brachyuran crabs.

Crab species	Removal percentages/ mangrove types	Location	Reference
Ucides cordatus	81.3% of *Rhizophora mangle* leaf production	Brazilian mangrove forest	Schories et al. (2003)
	90% *Avicennia* leaves 86% *Laguncularia* leaves 54% *Rhizophora* leaves 68% *Rhizophora* propagules	North Brazil mangrove	Nordhaus et al. (2006)
Parasesarma spp., *Neosarmatium* spp.	71% of *Ceriops* leaf production 79% of *Bruguiera* leaf production 33% of *Avicennia marina* leaf production 37.5%–85% mangrove leaf material	Northern Australia	Robertson and Daniel (1989)
Parasesarma messa	28% of leaf production of *Rhizophora* spp.	North Australia	Robertson (1986)
Neosarmatium trispinosum	50% of fresh *Rhizophora stylosa* leaf litter	Queensland, Australia	Harada and Lee (2016)
Parasesarma onychophorum, *Parasesarma eumolpe*	79±3% mixed *Rhizophora, Avicennia,* and *Bruguiera* 9%–30% *Rhizophora apiculata, R. stylosa*	Peninsular Malaysia	Ashton (2002) Leh and Sasekumar (1985)

Table 5.1 Consumption of mangrove litter production by brachyuran crabs—cont'd

Crab species	Removal percentages/ mangrove types	Location	Reference
Parasesarma bidens, Parasesarma affine	>57% Kandelia obovata leaf litter production	Hong Kong	Lee (1989a, b)
Neosarmatium meinerti	44% of Avicennia marina	South Africa	Emmerson and McGwynne (1992)
	>100% of Avicennia marina		Ólafsson et al. (2002)
Neosarmatium malabaricum	80% of leaves (Excoecaria agallocha, Avicennia officinalis, Aegiceras corniculatum, R. mucronata) collected by crabs		Shanij et al. (2016)
Parasesarma plicatum	88.7% of daily K. obovata leaf litter fall		Chen and Ye (2008)
Neosarmatium meinerti	99.1% of Bruguiera gymnorhiza leaves		Steinke et al. (1993)
Guinearma huzardi, Guinearma alberti, Goniopsis pelii, Armases elegans, Metopograpsus curvatus	40%–80% (Avicennia germinans, Laguncularia racemosa, R. mangle)		Simon (2018)

Where appropriate, the most current taxonomic names are used in place of the original names.

The lack of understanding of the physiological mechanism that enables direct macrofaunal consumption of vascular plant detritus in coastal wetlands fuels the continued debate on the significance of this mineralization pathway. Data from studies using chemical tracers (e.g., stable isotopes, fatty acid biomarkers) suggest equivocal results from within even the same macrofaunal groups in terms of their dependence on vascular plant detritus as opposed to presumably more easily digestible C sources, e.g., the MPB or imported phytoplankton (e.g., Bouillon et al., 2002; Oakes et al., 2010; Raw et al., 2017). Despite their expected ease in assimilation by macrofauna, the MPB assemblage and their role in coastal wetland C dynamics are still obscure. Early reviews of coastal vegetated wetland C budgets suggested a minor contribution to the total C pool due to heavy shading under closed mangrove canopies (Alongi, 1994, 2009a), but this limitation may be less in incomplete canopies or saltmarshes. The notion that mangrove forest sediment does not support a productive MPB community has recently been challenged (Kwon et al., 2020). While this study reported reduced (up to 50%) MPB primary productivity in vegetated wetlands, MPB biomass in an Australian mangrove was almost double that of unvegetated tidal flats.

Another missing piece in the macrofauna-mediated mineralization pathways is the assimilation of dissolved organic C (DOC), which is ready available in vegetated coastal wetlands (Dittmar et al., 2006; Marchand et al., 2006; Kristensen et al., 2008). Significant amounts of free organic carbon are also produced by the MPB as extracellular polymeric substances (EPS) on open tidal flats and also on the substrate of vegetated wetlands (Underwood, 2002). While bioturbation by macrofauna, e.g., burrow construction, may strongly influence DOC movement and mineralization (Call et al., 2019), few studies have investigated the consumption or uptake of DOC by macrofauna.

This chapter will review the main C mineralization pathways mediated by the macrofauna in coastal wetlands, across the range of production sources as well as habitat types. Mineralization of wetland primary production by macrofaunal consumption is mainly mediated by animal respiration and increased rate of nutrient release from processed organic matter. As organic matter from coastal wetland production varies widely in quality and quantity, the physiological mechanisms by which low-quality vascular plant production is assimilated by wetland macrofauna will be discussed. Indirect nonconsumptive effects, such as bioturbation, will be discussed separately elsewhere in this book.

5.2 The C pool in coastal wetlands

Despite the dynamic and harsh physical environment, coastal wetlands are among the most productive ecosystems globally. Net aerial primary productivity (NPP) of macrophytes in vegetated coastal wetlands, such as mangrove forests and salt marshes, varies widely according to factors like latitude, stand/patch characteristics (species, tidal position, plant density), and salinity, but can reach $1000\,g\,C\,m^{-2}$ $year^{-1}$ (Alongi, 2009a; Mendelssohn and Morris, 2000). It should be noted, however, that estimates of macrophyte production vary widely with the methods used, and most may not represent biomass production that will be subjected to macrofauna mineralization, e.g., the "light attenuation" method for mangrove productivity (Bunt et al., 1979).

Macroalgae occurrence in mangroves and saltmarshes is often seasonal and dependent on local nutrient regimes. Eutrophication usually leads to significant macroalgal bloom, especially of chlorophytes, e.g., *Enteromorpha* spp. Epiphytic macroalgae are commonly associated with certain mangrove structures, e.g., the pneumatophore of *Avicennia marina*, which are relatively well studied (e.g., Saifullah and Ahmed, 2007; Steinke et al., 2003). Macroalgae usually only contribute a relatively low percentage of the overall C pool in coastal wetlands. Lee (1990b) assessed that the macroalgal assemblage in a mangrove wetland in Hong Kong (mainly *Enteromorpha* spp.) reached a peak biomass of $\sim7\,g\,C\,m^{-2}$ during the winter, with a net annual primary productivity estimated at 2.3% of the overall primary production of a 10ha habitat area.

MPB production is less studied in coastal wetlands, particularly in mangroves. MPB production is usually an order of magnitude smaller than their associated vascular plants in salt marshes (Sullivan and Currin, 2000), with light availability (e.g., in stands of tall/short forms of *Spartina*) and aridity being influential factors on the relative levels of the two. Some coastal wetlands, where canopy shading is significant, e.g., mangroves, were assumed to have suppressed MPB production because of light limitation and high concentrations of sediment tannins (Alongi, 1994). Studies of the MPB assemblages of salt marshes, however, recorded highly productive communities shaped by factors such as the tidal position and height of the associated vegetation (e.g., short vs long *Spartina* grass) (Sullivan and Currin, 2000). Kwon et al. (2020) reported MPB primary productivity under a mangrove canopy in Australia at $81\,mg\,C\,m^{-2}\,h^{-1}$, which is higher than the bare flat or salt marsh sites measured in the same study. However, the relatively higher Chl-a concentration ($102\,mg\,Chl\text{-}a\,m^{-2}$) resulted in a lower primary productivity ($0.79\,mg\,C\,mg\,Chl\text{-}a^{-1}\,m^{-2}\,h^{-1}$) than in the open tidal flats at the same location. A similar pattern was found at a location in Cambodia (Kwon et al., 2020). These data suggest that mangrove MPB has the potential as an important food source for wetland macrofauna (Kwon et al., 2020). A significant proportion of the production from MPB is in the form of extracellular polymeric substances (EPS) comprising relatively labile C (Underwood and Paterson, 2003), which increases the mineralization rate of MPB production.

Shallow-water seagrasses, especially fast-growing species with relatively lower structural carbon, are important sources of organic matter to coastal macrofauna such as fish, sea urchins, and some megafauna, e.g., turtles and dugongs. With primary productivity comparable to mangroves and salt marshes (Hemminga and Duarte, 2000), and their capacity for trapping also allochthonous organic matter, the mineralization of seagrass production is an important part of C dynamics in shallow coastal ecosystems.

Apart from autochthonous sources, allochthonous C from imported and settled phytoplankton, macroalgae, or terrigenous material also contribute to the reduced C pool in coastal wetlands. Sediments of coastal wetlands are highly C-dense, e.g., $33\pm6\,kg\,C_{org}\,m^{-2}$ in the top 1-m of seagrass meadows (Fourqurean et al., 2012), making them the largest C storage compartment ($>90\%$ of total C stock in estuarine mangroves) in coastal vegetated wetlands (Donato et al., 2011; Fourqurean et al., 2012; Ouyang and Lee, 2014). Bidirectional movement (outwelling and/or in-welling) of C may occur among connected wetlands. Holistic analyses are few, but the relative contribution of the various production sources to the C pool available to macrofaunal consumers seems to vary significantly according to local hydrological and edaphic conditions (e.g., Dunn et al., 2008).

Consumption and subsequent mineralization of organic carbon in coastal wetlands depends on both the availability and nutritional quality of the carbon sources. Several indicators, such as the concentrations of secondary chemicals (e.g., tannins), carbon and nitrogen content, have been used in assessing nutritional quality of plant production. Wide variations exist among the various producer groups as well as at the species level in coastal wetlands (Table 5.2). Vascular plants generally offer lower quality material (as reflected by their lower %N and higher C/N ratio) with higher feeding deterrents (high % tannins).

Table 5.2 Examples of difference in quality of organic production from various producer sources in coastal wetlands.

Organic matter source	Taxon	Tannins (%)	Carbon (%)	Nitrogen (%)	C/N ratio	Reference
Microphytobenthos	Mixed, mainly diatoms	ND	17.3	1.0	17.3	Lee unpubl. data
Phytoplankton	96 species	ND	34.1	5.5	6.2	Duarte (1992)
			29.4	2.9	10.1	Lee (unpubl. data)
Macroalgae	46 species		24.8	1.9	13.1	Duarte (1992)
	Enteromorpha intestinalis		54	1.4	38.0	Shafique et al. (2013)
	Enteromorpha clathrata		37	2.6	14.2	
	Ulva reticulata		59	3.2	18.4	
	Chlorophyta (12 spp.)	8.2–29.4				Petchidurai et al. (2019)
	Rhodophyta (9 spp.)	6.0–24.9				
Seagrass	Posidonia oceanica[g]	5–10				Pergent et al. (2008) and Agostini et al. (1998)
	27 species		33.5	1.9	17.6	Duarte (1992)
	Thalassia testudinum[g]		36.9	1.82	24.6	Fourqurean and Zieman (2002) and Arnold and Targett (2002)
	Halodule uninervis[g]	3–11		2.4		Udy et al. (1999)
	Syringodium isoetifolium[g]		34.9	1.6	12.4	Holmer and Olssen (2002)
	Enhalus acoroides[d]			3.28		

						References
Saltmarsh plants	Spartina alterniflora[d]	0	48	0.95	50.5	Bärlocher and Moulton (1999) and Marinucci et al. (1983)
	Scirpus triqueter[g]		40	<1	>40	Valiela et al. (1985)
	Scirpus mariquete[g]		45.1	0.76	59.3	Zhou et al. (2007)
	Limonium vulgare	43.7	43.0	1.10	39.3	Summers et al. (1993)
	Salicornia europaea	8.2				
	Triglochin maritima	8.8				
	Plantago maritima	17.4				
	Aster tripolium	22.2				
	Armeria maritima	15.0				
	Puccinellia maritima	5.4				
	Halimione portulacoides	7.4				
Mangroves	Kandelia obovata[d]	2.35	36.7	0.75	48.9	Lee (1993)
	Aegiceras corniculatum[d]	1.95	54.6	0.79	69.1	Lee (1993)
	Avicennia marina[d]	0.86	34.5	1.26	27.4	Tong et al. (2006)
	K. obovata[g]	9.2		1.80		Hernes et al. (2001)
	Rhizophora mangle[d]	10				
		5.2				Hernes and Hedges (2004)
	9 species of Bruguiera[g] Ceriops[g], Kandelia[g], and Rhizophora[g]	8–44*				Basak et al. (1998)
	Avicennia germinans[g]		46	2.06	22.3	McKee and Faulkner (2000)
	Laguncularia racemosa[g]		46	1.02	45.1	Holmer and Olssen (2002)
	R. mangle[g]		46	1.44	31.9	

Continued

Table 5.2 Examples of difference in quality of organic production from various producer sources in coastal wetlands—cont'd

Organic matter source	Taxon	Tannins (%)	Carbon (%)	Nitrogen (%)	C/N ratio	Reference
	R. apiculata[d]		34.8	0.47	86.4	Werry and Lee (2005)
	R. stylosa[d]		47.4	0.41	115.6	Tietjen and Alongi (1990)
	Avicennia marina[d]		45.8	0.65	70.4	
	R. stylosa[d]	11.5	40.2	1.04	38.7	Wang et al. (2014)
	Avicennia marina[d]	3.6	41.0	1.98	20.7	
	Heritiera littoralis[g], Lumnitzera racemosa[g], Bruguiera gymnorhiza[g], Aegiceras corniculatum[g], K. obovata[g]	6–15				
	Excoecaria agallocha[g], Acanthus ilicifolius[g] Avicennia marina[g]	<1.5				Wang et al. (2014)

Data on salt marsh, seagrass, and mangroves are based on dry weight of either green or senescent leaf tissues, respectively, indicated by the superscripts "g" and "d." Tannin values include condensed, hydrolysable, and total tannin measurements. * The values reported by this study are particularly high.

5.3 Herbivory
5.3.1 Macrophyte production

Grazing on seagrass is highly variable among species and herbivore combinations, but consumption is generally <15% of leaf production (Cebrián and Duarte, 1998). Macrofaunal consumption of live biomass of mangrove and salt marsh macrophytes is usually limited to <10% of standing biomass (Teal, 1962; Lee, 1991), with leaf quality such as tannin levels and toughness being important in determining the level of attack as well as the consumer identity. Aerial consumers are predominant on mangrove and salt marsh macrophytes, dominated by insects (Tong et al., 2006). Mass defoliations of mangroves, however, are not uncommon and are usually linked to stress on the trees, exacerbated by factors like eutrophication (e.g., Anderson and Lee, 1995). Indirect impact of macrofauna consumption on the mineralization rate of wetland macrophyte production may occur as herbivore attack may alter leaf chemistry (Tong et al., 2003) or longevity (Onuf et al., 1977; Lee, 1990a, 1991; Reef et al., 2012). Feller (2002) reported that the species of xylem- and phloem-feeding insects killed over 50% of the *Rhizophora mangle* canopy, which is much more than the 6% consumed by other herbivores in Belize. Massive defoliation, whether resulting in completely consumed or precocious abscission of heavily damaged foliage, results in concentrated fluxes of relatively high-quality material in the form of insect frass and partially consumed leaves of higher N content available to macrodetritivores or exported (Zimmer and Topp, 2002). Defoliation through abiotic processes may also influence coastal wetland carbon mineralization. Massive increases in substrate availability through cyclone defoliation result in sharp increases in soil respiration in mangroves (Ouyang et al., 2021). Although not previously measured, similar spikes in mineralization rate are likely to follow defoliation events mediated by macrofauna.

Consumption by invertebrate macrofauna has traditionally been considered minimal in salt marshes, but recent studies have implicated crab consumption in massive die-off of *Spartina* salt marshes (Bertness et al., 2014). Belowground herbivory by the marsh crab *Sesarma reticulatum* can cause mortality of *Spartina* even after the plant has attained size refuge to aboveground herbivory (Coverdale et al., 2012). Such massive die-off is also expected to result in concentrated fluxes of marsh C for mineralization.

Live macrophyte biomass in coastal wetlands is also used by some megafauna (Lee et al., 2017), but consumption in intertidal wetlands such as saltmarsh and mangroves is occasional in most cases and use as staple food is rare. Terrestrial grazing mammals such as goats and camels and arboreal species such as monkeys may mediate significant biomass mineralization (Moore, 2002), but their impact is poorly known, although a recent metaanalysis suggests potential influence on C storage (Davidson et al., 2017). Seagrass meadows are heavily used by megafauna such as dugong, manatees, and sea turtles as staple food. The effect of grazing on production through processes such as compensatory growth is, however, equivocal (e.g.,

Kathleen and Karen, 2005; Johnson et al., 2020). Johnson et al. (2020) investigated system carbon metabolism in turtle grazing areas at five locations in the Greater Caribbean and the Gulf of Mexico but failed to find any significant increase in carbon remineralization via heterotrophic respiration caused by grazing. However, how mineralization may be modified after seagrass biomass has passed through the mega-herbivore's gut has not been investigated.

Grazing by herbivores may have indirect consequences on C mineralization. Seagrass meadows (*Posidonia australis* and *Zostera nigracaulis*) that are heavily grazed by sea urchins (*Heliocidaris erythrogramma*) have been found to promote erosion and loss of the C pool to remineralization (Carnell et al., 2020). 35% of C stock was recorded in 6 months compared to ungrazed meadows while this loss continued even after cessation of urchin grazing, culminating in 46% loss after 3 years. This indirect effect is, however, expected to be less significant for emergent macrophytes such as mangrove and salt marsh plants, although bioturbation that result in sediment loss may also promote C loss and remineralization.

5.3.2 Microalgal production

Consumption of microalgal production by the macrofauna has been investigated to different extents among coastal wetlands. The predominance of a vascular plant detritus fueled food web in coastal wetlands has been challenged in the last two decades, partly due to the availability of techniques that enabled the assessment of the significance of microalgae in the C dynamics of coastal wetlands. While microalgae offer a more readily digestible C source to macroconsumers, their biomass and productivity levels were poorly known until the 1990s. Data available are, however, still largely limited to unvegetated tidal flats or salt marshes, while the MPB communities and their role in ecosystem C metabolism are yet to be explored in mangrove forests.

EPS production ranges from 0 to 18.2 μg glucose equivalent per μg Chl-a^{-1} h^{-1}, with factors such as light intensity and nature of the MPB assemblage contributing to the level (Underwood and Paterson, 2003). The majority of this production is in the form of carbohydrates, which are shown by pulse-chase experiments to be readily utilized by heterotrophic bacteria and protists, e.g., foraminiferans (Oakes et al., 2012; Miyatake et al., 2014). Oakes et al. (2012) further reported that 63% of the ^{13}C pulse enrichment was lost through respiration within 33 days. The rise of tracer techniques using natural abundance as well as compound-specific stable isotopes and fatty acid biomarkers suggests an important potential trophic role of the microphytobenthos to both the meiofauna (van der Heijden et al., 2019) and macrofauna (e.g., Oakes et al., 2010) in coastal wetlands. More direct evidence of the trophic link between MPB and macrofaunal production will be necessary to assess the importance of this C mineralization pathway in coastal wetlands.

5.4 **Detritivory**

Contrary to the earlier view that dead organic production in coastal wetlands is pre-
dominantly exported to drive nearshore detritus-based food chains (e.g., Teal, 1962;
Odum, 1980), exchange of organic matter is bidirectional. Geomorphology with reference to the hydrological regime (tidal/riverine influence) determines if net outwelling or in-welling occurs. Recent work on the C stock of coastal wetlands point to the
high density of C, dominated by dead organic matter accumulated in the sediment.
Microbial decomposition of this detritus pool is suppressed by the general lack of
oxygen as electron acceptor for aerobic respiration (see Chapter 2). Where export
and in situ macrofaunal consumption of the detritus are low, the slow decomposition
rate will promote C accumulation, leading to the majority of intertidal wetlands
being highly important C storages (Mcleod et al., 2011). The structural complexity
of vegetated coastal wetlands to trap allochthonous sediment and the associated C
adds to the high sediment C accumulation rate (Ouyang and Lee, 2014; Kamal
et al., 2017).

This fraction of the wetland production has long been regarded as the basis of
secondary production in coastal wetlands (Odum and Heald, 1975). Low-quality vascular plant detritus is thought to undergo gradual microbial transformation including
nitrogen immobilization to result in improved nutritional value to the macrofauna.
The contribution of other C sources, such as settled phytoplankton or MPB, was
regarded as secondary. This view was bolstered by the discovery that some macro-
detritivores, e.g., sesarmid crabs, can consume large quantities of litter before significant microbial enrichment (Robertson, 1986; Robertson and Daniel, 1989; Lee,
1997; Werry and Lee, 2005). Leaf-caching to allow microbial enrichment of the
stored mangrove leaf litter was proposed as a behavioral adaptation to the consumption of this low-quality C source (Giddins et al., 1986). Later investigations, however, suggested that microbial enrichment was negligible during the storage
period in *Ucides cordatus* (Nordhaus and Wolff, 2007), and some species actually
consume freshly senescent leaves as soon as they become available (Harada and
Lee, 2016).

This new trophic link suggested that a large proportion of the mangrove production that was previously assumed to be exported was actually retained and mineralized within the coastal wetland, driving further trophic linkages through coprophagy
of the mangrove C processed by these initial processors (Lee, 1997; Werry and Lee,
2005) and their predation up the food chain (Sheaves and Molony, 2000). The latter
process suggests a significant role of predators in regulating the C storage potential of
coastal wetlands (Atwood et al., 2015). The large ucidid crab *U. cordatus* was able to
consume 81.3% of the daily leaf and propagule production in a high intertidal mangrove forest in northern Brazil, with gut evacuation rates at $>0.31\,h^{-1}$. At their natural density, *U. cordatus* produces $\sim7.1\,t\,dwt\,ha^{-1}\,year^{-1}$ of fecal material
comprising largely partially digested mangrove production (Nordhaus and Wolff,
2007), with physical and chemical properties significantly more amenable to mineralization than unprocessed material.

The physically shredded and chemically processed detritus remaining in the crab's feces allows rapid microbial colonization and N immobilization, resulting in improved nutritive value to coprophagous invertebrates including deposit-feeders. The processed material may also contribute to the water column as well as benthic C pool in nearshore environments. Bouillon et al. (2007) reported significant trapping of mangrove-derived particulate organic carbon (POC) by seaward seagrass meadows in Gazi Bay, Kenya. Werry and Lee (2005) reported increase of the δ^{15}N of shredded mangrove leaf fragments in the feces of *Parasesarma erythodactyla* at 0.09‰ day^{-1} upon decomposition, likely a result of immobilization of atmospheric nitrogen (0‰), making the feces complex more nutritionally attractive to macrofauna. This processing by shredders of vascular plant production contributes significantly to the sedimentary C pool and could partly explain the generally low C/N ratios of wetland sediments and micro-POM in estuaries compared to senescent litter (e.g., <20 and 50–70, respectively, Koch et al., 2010; Werry and Lee, 2005). These links, however, also stimulated debates on how these macroconsumers are physiologically able to assimilate and sustain on a staple food of such low quality in coastal wetland communities, particularly in terms of N availability (Kristensen et al., 2017).

5.5 Physiological adaptations for assimilation of vascular plant detritus

5.5.1 Endogenous versus microbial-assisted capacity for cellulose digestion

For macrofauna to contribute to mineralization of coastal wetland production through consumption, cellulases are crucial in breaking down cellulose, which is glucose units linked by β-1,4-glycosidic bonds. Three major groups of enzymes: endo-β-1,4-glucanase, exo-1,4-β-glucanase, and β-glucosidase are required for complete breakdown of cellulose into glucose (Allardyce and Linton, 2008). Before the discovery of the genomic origin of cellulase in termites (Fig. 5.1), all herbivorous animals were assumed to be assisted by symbionts in cellulose digestion, despite a fair amount of indirect evidence demonstrating endogenous cellulase production in invertebrates. The distribution of cellulase in more than 60 invertebrates including annelids, crustaceans, mollusks, and brachiopods was reviewed by Yokoe and Yasumasu (1964). The absence of wetland-associated species in the review shows that mangrove fauna were largely excluded from the discussion. Recent molecular biological techniques such as transcriptomics, proteomics, and bioinformatics enable a more comprehensive understanding of cellulases. For example, the functions of different families of glycosyl hydrolase (GH) are now recorded in well-established databases such as carbohydrate-active enzyme (CAZy), the structures of different cellulases, and the evolution of the genes have now been clarified. GH1, 3, 5–10, 12, 19, 26, 30, 44, 48, 51, 61, 74, 116, and 124 are cellulases out of 130 GH families (Cantarel et al., 2009). The structures of the cellulases in termites and nematodes

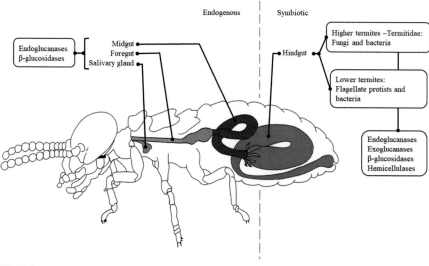

FIG. 5.1

The cellulolytic system of termites. Endogenous cellulases of lower termites is secreted by salivary gland (Watanabe et al. 1997; Brune 2014; Talia and Arneodo 2018), while secretion is attributed to midgut epithelium in higher termites (Tokuda et al. 1999). Hindgut of termites is colonized with dense microbial community, cellulolytic protistan flagellates, and bacteria in lower termites, while higher termites lost the flagellates and acquired symbiotic bacteria and fungi such as Basidiomycetes for the provision of cellulases and hemicellulase (i.e., Laccase and xylanase) in the hindgut (Johjima et al. 2006; Liu et al. 2013).

The termite diagram is modified from Brune, A., Ohkuma, M., 2010. Diversity, structure, and evolution of the termite gut microbial community. Biology of Termites: A Modern Synthesis. Springer, Dordrecht, pp. 413–438.

were the first to be reported in 1998 (Smant et al., 1998; Watanabe et al., 1998). The following discussion focuses on the most abundant and diverse detritivore taxa, namely, the mollsucs and crustaceans, in coastal wetlands. Detritivory by insects has been studied intensively in terrestrial systems and will not be addressed here.

5.5.1.1 Mollusks

The presence of cellulase in the shipworm *Teredo navalis* was discovered more than 90 years ago (Boynton and Miller, 1927). Since then the ability of many bivalves to assimilate plant materials has been confirmed, prompting even greater interest in their capacity for cellulase production. However, there have been limited studies on cellulase production among coastal wetland macrofauna. Payne et al. (1972) discovered cellulase activities in the crystalline style, digestive diverticula and midgut of the estuarine bivalve *Scrobicularia plana,* and the hydrolysis of cellulose to glucose can be fully processed in the digestive diverticula but not in the style. For shipworms, cellulase activity was detected in the appendix, anal canal, intestine, digestive diverticula, gills, and stomach (de Moraes Akamine et al., 2018). In the last two decades, more detailed investigations on the characterization (e.g., optimum temperature and pH) of cellulase emerged, with a wider coverage of the group including nonshipworm bivalves and gastropods (Table 5.3). The types and activities of cellulases discovered were greatly influenced by the methods used, making it difficult to draw conclusions by comparing results from separate studies. The GH groups screened from mollusks include 5, 6, 9, 10, 11, 18, 45, and 53 (Badariotti et al., 2011; de Moraes Akamine et al., 2018; O'Connor et al., 2014; Sakamoto and Toyohara, 2009). For example, the cellulase from the mangrove oyster *Crassostrea rivularis* was deduced to be GHF45, an endoglucanase functional group (An et al., 2015).

Strong evidence also confirms cellulase production in gastropods. Various types of carbohydrate-hydrolyzing enzymes were found in the saltmarsh periwinkles *Littoraria irrorata* and *Littoraria saxatalis* (Table 5.3), including endoglucanase and β-glucosidases, indicating that it can mineralize carbon by hydrolyzing native cellulose to glucose (Bärlocher and Pitcher, 1999). Initially, only cellulase was investigated, hemicellulase was later found to be equally important in breaking down cellulose. Niiyama and Toyohara (2011) showed that cellulase and hemicellulase activities were widespread among aquatic invertebrates. High cellulase activities were detected in the mangrove whelk, *Terebralia palustris*, which feeds on mangrove leaves (Houbrick, 1991). Their high xyloglucanase activity further indicated their ability in breaking down cell walls, as xyloglucanase hydrolyzes the hemicellulose xyloglucan, which connects cellulose chains. The cellulase isolated from the mangrove oyster *Crassostrea rivularis* is salt-tolerant, well adapted to the brackish environment (An et al., 2015).

5.5.1.2 Crustaceans

Linton et al. (2006) investigated the potential for endogenous cellulase enzyme production in six decapods, putative partial endo-β-1,4-glucanase amino acids were sequenced. Adachi et al. (2012) indicated that endo-β-1,4-glucanase activities of three mangrove sesarmid crabs (*Episesarma versicolor, Perisesarma indiarum, and Episesarma palawanense*) were higher than those in marsh species in Japan (*Chiromantes dehanni, Chiromantes haematocheir,* and *Parasesarma pictum*), where availability of cellulosic material was lower. The latter group of crabs, however, is also common inhabitants of East Asian mangroves. It would be interesting to note if cellulase levels is habitat-dependent, driven by vascular plant detritus availability. Bui and Lee (2015a) demonstrated endogenous cellulase production by the mangrove crab *Parasesarma erythodactyla* with strong evidence of obtaining the cDNA of the GHF9 family of glycosyl hydrolase from the hepatopancreas and cellulase activities in the gastric juice. By comparing the abilities of cellulose digestion of six crabs in coastal wetlands, Kawaida et al. (2019) discovered that the mangrove crab *Parasesarma bidens* is equipped with high cellulose digesting ability, which is correlated with the significant proportion of plant materials in their diet.

Table 5.3 Cellulases detected from various animal taxa in coastal wetlands.

Taxa	Habitats	Enzyme types	References
Mollusks			
Scrobicularia plana	Estuary	Cellulase	Payne et al. (1972)
Tegillarca granosa	Mangrove	Endo-β-1,4-glucanase	Niiyama et al. (2012a,b)
Neoteredo reynei		Endoglucanase	de Moraes Akamine et al. (2018)
Littoraria irrorata		Cellulase, endoglucanase, xylanase, and laminarinase	Bärlocher et al. (1989)
Terebralia palustris		Cellulase and xyloglucanase	Niiyama and Toyohara (2011)
Magallana rivularis		Endoglucanase β-1,4 glucan and β-1,4 xylose linkage breaking enzymes	An et al. (2015)
Cerithideopsilla cingulata		Exo-β-1,4-glucanase endoglucanase,	An et al. (2014)
Batillaria zonalis	Mudflat	β-1,4-glucosidase, xylanase, and laminarinase	
Cyclina sinensis			
Solen linearis			
Crassostrea virginica	Salt marsh	β-1,4-glucanase and β-glucosidases	Langdon and Newell (1990)
Geukensia demissa			
Littoraria saxatalis		Cellulase, endoglucanse, and xylanase	Bärlocher and Pitcher (1999)
Cipangopaludina japonica	Estuary	Cellulase	Antonio et al. (2010)
Clithon retropictum			
Semisulcospira libertina			
Crustaceans			
Isopods			
Sphaeroma terebrans	Mangrove	Endoglucanase	Benson et al. (1999)
Limnoria spp.			
Copepods			
Acetes indicus	Estuary	Cellulase	Niiyama et al. (2012a)
Acetes japonicus			
Acetes sibogae			

Continued

Table 5.3 Cellulases detected from various animal taxa in coastal wetlands—cont'd

Taxa	Habitats	Enzyme types	References
Brachyuran crabs			
Callinectes sapidus	Shallow estuarine, bay and coastal waters	Maltose-hydrolyzing enzyme Cellulase and chitinase	McClintock et al. (1991) Allman et al. (2017)
Parasesarma erythodactyla	Mangrove	Endo-β-1,4-glucanase	Bui and Lee (2015a,b)
Episesarma versicolor			Adachi et al. (2012)
Episesarma palawanense			
Parasesarma indiarum			
Gelasimus borealis		Exo-β-1,4-glucanase, endoglucanase, β-1,4-glucosidase, xylanase, and laminarinase	An et al. (2014)
Parasesarma plicatum			
Mictyris longicarpus	Tidal mudflat		
Scylla serrata		Amylase, cellulase, and xylanase	Pavasovic et al. (2004)
Prawns/shrimps			
Palaemon styliferus	Estuary	Cellulase	Niiyama et al. (2012a)
Fenneropenaeus merguiensis			
Metapenaeus ensis			
Metapenaeus lysianassa			
Merispenaeopsis sculptilis			
Penaeus monodon			
Palaemon semmklinkii			

Where appropriate, the most current taxonomic names are used instead of those in the original reports.

Isopods from the genera *Limnoria* and *Sphaeroma* are abundant in mangrove forests (Lee et al., 2017), with some, such as *Limnoria sellifera,* being wood-boring specialists (Cookson et al., 2012) The research on isopod cellulase started in 1950s, mainly on *Limnoria* spp. Ray (1959) indirectly showed the production of cellulase in the midgut diverticula of isopods by the removal of cellulose and hemicellulose by *Limnoria* sp. In the 1970s, the absence or impoverished microbiota in the alimentary canal indirectly proved the endogenous production of cellulase by *Limnoria* sp. (Boyle and Mitchell, 1978; Sleeter et al., 1978). Zimmer and Bartholmé (2003) demonstrated that sympatric freshwater detritivores may assimilate vascular plant leaf litter in different ways: the isopod *Asellus aquaticus* hosts bacterial endosymbionts whereas the amphipod *Gammarus pulex* relied on endogenous enzymes for assimilation. With advanced sequencing techniques, β-1,4-glucanases are now proven to be widespread in isopods, mainly classified as GH9 family14, which is also commonly found in crustaceans such as the land crab *Gecarcoidea natalis* and the mangrove crab *Parasesarma erythodactyla*. GH9 sequences are highly conserved among isopods and multiple copies of GH9 were present (Bredon et al., 2019), which have strong expression and result in high cellulase production, reflecting the importance of GH9 in these detritivores. β-glucosidases are also present in the transcriptome of marine isopods including *Sphaeroma terebrans* and *Limnoria tripunctata*, which are pests on mangroves. Some marine isopods could use hemocyanins to modify lignin during digestion of lignocellulose (Besser et al., 2018; King et al., 2010).

There are relatively less data on cellulase production by coastal wetland crustaceans other than brachyuran crabs and isopods, but Niiyama et al. (2012a) reported that various cellulases can be found in mangrove and estuarine copepods and decapods (Fig. 5.2). The widespread macrofaunal utilization of cellulases, or GH, was indicated by a review of lignocellulose degradation across Tree of Life (Cragg et al., 2015), that presence of GH is common, irrespective of whether the GH is endogenously produced or from symbionts. Recently, Linton (2020) made a detailed review on the structure and function of cellulases and hemicellulases, pointing out that some of the "cellulases," especially endoglucanases, do not always function as cellulase, but could actually function in the immune system. Therefore, the widespread occurrence of true functional cellulases may have been exaggerated. The origin of the cellulase detected in most studies, i.e., endogenous or assisted to various degrees by gut endosymbionts, is still obscure. However, this uncertainty does not change the fact that mangrove carbon can be mineralized through consumption by these animals.

5.5.2 Cellulolytic activities from endosymbionts

Cellulose-decomposing microorganisms, especially bacteria and fungi, are cosmopolitan in marine habitats and play an important role as organic matter mineralizers and modulate their productivity. These groups are highly diverse in mangroves (Lee et al., 2017), potentially also in other coastal wetlands where cellulose and similar recalcitrant polysaccharides are an important component of the organic production.

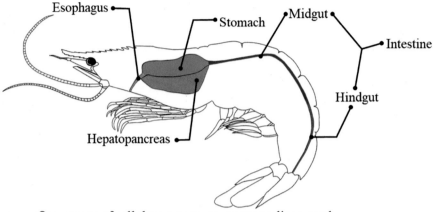

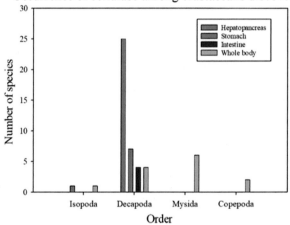

FIG. 5.2

The occurrence of cellulases in crustaceans. The shrimp digestive system is used as a representative crustacean digestive system. The mangrove, wetland, or saltmarsh crustacean cellulolytic system is still understudied. Most of the knowledge on the topic is from the decapods. As the hepatopancreas is perceived as the main digestive enzyme secreting organ, most of the studies only included the hepatopancreas, while the possible role of the intestine in enzyme secretion is still not explored. There is no description of amphipod cellulolytic system yet. Crustaceans are generalized to be depending on endogenous cellulases only because of the case of isopod (King et al. 2010), but there is still debate about the system. Studies of the symbionts of crustaceans only demonstrate the presence of cellulolytic bacteria indirectly or described overall enzyme activity, with no isolation of cellulolytic symbionts made to date.

Cellulolytic symbiotic bacteria are important to ruminant nutrition while the necessity for symbionts is quelled in detritivorous macrofauna with the capacity for endogenous cellulase production.

5.5.2.1 Mollusks

The data on symbionts of mollusks are mainly about shipworms. Kohlmeyer and Bebout (1986) observed that the gastropod *Littoraria angulifera* regularly ingests fungi and suggested that the cellulase produced by the fungi might help break down cellulose in its diet. As vascular plant detritus has a high C/N ratio (%N is usually <0.5% and C/N > 50 in most mangrove detritus), focus has been on nitrogen-fixing bacteria. While C is potentially unlimited, the recalcitrant nature of this C source suggests that cellulolytic symbionts are essential for detritivores utilizing high C/N food sources, e.g., shipworms. However, there is limited information about cellulolytic symbionts of mollusks in general (Rosenberg and Breiter, 1969).

Cellulolytic symbionts from digestive tract of mangrove macrofauna were first discover as early as 1979, three cellulolytic streptomycetes were isolated from the gut of a mangrove bivalve borer, *Barnea birmanica* (Balaasubramanin et al., 1979). Betcher et al. (2012) surveyed the common distribution and abundance of symbionts in five shipworm species from the genera of *Bankia*, *Lyrodus*, and *Teredo* and discovered that bacteria mainly colonized the surface of fecal pellets in the intestine. Only small populations of symbionts can be found in the cecum, the organ proposed to be the location for lignocellulose digestion and absorption (Bazylinski and Rosenberg, 1983). Therefore, there is no strong evidence to show that shipworms depend on the gut cellulolytic bacteria for cellulose or lignocellulose digestion. Instead, the gills of the shipworm are colonized by abundant endosymbionts (Betcher et al., 2012). Intriguingly, these endosymbionts include some cellulolytic and nitrogen-fixing taxa (Distel et al., 2002a,b; Lechene et al., 2007; Waterbury et al., 1983). The gill endosymbiont community of the mangrove shipworm *Lyrodus pedicellatus* comprises microbes that are closely related to polysaccharide-degrading gamma-proteobacteria (Distel et al., 2002a,b; Luyten et al., 2006). For mangrove-associated *Neoteredo reynei*, highest endoglucanase activity was detected in the gill extract among the tested organs (de Moraes Akamine et al., 2018). This study attributed the high endoglucanase activity to endosymbionts and suggested that the wood-degrading enzymes produced by endosymbionts in the marine shipworm gill can be selectively translocated to the host's bacteria-free gut as suggested by O'Connor et al. (2014). This hypothesis is supported by the resemblance between the protein identified in host gut cells and gill endosymbionts, whose genomes were proven to encode 83–128 predicted GH genes.

Ingesting predigested carbon could compensate for the absence of gut symbionts. Two mussels, *Geukensia demissa* and *Mytilus edulis*, respectively dominant on protected salt marshes and exposed rocky shores, have been reported to ingest and assimilate carbon from cellulolytic bacteria (Kreeger and Newell, 1996). There is a paucity of studies on gastropod gut cellulolytic bacteria, with a general lack of data on coastal species. The activities of cellulase and chitinase in the terrestrial snail *Helix pomatia* were not related to the population size of bacteria, implying the animal only depends on endogenous cellulase (Stradine and Whitaker, 1963). However, it is unclear if this applies to coastal herbivorous gastropods. How the relative significance of endogenous versus symbiont cellulases vary along the land-sea gradient, where the abundance of vascular plant organic C varies significantly, deserves further attention. The pattern may be intimately related to the evolution of detritivory in coastal wetland macrofauna and their role in C mineralization.

5.5.2.2 Crustaceans

Early studies of crustacean gut symbionts focused on mangrove-associated isopods and amphipods. Unlike their terrestrial counterparts (Bouchon et al., 2016; Kostanjšek et al., 2004, 2005; Zimmer, 2002), there is no record of gut bacteria in marine wood-boring isopods *Limnoria tripunctata* or amphipods *Chelura terebrans*, as revealed by scanning electron microscopy (Boyle and Mitchell, 1978). The exoskeleton of marine or estuarine isopods and the surface of their tunnels in decomposing wood are colonized by bacteria (Sleeter et al., 1978). It was suggested that certain cellulase genes have long been evolutionarily adopted by the isopod lineage, enabling some, especially marine, species to digest cellulose without the assistance of gut bacteria (King et al., 2010). Harris (1993) reviewed and investigated 16 estuarine crustacean species and found that except the carnivorous *Scylla serrata*, all the hindguts were colonized by bacteria. Harris suggested that the extensive colonization of epimural rod bacteria in detritivores resembles bacteria in the termite hindgut, which are capable of cellulose digestion (Breznak and Pankratz, 1977).

In the last decade, assessing the gut bacteria community and diversity was made easy because of the development of molecular biological and bioinformatic techniques. There are already numerous studies on crustaceans (e.g., Mattila et al. 2014; Dong et al., 2018), but none of them focus on cellulolytic bacteria. This is because most of the bacterial communities were identified solely through 16S rRNA genes, which fail to fully confirm the functions of the bacteria, although this could be deduced using bioinformatics tools such as QIIME 2 (Bolyen et al., 2019). While the data on crustacean symbionts are growing, a recent study by Martin (2020) suggested that the midgut of marine crustaceans lacks resident symbionts (cf their terrestrial and freshwater counterparts). It was suggested that bacteria from the environment were ingested during feeding, and only transiently lived in and passed through the gut of the "host." Despite the presence or absence of resident bacteria in guts, it is reasonable that the animals in coastal wetlands regularly ingest environmental bacteria and fungi from their habitats, which support diverse and abundant bacteria, including cellulolytic forms (Behera et al., 2017). Although there is a diverse assemblage of cellulolytic endosymbionts isolated from the macrofauna, the critical question of how much they contribute to the cellulose digestion and carbon mineralization mediated by their hosts is still obscure.

5.6 **Implications for C fluxes in coastal wetlands**

Macrofaunal consumption significantly increases carbon mineralization in coastal wetlands (Fig. 5.3). Consumption by initial detritus processors, e.g., sesarmid and ucidid crabs, can remove up to >80% of litterfall in tropical mangrove forests (Robertson and Daniel, 1989; Nordhaus et al., 2006). Despite some variations, these shredders consume >50% of daily mangrove litter production (Table 5.1). Lee (1989a) measured crab consumption rate and microbial decomposition rate of leaf litter of the mangrove *Kandelia obovata*. At an average density of 4 individuals m^{-2}, the sesarmid crabs were able to consume $\sim$0.35 g C m^{-2} day^{-1} (>57% of annual leaf litter production) whereas microbial decay would result in $\sim$0.010 g C m^{-2} day^{-1}. While not all ingested leaf material will be directly mineralized, this large difference in leaf processing rate means that crab-mediated mineralization of *K. obovata* leaf litter will still be many times faster than that through microbial decay during the crab-active season. While this factor may be an overestimate because of incomplete assimilation as well as biomass accumulation by the crabs, the accelerated mineralization of the processed litter and its subsequent consumption by coprophagous invertebrates (Lee, 1997; Werry and Lee, 2005;

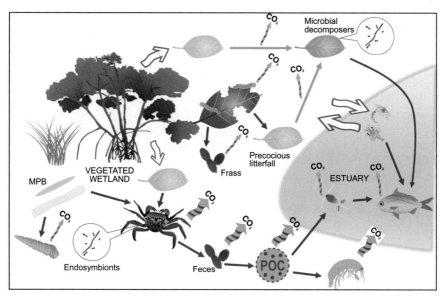

FIG. 5.3

Schematic diagram illustrating the various mineralization processes mediated by the macrofauna in a mangrove wetland. *Red arrows*, macrofauna-mediated trophic processes; *green arrows*, trophic processes not mediated by macrofauna; *wavy blue arrows*, translocation of organic C; *wavy green/red arrows*, CO_2 emission, with relative significance as a pathway of mineralization of wetland C indicated by thickness.

Nordhaus and Wolff, 2007) would ensure a large difference in mineralization rate is achieved. Fecal material derived from crab processing also decomposes much faster than litter undergoing only microbial decomposition. While the atomic C/N of the leaf litter of *Avicennia marina* decreased at $0.28\,day^{-1}$ through microbial decomposition over a 168-day period, the feces produced by the shredder crab *Parasesarma erythodactyla* had a rate of $0.50\,day^{-1}$ over 28 days (Werry and Lee, 2005). This much higher rate of decrease is a result of both increased C mineralization and N immobilization.

Macrofauna can also promote mineralization of wetland C in contiguous habitats via their own movement or translocation of processed material. C exchange between hydrologically connected coastal habitats is often bidirectional, but most mangroves/ salt marshes tend to export some proportion of their production (Lee, 1995; Hübner et al., 2015). POC in the form of macrofauna processed and partially mineralized material contributes to the suspended organic matter in estuarine creeks (Werry and Lee, 2005). This "micro-POC" may also contribute significantly to mangrove-derived C trapped significantly in contiguous seagrass meadows (e.g., Bouillon et al., 2007), where rapid mineralization of the C then occurs. This efficient trapping and mineralization of mangrove C by the seagrass meadows, however, reduces export further offshore. Dissolved organic carbon (DOC) usually makes up the bulk of total organic C pool in waters around coastal wetlands, with mangroves estimated to contribute >10% of terrigenous refractory DOC exported to the ocean (Dittmar et al., 2006). Physical processing of litter by macrofauna will also facilitate release of DOC, although it is not known if subsequent mineralization will be significantly affected as this material is mostly recalcitrant.

Consumption of organic production by macrofauna with ranges beyond coastal wetlands also enables C exchange far beyond what can be accomplished by microbial means in terms of spatial extent as well as C flux. The significant range of megafauna such as dugongs and turtles grazing on seagrass can mediate transfer of nutrients over spatial scales of 100 km or more, as well as from the ocean back to the terrestrial food chain. Female green turtles (*Chelonia mydas*), which specializes in grazing seagrass, can transfer significant biomass supported by seagrass production in the form of eggs that support terrestrial predators on the eggs or hatchlings, e.g., mammals, reptiles, and birds. A similar effect is evident also when insect herbivores consume vascular plant biomass in coastal wetlands and release as mineralized products in adjacent habitats.

Many of the most important macrofaunal consumers (e.g., brachyuran crabs) of coastal wetland production have obligatory euhaline planktonic larval stages. Reproductive output contributes a significant percentage to the overall production in these crustaceans, e.g., gonosomatic index in female individuals of the mangrove sesarmids *Parasesarma bidens* and *Parasesarma affine* reach 5%–6% at peak reproductive season (Lee and Kwok, 2002). This output was, however, dependent on species of mangrove associated with the crabs, hypothesized to be mediated through the difference in quality and quantity of food available. The flux of larvae of macrofauna dependent on coastal wetland C between the estuary and offshore areas can be significant, but often ignored in coastal C dynamics. Dittel et al. (1991) estimated that a net export of 3.3×10^9 and 1.4×10^9 zoea I larvae over a 5-day period for fiddler and grapsid crabs, respectively, in a Costa Rican mangrove creek. Wetland macrofauna could therefore drive C mineralization also in distant but connected habitats.

Acknowledgments

CYL thanks The Chinese University of Hong Kong for financial support toward her PhD degree.

References

Adachi, K., Toriyama, K., Azekura, T., Morioka, K., Tongnunui, P., Ikejima, K., 2012. Potent cellulase activity in the hepatopancreas of mangrove crabs. Fish. Sci. 78, 1309–1314.

Agostini, S., Desjobert, J.M., Pergent, G., 1998. Distribution of phenolic compounds in the seagrass *Posidonia oceanica*. Phytochemistry 48, 611–617.

Allardyce, B.J., Linton, S.M., 2008. Purification and characterisation of endo-β-1,4-glucanase and laminarinase enzymes from the gecarcinid land crab *Gecarcoidea natalis* and the aquatic crayfish *Cherax destructor*. J. Exp. Biol. 211, 2275–2287.

Allman, A.L., Williams, E.P., Place, A.R., 2017. Growth and enzyme production in blue crabs (*Callinectes sapidus*) fed cellulose and chitin supplemented diets. J. Shellfish Res. 36, 283–291.

Alongi, D.M., 1994. Zonation and seasonality of benthic primary production and community respiration in tropical mangrove forests. Oecologia 98, 320–327.

Alongi, D.M., 2009a. The Energetics of Mangrove Forests. Springer, New York.

Alongi, D.M., 2009b. Paradigm shifts in mangrove biology. In: Perillo, G.M.E., Wolanski, E., Cahoon, D.R., Brinson, M.M. (Eds.), Coastal Wetlands: An Integrated Ecosystem Approach. Elsevier, The Netherlands.

An, T., Lyu, J., Jia, W., Wang, M., Wei, S., Zhang, Y., 2014. Role of macrobenthic fauna in mangrove carbon fluxes indicated by their cellulase and hemicellulase activities. Mar. Biol. Res. 10, 934–940.

An, T., Dong, Z., Lv, J., Liu, Y., Wang, M., Wei, S., Song, Y., Zhang, Y., Deng, S., 2015. Purification and characterization of a salt-tolerant cellulase from the mangrove oyster, *Crassostrea rivularis*. Acta Biochim. Biophys. Sin. 47, 299–305.

Anderson, C., Lee, S.Y., 1995. Defoliation of the mangrove *Avicennia marina* in Hong-Kong—cause and consequences. Biotropica 27, 218–226.

Antonio, E.S., Kasai, A., Ueno, M., Kurikawa, Y., Tsuchiya, K., Toyohara, H., Ishihi, Y., Yokoyama, H., Yamashita, Y., 2010. Consumption of terrestrial organic matter by estuarine molluscs determined by analysis of their stable isotopes and cellulase activity. Estuar. Coast. Shelf Sci. 86, 401–407.

Arnold, T.M., Targett, N.M., 2002. Marine tannins: the importance of a mechanistic framework for predicting ecological roles. J. Chem. Ecol. 28, 1919–1934.

Ashton, E.C., 2002. Mangrove sesarmid crab feeding experiments in peninsular Malaysia. J. Exp. Mar. Biol. Ecol. 273, 97–119.

Atwood, T.B., Connolly, R.M., Ritchie, E.G., Lovelock, C.E., Heithaus, M.R., Hays, G.C., Fourqurean, J.W., Macreadie, P.I., 2015. Predators help protect carbon stocks in blue carbon ecosystems. Nat. Climate Change 5, 1038–1045.

Badariotti, F., Lelong, C., Dubos, M.P., Favrel, P., 2011. Identification of three singular glycosyl hydrolase family 18 members from the oyster *Crassostrea gigas*: structural characterization, phylogenetic analysis and gene expression. Comp. Biochem. Physiol. B Biochem. Mol. Biol. 158, 56–63.

Balaasubramanin, T., Lakshmanaperumalsamy, P., Chandramohan, D., Natarajan, R., 1979. Cellulolytic activity of streptomycetes isolated from the digestive tract of a marine borer. Int. J. Mol. Sci. 8, 111–114.

Bärlocher, F., Moulton, V.D., 1999. *Spartina alterniflora* in two New Brunswick salt marshes. I. Growth and decomposition. Bull. Mar. Sci. 64, 299–305.

Bärlocher, F., Pitcher, P.A., 1999. *Spartina Alterniflora* in two New Brunswick salt marshes. II. Potential use by *Littorina saxatalis*. Bull. Mar. Sci. 64, 307–313.

Bärlocher, F., Arsuffi, T.L., Newell, S.Y., 1989. Digestive enzymes of the salt marsh periwinkle *Littorina irrorata* (Mollusca: Gastropoda). Oecologia 80, 39–43.

Basak, U.C., Das, A.B., Das, P., 1998. Seasonal changes in organic constituents in leaves of nine mangrove species. Mar. Freshw. Res. 49, 369–372.

Bazylinski, D.A., Rosenberg, F.A., 1983. Occurrence of a brush border in the caecum (appendix) of several *Teredo* and *Bankia* species (Teredinidae: Bivalvia: Mollusca). Veliger 25, 251–254.

Behera, B.C., Sethi, B.K., Mishra, R.R., Dutta, S.K., Thatoi, H.N., 2017. Microbial cellulases–diversity & biotechnology with reference to mangrove environment: a review. J. Genet. Eng. Biotech. 15, 197–210.

Benson, L.K., Rice, S.A., Johnson, B.R., 1999. Evidence of cellulose digestion in the wood boring isopod *Sphaeroma terebrans*. Florida Sci. 62, 128–144.

Bertness, M.D., Brisson, C.P., Coverdale, T.C., Bevil, M.C., Crotty, S.M., Suglia, E.R., 2014. Experimental predator removal causes rapid salt marsh die-off. Ecol. Lett. 17, 830–835.

Besser, K., Malyon, G.P., Eborall, W.S., da Cunha, G.P., Filgueiras, J.G., Dowle, A., Garcia, L.C., Page, S.J., Dupree, R., Kern, M., Gomez, L.D., 2018. Hemocyanin facilitates lignocellulose digestion by wood-boring marine crustaceans. Nat. Commun. 9, 1–14.

Betcher, M.A., Fung, J.M., Han, A.W., O'Connor, R., Seronay, R., Concepcion, G.P., Distel, D.L., Haygood, M.G., 2012. Microbial distribution and abundance in the digestive system of five shipworm species (Bivalvia: Teredinidae). PLoS One 7, e45309.

Bolyen, E., Rideout, J.R., Dillon, M.R., Bokulich, N.A., Abnet, C.C., Al-Ghalith, G.A., Alexander, H., Alm, E.J., Arumugam, M., Asnicar, F., Bai, Y., 2019. Reproducible, interactive, scalable and extensible microbiome data science using QIIME 2. Nat. Biotech. 37, 852–857.

Bouchon, D., Zimmer, M., Dittmer, J., 2016. The terrestrial isopod microbiome: an all-in-one toolbox for animal-microbe interactions of ecological relevance. Front. Microbiol. 7, 1472.

Bouillon, S., Koedam, N., Raman, A.V., Dehairs, F., 2002. Primary producers sustaining macro-invertebrate communities in intertidal mangrove forests. Oecologia 130, 441–448.

Bouillon, S., Dehairs, F., Velimirov, B., Abril, G., Borges, A.V., 2007. Dynamics of organic and inorganic carbon sources across contiguous mangrove and seagrass systems (Gazi Bay, Kenya). J. Geophys. Res. 112, G02018.

Boyle, P.J., Mitchell, R., 1978. Absence of microorganisms in crustacean digestive tracts. Science 200, 1157–1159.

Boynton, C.L., Miller, R.C., 1927. The occurrence of a cellulase in the ship-worm. J. Biol. Chem. 75, 219–287.

Bredon, M., Herran, B., Lheraud, B., Bertaux, J., Grève, P., Moumen, B., Bouchon, D., 2019. Lignocellulose degradation in isopods: new insights into the adaptation to terrestrial life. BMC Genomics 20 (462).

Breznak, J.A., Pankratz, H.S., 1977. In situ morphology of the gut microbiota of wood-eating termites [*Reticulitermes flavipes* (Kollar) and *Coptotermes formosanus* Shiraki]. Appl. Environ. Microbiol. 33, 406–426.

Brune, A., 2014. Symbiotic digestion of lignocellulose in termite guts. Nat. Rev. Microbiol. 12, 168–180.

Bui, T.H.H., Lee, S.Y., 2015a. Endogenous cellulase production in the leaf litter foraging mangrove crab *Parasesarma erythodactyla*. Comp. Biochem. Phys. B 179, 27–36.

Bui, T.H.H., Lee, S.Y., 2015b. Potential contributions of gut microbiota to the nutrition of the detritivorous sesarmid crab *Parasesarma erythodactyla*. Mar. Biol. 162, 1969–1981.

Bunt, J.S., Boto, K.G., Boto, G., 1979. A survey method for estimating potential levels of mangrove forest primary production. Mar. Biol. 52, 123–128.

Call, M., Santos, I.R., Dittmar, T., de Rezende, C.E., Asp, N.E., Maher, D.T., 2019. High porewater derived CO_2 and CH_4 emissions from a macro-tidal mangrove creek in the Amazon region. Geochim. Cosmochim. Acta 247, 106–120.

Cantarel, B.I., Coutinho, P.M., Rancurel, C., Bernard, T., Lombard, V., Henrissat, B., 2009. The Carbohydrate-Active EnZymes database (CAZy): an expert resource for glycogenomics. Nucleic Acids Res. 37, 233–238.

Carnell, P.E., Ierodiaconou, D., Atwood, T.B., Macreadie, P.I., 2020. Overgrazing of seagrass by sea urchins diminishes blue carbon stocks. Ecosystems 23, 1437–1448.

Cebrián, J., 2002. Variability and control of carbon consumption, export, and accumulation in marine communities. Limnol. Oceanogr. 47, 11–22.

Cebrián, J., Duarte, C.M., 1998. Patterns in leaf herbivory on seagrasses. Aquat. Bot. 60, 67–82.

Chen, G.-C., Ye, Y., 2008. Leaf consumption by *Sesarma plicata* in a mangrove forestat Jiulongjiang estuary, China. Mar. Biol. 154, 997–1007.

Cookson, L.J., Cragg, S.M., Hendy, I.W., 2012. Wood-boring limnoriids (Crustacea, Isopoda) including a new species from mangrove forests of the Tukang Besi Archipelago, Indonesia. Zootaxa 3248, 25–34.

Coverdale, T.C., Altieri, A.H., Bertness, M.D., 2012. Belowground herbivory increases vulnerability of New England salt marshes to die-off. Ecology 93, 2085–2094.

Cragg, S.M., Beckham, G.T., Bruce, N.C., Bugg, T.D., Distel, D.L., Dupree, P., Etxabe, A.G., Goodell, B.S., Jellison, J., McGeehan, J.E., McQueen-Mason, S.J., 2015. Lignocellulose degradation mechanisms across the tree of life. Curr. Opin. Chem. Biol. 29, 108–119.

Currin, C.A., Newell, S.Y., Paerl, H.W., 1995. The role of standing dead spartina-alterniflora and benthic microalgae in salt-marsh food webs—considerations based on multiple stable-isotope analysis. Mar. Ecol. Prog. Ser. 121, 99–116.

Davidson, K.E., Fowler, M.S., Skov, M.W., Doerr, S.H., Beaumont, N., Griffin, J., 2017. Livestock grazing alters multiple ecosystem properties and services in salt marshes: a meta-analysis. J. Appl. Ecol. 54, 1395–1405.

de Moraes Akamine, D.T., da Silva, D.D.A.C., de Lima, C.G., Carvalho, T.V., Brienzo, M., 2018. Endoglucanase activity in *Neoteredo reynei* (Bivalvia, Teredinidae) digestive organs and its content. World J. Microbiol. Biotech. 34, 84.

Distel, D.L., Beaudoin, D.J., Morrill, W., 2002a. Coexistence of multiple proteobacterial endosymbionts in the gills of the wood-boring bivalve *Lyrodus pedicellatus* (Bivalvia: Teredinidae). Appl. Environ. Microbiol. 68, 6292–6299.

Distel, D.L., Morrill, W., MacLaren-Toussaint, N., Franks, D., Waterbury, J., 2002b. *Teredinibacter turnerae* gen. nov., sp. nov., a dinitrogen-fixing, cellulolytic, endosymbiotic gamma-proteobacterium isolated from the gills of wood-boring molluscs (Bivalvia: Teredinidae). Int. J. Syst. Evol. Microbiol. 52, 2261–2269.

Dittel, A.I., Epifanio, C.D., Lizano, O., 1991. Flux of crab larvae in a mangrove creek in the Gulf of Nicoya, Costa Rica. Estuar. Coast. Shelf Sci. 32, 129–140.

Dittmar, T., Hertkorn, N., Kattner, G., Lara, R.J., 2006. Mangroves, a major source of dissolved organic carbon to the oceans. Global Biogeochem. Cy. 20.

Donato, D.C., Kauffman, J.B., Murdiyarso, D., Kurnianto, S., Stidham, M., Kanninen, M., 2011. Mangroves among the most carbon-rich forests in the tropics. Nat. Geosci. 4, 293–297.

Dong, J., Li, X., Zhang, R., Zhao, Y., Wu, G., Liu, J., Zhu, X., Li, L., 2018. Comparative analysis of the intestinal bacterial community and expression of gut immunity genes in the Chinese Mitten Crab (*Eriocheir sinensis*). AMB Express 8, 192.

Duarte, C.M., 1992. Nutrient concentration of aquatic plants: patterns across species. Limnol. Oceanogr. 37, 882–889.

Duarte, C.M., Cebrian, J., 1996. The fate of marine autotrophic production. Limnol. Oceanogr. 41, 1758–1766.

Dunn, R.J.K., Welsh, D.T., Teasdale, P.R., Lee, S.Y., Lemckert, C.J., Meziane, T., 2008. Investigating the distribution and sources of organic matter in surface sediment of Coombabah Lake (Australia) using elemental, isotopic and fatty acid biomarkers. Cont. Shelf Res. 28, 2535–2549.

Emmerson, W.D., McGwynne, L.E., 1992. Feeding and assimilation of mangrove leaves by the crab *Sesarma meinerti* de Man in relation to leaf-litter production in Mgazana, a warm-temperate southern African mangrove swamp. J. Exp. Mar. Biol. Ecol. 157 (1), 41–53.

Feller, I.C., 2002. The role of herbivory by wood-boring insects in mangrove ecosystems in Belize. Oikos 97, 167–176.

Fourqurean, J.W., Zieman, J.C., 2002. Nutrient content of the seagrass *Thalassia testudinum* reveals regional patterns of relative availability of nitrogen and phosphorus in the Florida Keys USA. Biogeochem 61, 229–245.

Fourqurean, J.W., Duarte, C.M., Kennedy, H., Marba, N., Holmer, M., Mateo, M.A., Apostolaki, E.T., Kendrick, G.A., Krause-Jensen, D., McGlathery, K.J., Serrano, O., 2012. Seagrass ecosystems as a globally significant carbon stock. Nat. Geosci. 5, 505–509.

Giddins, R.L., Lucas, J.S., Neilson, M.J., Richards, G.N., 1986. Feeding ecology of the mangrove crab *Neosarmatium smithi* (Crustacea, Decapoda, Sesarmidae). Mar. Ecol. Prog. Ser. 33, 147–155.

Harada, Y., Lee, S.Y., 2016. Foraging behavior of the mangrove sesarmid crab *Neosarmatium trispinosum* enhances food intake and nutrient retention in a low-quality food environment. Estuar Coast Shelf S 174, 41–48.

Harris, J.M., 1993. Widespread occurrence of extensive epimural rod bacteria in the hindguts of marine Thalassinidae and Brachyura (Crustacea: Decapoda). Mar. Biol. 116, 615–629.

Hemminga, M., Duarte, C., 2000. Seagrass Ecology. Cambridge University Press, Cambridge.

Hernes, P.J., Hedges, J.I., 2004. Tannin signatures of barks, needles, leaves, cones, and wood at the molecular level. Geochim. Cosmochim. Acta 68, 1293–1307.

Hernes, P.J., Benner, R., Cowie, G.L., Goñi, M.A., Bergamaschi, B.A., Hedges, J.I., 2001. Tannin diagenesis in mangrove leaves from a tropical estuary: a novel molecular approach. Geochim. Cosmochim. Acta 18, 3109–3122.

Holmer, M., Olssen, A.B., 2002. Role of decomposition of mangrove and seagrass detritus in sediment carbon and nitrogen cycling in a tropical mangrove forest. Mar. Ecol. Prog. Ser. 230, 87–101.

Houbrick, R.S., 1991. Systematic review and functional morphology of the mangrove snails *Terebralia* and *Telescopium* (Potamididae; Prosobranchia). Malacologia 33, 289–338.

Hübner, L., Pennings, S.C., Zimmer, M., 2015. Sex- and habitat-specific movement of an omnivorous semi-terrestrial crab controls habitat connectivity and subsidies: a multi-parameter approach. Oecologia 178, 999–1015.

Johjima, T., Taprab, Y., Noparatnaraporn, N., Kudo, T., Ohkuma, M., 2006. Large-scale identification of transcripts expressed in a symbiotic fungus (*Termitomyces*) during plant biomass degradation. Appl. Microbiol. Biotechnol. 73, 195–203.

Johnson, R.A., Gulick, A.G., Constant, N., Bolten, A.B., Smulders, F.O.H., Christianen, M.J.A., Nava, M.I., Kolasa, K., Bjorndal, K.A., 2020. Seagrass ecosystem metabolic carbon capture in response to green turtle grazing across Caribbean meadows. J. Ecol. 108, 1101–1114.

Kamal, S., Warnken, J., Bakhtiyari, M., Lee, S.Y., 2017. Sediment distribution in shallow estuaries at fine scale: in situ evidence of the effects of three-dimensional structural complexity of mangrove pneumatophores. Hydrobiologia 803, 121–132.

Kathleen, L.M., Karen, A.B., 2005. Simulated green turtle grazing affects structure and productivity of seagrass pastures. Mar. Ecol. Prog. Ser. 305, 235–247.

Kawaida, S., Nanjo, K., Ohtsuchi, N., Kohno, H., Sano, M., 2019. Cellulose digestion abilities determine the food utilization of mangrove estuarine crabs. Estuar. Coast. Shelf Sci. 222, 43–52.

King, A.J., Cragg, S.M., Li, Y., Dymond, J., Guille, M.J., Bowles, D.J., Bruce, N.C., Graham, I.A., McQueen-Mason, S.J., 2010. Molecular insight into lignocellulose digestion by a marine isopod in the absence of gut microbes. Proc. Natl. Acad. Sci. U. S. A. 107, 5345–5350.

Koch, B.P., Dittmar, T., Lara, R.J., 2010. The biogeochemistry of the Caeté mangrove-shelf system. In: Saint-Paul, U., Schnieder, H. (Eds.), Mangrove Dynamics and Management in North Brazil. Springer-Verlag, Berlin, pp. 45–67.

Kohlmeyer, J., Bebout, B., 1986. On the occurrence of marine fungi in the diet of *Littorina angulifera* and observations on the behavior of the periwinkle. Mar. Ecol. 7, 333–343.

Kostanjšek, R., Lapanje, A., Rupnik, M., Štrus, J., Drobne, D., Avguštin, G., 2004. Anaerobic bacteria in the gut of terrestrial isopod crustacean *Porcellio scaber*. Folia Microbiol 49, 179–182.

Kostanjšek, R., Štrus, J., Lapanje, A., Avguštin, G., Rupnik, M., Drobne, D., 2005. Intestinal microbiota of terrestrial isopods. In: Intestinal Microorganisms of Termites and Other Invertebrates. vol. 6, pp. 115–131.

Kreeger, D.A., Newell, R.I.E., 1996. Ingestion and assimilation of carbon from cellulolytic bacteria and heterotrophic flagellates by the mussels *Geukensia demissa* and *Mytilus edulis* (Bivalvia, Mollusca). Aquat. Microb. Ecol. 11, 205–214.

Kristensen, E., Bouillon, S., Dittmar, T., Marchand, C., 2008. Organic carbon dynamics in mangrove ecosystems: a review. Aquat. Bot. 89, 201–219.

Kristensen, E., Lee, S.Y., Mangion, P., Quintana, C.O., Valdemarsen, T., 2017. Trophic fractionation of stable isotopes and food source partitioning by leaf-eating crabs in mangrove environments. Limnol. Oceanogr. 62, 2097–2112.

Kwon, B.O., Kim, H., Noh, J., Lee, S.Y., Nam, J., Khim, J.S., 2020. Spatiotemporal variability in microphytobenthic primary production across bare intertidal flat, saltmarsh, and mangrove forest of Asia and Australia. Mar. Pollut. Bull. 151.

Langdon, C.J., Newell, R.I., 1990. Utilization of detritus and bacteria as food sources by two bivalve suspension-feeders, the oyster *Crassostrea virginica* and the mussel *Geukensia demissa*. Mar. Ecol. Prog. Ser. 58, 299–310.

Lechene, C.P., Luyten, Y., McMahon, G., Distel, D.L., 2007. Quantitative imaging of nitrogen fixation by individual bacteria within animal cells. Science 317, 1563–1566.

Lee, S.Y., 1989a. The importance of Sesarminae crabs *Chiromanthes* spp. and inundation frequency on mangrove (*Kandelia candel* (L) Druce) leaf litter turnover in a Hong-Kong tidal shrimp pond. J. Exp. Mar. Biol. Ecol. 131, 23–43.

Lee, S.Y., 1989b. Litter production and turnover of the mangrove *Kandelia candel* (L.) Druce in a Hong-Kong tidal shrimp pond. Estuar. Coast. Shelf Sci. 29, 75–87.

Lee, S.Y., 1990a. The intensity and consequences of herbivory on *Kandelia candel* (L.) Druce leaves at the Mai Po Marshes, Hong Kong. In: Morton, B. (Ed.), The Marine Flora and Fauna of Hong Kong and Southern China II. Hong Kong University Press, Hong Kong.

Lee, S.Y., 1990b. Primary productivity and particulate organic-matter flow in an estuarine mangrove-wetland in Hong Kong. Mar. Biol. 106, 453–463.

Lee, S.Y., 1991. Herbivory as an Ecological Process in a *Kandelia-candel* (Rhizophoraceae) mangal in Hong-Kong. J. Trop. Ecol. 7, 337–348.

Lee, S.Y., 1993. Leaf choice of the sesarmid crabs Chiromanthes bidens and C. plicata in a Hong Kong mangal. In: Morton, B. (Ed.), Proceedings of the International Conference on Marine Biology of Hong Kong and the South China Sea, University of Hong Kong, October 1990. Hong Kong University Press, Hong Kong, pp. 597–604.

Lee, S.Y., 1995. Mangrove outwelling—a review. Hydrobiologia 295, 203–212.

Lee, S.Y., 1997. Potential trophic importance of the faecal material of the mangrove sesarmine crab *Sesarma messa*. Mar. Ecol. Prog. Ser. 159, 275–284.

Lee, S.Y., Kwok, P.W., 2002. The importance of mangrove species association to the population biology of two sesarmine crabs, *Perisesarma bidens* and *Parasesarma affinis*. Wetl. Ecol. Manag. 10, 215–226.

Lee, S.Y., Jones, E.B.G., Diele, K., Castellanos-Galindo, G.A., Nordhaus, I., 2017. Biodiversity. In: Rivera-Monroy, V.H., Lee, S.Y., Kristensen, E., Twilley, R.R. (Eds.), Mangrove Ecosystems: A Global Biogeographic Perspective: Structure, Function, and Services. Springer International Publishing, New York.

Leh, C.M.U., Sasekumar, A., 1985. The food of sesarmid crabs in Malaysia mangrove forests. Malay. Nat. J. 39, 135–145.

Linton, S.M., 2020. The structure and function of cellulase (endo-β-1, 4-glucanase) and hemicellulase (β-1, 3-glucanase and endo-β-1, 4-mannase) enzymes in invertebrates that consume materials ranging from microbes, algae to leaf litter. Comp. Biochem. Physiol. B Biochem. Mol. Biol. 240, 110354.

Linton, S.M., Greenaway, P., Towle, D.W., 2006. Endogenous production of endo-β-1, 4-glucanase by decapod crustaceans. J. Comp. Physiol. B. 176, 339–348.

Liu, N., Zhang, L., Zhou, H., Zhang, M., Yan, X., Wang, Q., Long, Y., Xie, L., Wang, S., Huang, Y., Zhou, Z., 2013. Metagenomic insights into metabolic capacities of the gut microbiota in a fungus-cultivating termite (*Odontotermes yunnanensis*). PLoS One 8, e69184.

Luyten, Y.A., Thompson, J.R., Morrill, W., Polz, M.F., Distel, D.L., 2006. Extensive variation in intracellular symbiont community composition among members of a single population of the wood-boring bivalve *Lyrodus pedicellatus* (Bivalvia: Teredinidae). Appl. Environ. Microbiol. 72, 412–417.

Marchand, C., Alberic, P., Lallier-Verges, E., Baltzer, F., 2006. Distribution and characteristics of dissolved organic matter in mangrove sediment pore waters along the coastline of French Guiana. Biogeochemistry 81, 59–75.

Marinucci, Hobbie, Helfrich, 1983. Effect of litter nitrogen on decomposition and microbial biomass in *Spartina alterniflora*. Microb. Ecol. 9, 27–40.

Martin, G.G., Natha, Z., Henderson, N., Bang, S., Hendry, H., Loera, Y., 2020. Absence of a microbiome in the midgut trunk of six representative Crustacea. J. Crustac. Biol. 40, 122–130.

Mattila, J.M., Zimmer, M., Vesakoski, O., Jormalainen, V., 2014. Habitat-specific gut microbiota of the marine herbivore *Idotea balthica* (Isopoda). J. Exp. Mar. Biol. Ecol. 455, 22–28.

McClintock, J.B., Klinger, T.S., Marion, K., Hsueh, P., 1991. Digestive carbohydrases of the blue crab *Callinectes sapidus* (Rathbun): implications in utilization of plant-derived detritus as a trophic resource. J. Exp. Mar. Biol. Ecol. 148, 233–239.

McKee, K.L., Faulkner, P.L., 2000. Restoration of biogeochemical function in mangrove forests. Restor. Ecol. 8, 247–259.

Mcleod, E., Chmura, G.L., Bouillon, S., Salm, R., Bjork, M., Duarte, C.M., Lovelock, C.E., Schlesinger, W.H., Silliman, B.R., 2011. A blueprint for blue carbon: toward an improved understanding of the role of vegetated coastal habitats in sequestering CO2. Front. Ecol. Environ. 9, 552–560.

Mendelssohn, I.A., Morris, J.T., 2000. Eco-physiological controls on the productivity of *Spartina alterniflora* Loisel. In: Weinstein, M.P., Kreeger, D.A. (Eds.), Concepts and Controversies in Tidal Marsh Ecology. Kluwer Academic Publishers, Dordrecht, The Netherlands.

Miyatake, T., Moerdijk-Poortvliet, T.C.W., Stal, L.J., Boschker, H.T.S., 2014. Tracing carbon flow from microphytobenthos to major bacterial groups in an intertidal marine sediment by using an in situ ^{13}C pulse-chase method. Limnol. Oceanogr. 59, 1275–1287.

Moore, P.G., 2002. Mammals in intertidal and maritime ecosystems: interactions, impacts and implications. Oceanogr. Mar. Biol. 40, 491–608.

Niiyama, T., Toyohara, H., 2011. Widespread distribution of cellulase and hemicellulase activities among aquatic invertebrates. Fish. Sci. 77, 649–655.

Niiyama, T., Hanamura, Y., Tanaka, K., Toyohara, H., 2012a. Occurrence of cellulase activities in mangrove estuarine mysids and Acetes shrimps. JIRCAS Working Rep. 75, 35–39.

Niiyama, T., Toyohara, H., Tanaka, K., 2012b. Cellulase activity in blood cockle (*Anadara granosa*) in the Matang Mangrove Forest Reserve, Malaysia. Japan Agric. Res. Quart. 46, 355–359.

Nordhaus, I., Wolff, M., 2007. Feeding ecology of the mangrove crab *Ucides cordatus* (Ocypodidae): food choice, food quality and assimilation efficiency. Mar. Biol. 151, 1665–1681.

Nordhaus, I., Wolff, M., Diele, K., 2006. Litter processing and population food intake of the mangrove crab *Ucides cordatus* in a high intertidal forest in northern Brazil. Estuar. Coast. Shelf Sci. 67, 239–250.

O'Connor, R.M., Fung, J.M., Sharp, K.H., Benner, J.S., McClung, C., Cushing, S., Lamkin, E. R., Fomenkov, A.I., Henrissat, B., Londer, Y.Y., Scholz, M.B., Posfai, J., Malfatti, S., Tringe, S.G., Woyke, T., Malmstrom, R.R., Coleman-Derr, D., Altamia, M.A., Dedrick, S., Kaluziak, S.T., Haygood, M.G., Distel, D.L., 2014. Gill bacteria enable a novel digestive strategy in a wood-feeding mollusk. Proc. Natl. Acad. Sci. U. S. A. 111, E5096–E5104.

Oakes, J.M., Connolly, R.M., Revill, A.T., 2010. Isotope enrichment in mangrove forests separates microphytobenthos and detritus as carbon sources for animals. Limnol. Oceanogr. 55, 393–402.

Oakes, J.M., Eyre, B.D., Middelburg, J.J., 2012. Transformation and fate of microphytobenthos carbon in subtropical shallow subtidal sands: a ^{13}C labeling study. Limnol. Oceanogr. 57, 1846–1856.

Odum, E.P., 1968. A Research Challenge: Evaluating the Productivity of Coastal and Estuarine Water. Proceedings of the Second Sea Grant Congress. University of Rhode Island, Kingston.

Odum, E.P., 1980. The status of three ecosystem level hypotheses regarding salt marshes: tidal subsidy, outwelling and the detritus based food chain. In: Kennedy, V.S. (Ed.), Estuarine Perspectives. Academic Press, New York.

Odum, W.E., Heald, E.J., 1975. The detritus-based food web of an estuarine mangrove community. In: Cronin, L.E. (Ed.), Estuarine Research. Academic Press, New York.

Ólafsson, E., Buchmayer, S., Skov, M.W., 2002. The East African decapod crab *Neosarmatium meinerti* (de Man) sweeps mangrove floors clean of leaf litter. AMBIO J. Hum. Environ. 31 (7–8), 569–573.

Onuf, C.P., Teal, J.M., Valiela, I., 1977. Interactions of nutrients, plant growth and herbivory in a mangrove ecosystem. Ecology 58 (3), 514–526.

Ouyang, X., Lee, S.Y., 2014. Updated estimates of carbon accumulation rates in coastal marsh sediments. Biogeosciences 11, 5057–5071.

Ouyang, X., Guo, F., Lee, S.Y., 2021. The impact of super-typhoon Mangkhut on sediment nutrient density and fluxes in a mangrove forest in Hong Kong. Sci. Total Environ. https://doi.org/10.1016/j.scitotenv.2020.142637.

Pavasovic, M., Richardson, N.A., Anderson, A.J., Mann, D., Mather, P.B., 2004. Effect of pH, temperature and diet on digestive enzyme profiles in the mud crab, *Scylla serrata*. Aquaculture 242, 641–654.

Payne, D.W., Thorpe, N.A., Donaldson, E.M., 1972. Cellulolytic activity and a study of the bacterial population in the digestive tract of *Scrobicularia plana* (Da Costa). Proc. Malac. Soc. Lond. 40, 147–160.

Pergent, G., Boudouresque, C.F., Dumay, D., Pergent-Martini, C., Wyllie-Echeverria, S., 2008. Competition between the invasive macrophyte *Caulerpa taxifolia* and the seagrass *Posidonia oceanica*: contrasting strategies. BMC Ecol. 8, 20.

Petchidurai, G., Nagoth, J.A., John, M.S., Sahayaraj, K., Murugesan, N., Pucciarelli, S., 2019. Standardization and quantification of total tannins, condensed tannin and soluble phlorotannins extracted from thirty-two drifted coastal macroalgae using high performance liquid chromatography. Biores. Technol. Rep. 7, 100273.

Raw, J.L., Perissinotto, R., Miranda, N.A.F., Peer, N., 2017. Feeding dynamics of *Terebralia palustris* (Gastropoda: Potamididae) from a subtropical mangrove ecosystem. Molluscan Res. 37, 258–267.

Ray, D.L., 1959. Nutritional Physiology of Limnoria. University of Washington Press.

Reef, R., Ball, M.C., Lovelock, C.E., 2012. The impact of a locust plague on mangroves of the arid Western Australia coast. J. Trop. Ecol. 28, 307–311.

Robertson, A.I., 1986. Leaf-burying crabs—their influence on energy-flow and export from mixed mangrove forests (*Rhizophora* spp.) in northeastern Australia. J. Exp. Mar. Biol. Ecol. 102, 237–248.

Robertson, A.I., Daniel, P.A., 1989. The influence of crabs on litter processing in high intertidal mangrove forests in tropical Australia. Oecologia 78, 191–198.

Rosenberg, F.A., Breiter, H., 1969. The role of cellulolytic bacteria in the digestive processes of the shipworm. I. Isolation of some cellulolytic microorganisms from the digestive system of teredine borers and associated waters. Mater. Org. 4, 147–159.

Saifullah, S.M., Ahmed, W., 2007. Epiphytic algal biomass on pneumatophores of mangroves of Karachi, Idus delta. Pak. J. Bot. 39, 2097–2102.

Sakamoto, K., Toyohara, H., 2009. Molecular cloning of glycoside hydrolase family 45 cellulase genes from brackish water clam *Corbicula japonica*. Comp. Biochem. Physiol. B Biochem. Mole. Biol. 152, 390–396.

Schories, D., Bergan, A.B., Barletta, M., Krumme, U., Mehlig, U., Rademaker, V., 2003. The keystone role of leaf-removing crabs in mangrove forests of North Brazil. Wetl. Ecol. Manage. 11, 243–255.

Shafique, S., Siddiqui, P.J.A., Aziz, R.A., Shoaib, N., 2013. Variations in carbon and nitrogen contents during decomposition of three macroalgae inhabiting sandspit backwater, Karachi. Pak. J. Bot. 45, 1115–1118.

Shanij, K., Praveen, V.P., Suresh, S., Oommen, M.M., Nayar, T.S., 2016. Leaf litter translocation and consumption in mangrove ecosystems: the key role played by the sesarmid crab *Neosarmatium malabaricum*. Curr. Sci. 110, 1969–1976.

Sheaves, M., Molony, B., 2000. Short-circuit in the mangrove food chain. Mar. Ecol. Prog. Ser. 199, 97–109.

Simon, N.L., 2018. Cameroon mangrove forest ecosystem: ecological and environmental dimensions. In: Sharma, S. (Ed.), Mangrove Ecosystems Ecology and Function. Intech Open, https:/doi.org/10.5772/intechopen.79021.

Sleeter, T.D., Boyle, P.J., Cundell, A.M., Mitchell, R., 1978. Relationships between marine microorganisms and the wood-boring isopod *Limnoria tripunctata*. Mar. Biol. 45, 329–336.

Smant, G., Stokkermans, J.P., Yan, Y., De Boer, J.M., Baum, T.J., Wang, X., Hussey, R.S., Gommers, F.J., Henrissat, B., Davis, E.L., Helder, J., 1998. Endogenous cellulases in animals: isolation of β-1, 4-endoglucanase genes from two species of plant-parasitic cyst nematodes. Proc. Natl. Acad. Sci. U. S. A. 95, 4906–4911.

Steinke, T.D., Lubke, R.A., Ward, C.J., 2003. The distribution of algae epiphytic on pneumatophores of the mangrove, Avicennia marina, at different salinities in the Kosi System. S. Afr. J. Bot. 69 (4), 546–554.

Steinke, T.D., Rajh, A., Holland, A.J., 1993. The feeding behaviour of the red mangrove crab *Sesarma meinerti* De Man, 1887 (Crustacea: Decapoda: Grapsidae) and its effect on the degradation of mangrove leaf litter. S. Afr. J. Mar. Sci. 13, 151–160.

Stradine, G.A., Whitaker, D.R., 1963. On the origin of the cellulase and chitinase of *Helix pomatia*. Biochem. Cell Biol. 41, 1621–1626.

Sullivan, M.J., Currin, C.A., 2000. Community structure and functional dynamics of benthic microalgae in salt marshes. In: Weinstein, M.P., Kreeger, D.A. (Eds.), Concepts and Controversies in Tidal Marsh Ecology. Kluwer Academic Publishers, Dordrecht, pp. 81–106.

Summers, S.W., Stansfield, J., Perry, S., Atkins, C., Bishop, J., 1993. Utilization, diet and diet selection by brent geese Branta berniclu berniclu on salt-marshes in Norfolk. J. Zool. Lond. 231, 249–273.

Talia, P., Arneodo, J., 2018. Lignocellulose degradation by termites. In: Termites and Sustainable Management. Springer International Publishing, pp. 101–117.

Teal, J.M., 1962. Energy-flow in salt-marsh ecosystem of Georgia. Ecology 43, 614.

Tenore, K.R., 1977. Utilization of aged detritus derived from different sources by polychaete *Capitella capitata*. Mar. Biol. 44, 51–55.

Tenore, K.R., Hanson, R.B., Mcclain, J., Maccubbin, A.E., Hodson, R.E., 1984. Changes in composition and nutritional-value to a benthic deposit feeder of decomposing detritus pools. Bull. Mar. Sci. 35, 299–311.

Tietjen, J.H., Alongi, D.M., 1990. Population growth and effects of nematodes on nutrient regeneration and bacteria associated with mangrove detritus from northeastern Queensland (Australia). Mar. Ecol. Prog. Ser. 68, 169–179.

Tokuda, G., Lo, N., Watanabe, H., Slaytor, M., Matsumoto, T., Noda, H., 1999. Metazoan cellulase genes from termites: intron/exon structures and sites of expression. Biochim. Biophys. Acta - Gene Struct. Expr. 1447, 146–159.

Tong, Y.F., Lee, S.Y., Morton, B., 2003. Effects of artificial defoliation on growth, reproduction and leaf chemistry of the mangrove *Kandelia candel*. J. Trop. Ecol. 19, 397–406.

Tong, Y., Lee, S.Y., Morton, B., 2006. The herbivore assemblage, herbivory and leaf chemistry of the mangrove *Kandelia obovata* in two contrasting forests in Hong Kong. Wetl. Ecol. Manage. 14, 39–52.

Twilley, R.R., Lugo, A.E., Pattersonzucca, C., 1986. Litter production and turnover in basin mangrove forests in southwest Florida. Ecology 67, 670–683.

Udy, J.W., Dennison, W.C., Lee Long, W.J., McKenzie, L.J., 1999. Responses of seagrass to nutrients in the Great Barrier Reef, Australia. Mar. Ecol. Prog. Ser. 185, 257–271.

Underwood, G.J.C., 2002. Adaptations of tropical marine microphytobenthic assemblages along a gradient of light and nutrient availability in Suva Lagoon, Fiji. Eur. J. Phycol. 37, 449–462.

Underwood, G.J.C., Paterson, D.M., 2003. The importance of extracellular carbohydrate production by marine epipelic diatoms. Adv. Bot. Res. 40, 183–240.

Valiela, I., Teal, J.M., Allen, S.D., Van Etten, R., Goehringer, D., Volkmann, S., 1985. Decomposition in salt marsh ecosystems: The phases and major factors affecting disappearance of above-ground organic matter. J. Exp. Mar. Biol. Ecol. 89 (1), 29–54.

van der Heijden, L.H., Graeve, M., Asmus, R., Rzeznik-Orignac, J., Niquil, N., Bernier, Q., Guillou, G., Asmus, H., Lebreton, B., 2019. Trophic importance of microphytobenthos and bacteria to meiofauna in soft-bottom intertidal habitats: a combined trophic marker approach. Mar. Environ. Res. 149, 50–66.

Wang, Y., Zhu, H., Tam, N.F., 2014. Polyphenols, tannins and antioxidant activities of eight true mangrove plant species in South China. Plant Soil 374, 549–563.

Watanabe, H., Nakamura, M., Tokuda, G., Yamaoka, I., Scrivener, A.M., Noda, H., 1997. Site of secretion and properties of endogenous endo-β-1,4-glucanase components from *Reticulitermes speratus* (Kolbe), a Japanese subterranean termite. Insect Biochem. Mol. Biol. 27, 305–313.

Watanabe, H., Noda, H., Tokuda, G., Lo, N., 1998. A cellulase gene of termite origin. Nature 394, 330–331.

Waterbury, J.B., Calloway, C.B., Turner, R.D., 1983. A cellulolytic nitrogen-fixing bacterium cultured from the gland of Deshayes in shipworms (Bivalvia: Teredinidae). Science 221, 1401–1403.

Werry, J., Lee, S.Y., 2005. Grapsid crabs mediate link between mangrove litter production and estuarine planktonic food chains. Mar. Ecol. Prog. Ser. 293, 165–176.

Yokoe, Y., Yasumasu, I., 1964. The distribution of cellulase in invertebrates. Comp. Biochem. Physiol. 13, 323–338.

Zhou, J., Wu, Y., Kang, Q., Zhang, J., 2007. Spatial variations of carbon, nitrogen, phosphorous and sulfur in the salt marsh sediments of the Yangtze Estuary in China. Estuar. Coast. Shelf Sci. 71, 47–59.

Zimmer, M., 2002. Nutrition in terrestrial isopods (Isopoda: Oniscidea) : an evolutionary-ecological approach. Biol. Rev. 77, 455–493.

Zimmer, M., Bartholmé, S., 2003. Bacterial endosymbionts in *Asellus aquaticus* (Isopoda) and *Gammarus pulex* (Amphipoda) and their contribution to digestion. Limnol. Oceanogr. 48, 2208–2213.

Zimmer, M., Topp, W., 2002. The role of coprophagy in nutrient release from the feces of phytophagous insects. Soil Biol. Biochem. 34, 1093–1099.

Water-air gas exchange of CO_2 and CH_4 in coastal wetlands

Judith A. Rosentreter[a,b]
[a]*Yale School of the Environment, Yale University, New Haven, CT, United States,*
[b]*Centre for Coastal Biogeochemistry, School of Environment, Science and Engineering,*
Southern Cross University, Lismore, NSW, Australia

6.1 Introduction

Anthropogenic carbon emissions to the atmosphere have increased significantly since preindustrial times, to the point that current emissions have reached their highest rates for the last 66 million years (Le Quéré et al., 2016; Zeebe et al., 2016; Friedlingstein et al., 2019). Following these recent changes, and in prediction of future climate scenarios (Myhre et al., 2013), global collaboration between governments, industries, and institutions has led to multidisciplinary research investigating the urgent mitigation of anthropogenic emissions (Höglund-Isaksson, 2012; Gallo et al., 2017; Lovelock and Duarte, 2019). Mitigation strategies focus on the reduction of carbon emissions (Hansen et al., 2013; Myhre et al., 2013; Wood and Roelich, 2019) and negative emission technology (NET) including afforestation/reforestation, bioenergy carbon capture, and storage, biochar, direct air capture, ocean fertilization, and enhanced weathering (Fuss et al., 2018; Minx et al., 2018). Another CO_2 removal strategy is to enhance natural carbon stores, in particular in highly productive coastal "blue carbon" wetlands (Nellemann et al., 2009; Mcleod et al., 2011; Duarte et al., 2013; Macreadie et al., 2019) that have carbon burial rates higher than tropical rainforests (Alongi, 2014). The carbon stored in coastal wetland sediments continues to accumulate over century to millennial time scales, thus is a long-term natural carbon sink. However, the role of natural carbon sinks in coastal wetlands may be counteracted by emissions of greenhouse gases such as carbon dioxide (CO_2), methane (CH_4), and nitrous oxide (N_2O) (Rosentreter et al., 2018c; Al-Haj and Fulweiler, 2020a,b; Oreska et al., 2020, Van Dam et al. in review). The radiative balance of a coastal wetland can be positive (net warming) or negative (net cooling), depending on the atmospheric CO_2 uptake (C sink) and emission (C source) to the atmosphere (Neubauer and Verhoeven, 2019). On global scale, blue carbon offsets by CH_4 emission remain uncertain, and estimates are sensitive to statistical assumptions because of the high variability associated with few available CH_4 flux

Carbon Mineralization in Coastal Wetlands. https://doi.org/10.1016/B978-0-12-819220-7.00003-0

measurements from mangrove, saltmarsh, and seagrass habitats (Rosentreter and Williamson, 2020). In the review by Al-Haj and Fulweiler (2020a,b), the authors concluded that high mean CH_4 fluxes result in a possible loss of climate benefits of mangroves and saltmarshes (offsets >100%), whereas other studies have highlighted the climate benefits and reduced but cooling effects of coastal wetlands (Rosentreter et al., 2018c; Taillardat et al., 2020; Oreska et al., 2020). As of today, the magnitude and direction of the climate change mitigation potential of blue carbon wetlands remains unclear but a highly discussed and urgent topic.

CO_2 and CH_4 are produced primarily in coastal sediments and reach the atmosphere via molecular diffusion through sediment-water-air interfaces (Kristensen et al., 2008; Call et al., 2015; Trifunovic et al., 2020; Oreska et al., 2020). In coastal wetlands, diffusive gas fluxes are often studied individually, i.e., from sediments to overlaying waters (benthic fluxes), diffusive fluxes from sediments to the air, and water-air gas exchange from coastal waters. Coastal wetlands are strongly influenced by tidal regimes (low-high tidal cycle, tidal elevation); therefore, all three interfaces of molecular gas diffusion should be considered when quantifying total CO_2 and CH_4 fluxes. Nevertheless, the majority of gas flux studies in mangrove forests and saltmarsh habitats have studied sediment-air fluxes (e.g., Magenheimer et al., 1996; Livesley and Andrusiak, 2012; Emery and Fulweiler, 2014; Olsson et al., 2015; Wilson et al., 2015; Nóbrega et al., 2016; Zheng et al., 2018; Cameron et al., 2019a,b), although in recent years there has been increased research activity to also quantifying water-air fluxes of CO_2 and CH_4 (e.g., Linto et al., 2014; Rosentreter et al., 2018a; Sea et al., 2018; Huertas et al., 2019; Trifunovic et al., 2020). In seagrass habitats, which are inundated most of the tidal cycle, the sediment-water flux using benthic chamber incubations or eddy covariance is the most commonly studied gas flux (Bahlmann et al., 2015; Garcias-Bonet and Duarte, 2017; Oreska et al., 2020).

The release of CH_4-containing gas bubbles from sediments (ebullition) (Purvaja et al., 2004; Padhy et al., 2020) and the transport through the aerenchyma of vascular plants (Purvaja et al., 2004; Krithika et al., 2008; He et al., 2019; Padhy et al., 2020) are additional pathways that need to be considered when estimating total CH_4 fluxes. For example, the total CH_4 flux in the freshwater Cattai wetland in Australia was dominated by plant-mediated fluxes (59%), while ebullition and diffusion each accounted for $\sim$20% of the CH_4 emission (Jeffrey et al., 2019a). Because CH_4 production is generally lower in marine compared to freshwater sediments, the ebullitive and plant-mediated flux has been widely ignored in coastal wetland studies. Although expected to be less important than in freshwater wetlands, the relative contribution of the ebullitive flux and the plant-mediated flux to total CH_4 emissions is mostly unknown and should be further explored in mangrove, saltmarsh, and seagrass ecosystems.

This chapter focusses on the diffusive water-air gas exchange of the two most potent greenhouse gases, CO_2, and CH_4, in mangrove, saltmarsh, and seagrasses waters. First, the range and magnitude of available CO_2 and CH_4 fluxes in coastal wetlands from the literature are reviewed and summarized. Then, the direct and indirect drivers that can control gas concentration gradients and fluxes across the water-air interface are discussed, and finally, commonly used methods that are available to estimate the gas exchange between the water phase and atmosphere are explained. A concluding section with implication of water-air fluxes on global estimates can be found at the end of this chapter.

6.2 Variability of water-air fluxes in coastal wetlands

Estimating CO_2 and CH_4 water-air gas exchange in coastal wetlands is fundamental for understanding their role in global coastal carbon budgets (Bauer et al., 2013). The diffusive flux is directly proportional to the concentration gradient of a specific gas and depends on the gas solubility as a function of temperature and salinity and the gas transfer velocity (Wanninkhof, 2014). There are different methods available to determine water-air fluxes, which are reviewed in Section 6.4 of this chapter. A positive flux implies an efflux from the water to the atmosphere, meaning a water body is a source of a gas to the atmosphere, which is also called "emission." In the case where gas concentrations are lower in the water compared to the atmosphere, this results in an uptake of a gas (influx) towards the water phase, which is indicated by a negative flux and means that the water body is a sink of a gas from the atmosphere.

In tidal wetlands, the water table fluctuates due to incoming and outgoing tides, depending on tidal elevation (high, mid, low tidal zone), tidal regimes (macro-, meso-, micro-tidal), tidal cycle (diurnal, semidiurnal), and lunar cycle (spring tides, neap tides). As such, the relative area of a tidal wetland being exposed (sediment-air flux) and inundated (water-air flux) is in constant change and also differs between marshes, mangroves. and seagrasses in the coastal ocean. For example, some high elevation saltmarshes may receive only occasional flooding (changes are also due to sea-level rise and available accommodation space, see in Rogers et al., 2019) (Mcowen et al., 2017), whereas most seagrass habitats that are primarily located in the outer estuarine zone or on the continental shelf are constantly submerged (McKenzie et al., 2020). Many mangrove areas are roughly inundated 50% and exposed 50% of the time (Rosentreter et al., 2018c; Bunting et al., 2018), although this general assumption may not be applicable to all coastlines of different climatic and tidal zones in the world. Salinity is one of the main drivers of surface water gas concentrations, and emissions are generally higher from oligohaline (low salinity) compared to polyhaline (high salinity) coastal wetlands (Poffenbarger et al., 2011; Martin and Moseman-Valtierra, 2015). Nevertheless, CO_2 and CH_4 can be emitted from high salinity coastal waters (Jacotot et al., 2018; Rosentreter et al., 2018c; Call et al., 2019; Huertas et al., 2019). Tidal and spatial variability of coastal wetland CO_2 and CH_4 fluxes is further closely linked to hydrogeology, the exchange of coastal groundwater discharge (CGD), and recirculated submarine groundwater discharge (SGD) or porewater (high in pCO_2 and CH_4) with surface water (Ovalle et al., 1990; Stieglitz et al., 2013; Maher et al., 2013; Call et al., 2019; Luijendijk et al., 2020). With regard to spatial variability, CO_2 and CH_4 fluxes may also vary with different water surface to wetland area ratios. For example, a tidal wetland creek (small ratio) may have a strong gas efflux (Barnes et al., 2006; Call et al., 2015; Rosentreter et al., 2018a), whereas a large estuary or bay fringed by coastal wetlands (large ratio) may show significantly lower emissions (Nirmal Rajkumar et al., 2008; Rosentreter et al., 2018b).

Because of regional-climatic, hydrogeological, geomorphological, physico-chemical, and biogeochemical drivers that control water-air gas exchange in dynamic coastal wetlands (for drivers see Section 6.3), fluxes show high spatial and temporal (tidal, diurnal, seasonal, and interannual) variability. Global estimates of mean, median, and ranges of CO_2 and CH_4 fluxes from coastal wetland waters are summarized in Table 6.1. Because of the high variability but also uncertainty associated with few water-air fluxes, data sets of coastal wetland fluxes are highly skewed, making global estimates sensitive to statistical assumptions.

6.2.1 Mangrove water-air-fluxes

The CO_2 water-air flux from mangrove tidal waters is generally positive, suggesting mangrove waters are a source of CO_2 to the atmosphere (Table 6.1). The CO_2 emissions can be attributed to the decomposition of high organic matter loads originated from either the mangrove system itself (autochthonous) or upstream rivers (allochthonous) or both. Aerobic respiration (Kristensen et al., 1994; Alongi et al., 2001), sulfate reduction (Kristensen et al., 1994; Bouillon et al., 2003, 2007), and iron reduction (Kristensen et al., 2000) have been suggested as the dominating pathways of microbial degradation of organic matter in mangrove sediments, affected by bioturbation, living root biomass, soil temperature and moisture, and tidal inundation (Kristensen et al., 1994; Kristensen and Alongi, 2006; Lovelock et al., 2006; Leopold et al., 2013; Cameron et al., 2019b). Calcium carbonate ($CaCO_3$) production in mangrove sediments represents an additional source of CO_2 to the atmosphere if $CaCO_3$ dissolution is lower than local calcification (Howard et al., 2017; Saderne et al., 2019).

CH_4 water-air fluxes are at least two magnitudes lower than CO_2 fluxes in mangroves because of the relatively lower concentrations in the water column. The total CH_4 flux to the atmosphere depends on the balance between production (methanogenesis) and consumption (methane oxidation) processes in sediments and water column and may include diffusive (Linto et al., 2014; Call et al., 2015, 2019; Rosentreter et al., 2018c), ebullitive (Purvaja et al., 2004; Padhy et al., 2020), and plant-mediated (Purvaja et al., 2004; Krithika et al., 2008; He et al., 2019; Padhy et al., 2020) fluxes. For example, gas bubble samples collected from mangrove sediments had higher CH_4 concentrations during the tide than before the tide in three locations at the Sundarbans mangroves in India (Padhy et al., 2020). Although, it may be noted that the gas bubble sampling method used by Purvaja et al. (2004) and Padhy et al. (2020) involved actively stirring up gas bubbles, which may not be comparable with naturally occurring gas bubble release. The same study in the Sundarbans mangroves further showed that CH_4 fluxes were higher from sediments

Table 6.1 Global mean, median, and range (min-max) of areal water-air fluxes of CO_2 and CH_4 from coastal wetlands.

Coastal wetland	CO_2 flux (mg CO_2 m^{-2} day^{-1})	CO_2 ref	CH_4 flux (mg CH_4 m^{-2} day^{-1})	CH_4 ref
Mangroves				
Mean	2487 3472	Rosentreter et al. (2018a) Alongi (2020a)	4.6 73[a] 8.5	Rosentreter et al. (2018b) Al-Haj and Fulweiler (2020a,b) Alongi (2020a)
Median	1914 2200	Rosentreter et al. (2018a) Alongi (2020a)	3.4 4.5[a] 4.2	Rosentreter et al. (2018b) Al-Haj and Fulweiler (2020a,b) Alongi (2020a)
(Min-max)	13–9726	Rosentreter et al. (2018a)	0.6–11 −1.1–1169[a]	Rosentreter et al. (2018b) Al-Haj and Fulweiler (2020a,b)
Saltmarshes				
Mean	2823±332	Alongi (2020b)	56.7[a] 8.4	Al-Haj and Fulweiler (2020a,b) Alongi (2020b)
Median	1557	Alongi (2020b)	3.6[a] 1.5	Al-Haj and Fulweiler (2020a,b) Alongi (2020b)
(Min-max)			−1.49–1510[a]	Al-Haj and Fulweiler (2020a,b)
Seagrasses				
Mean	−148[b]	This study	1.74[c] 3.9[d]	Al-Haj and Fulweiler (2020a,b) This study
Median	−132[b]	This study	1.04[c] 4.7[d]	Al-Haj and Fulweiler (2020a,b) This study
(Min-max)	−3168–3041[b]	This study	0.02–6.44[c] 1.9–4.9[d]	Al-Haj and Fulweiler (2020a,b) This study

Note, some global estimates present water-air combined with sediment-air or sediment-water fluxes as indicated by footnotes.

[a]Combined sediment-air and water-air fluxes.

[b]Based on the studies from Frankignoulle (1988), Gazeau et al. (2005), Polsenaere et al. (2012), Maher and Eyre (2012), Tokoro et al. (2014), Banerjee et al. (2018), Burkholz et al. (2019), and Van Dam et al. (2019). Seagrass studies that report NEM, NEE, or NCP based on O_2 measurements were excluded.

[c](Benthic) sediment-water fluxes or combined sediment-water-air fluxes.

[d]Based on the studies from Deborde et al. (2010) and Banerjee et al. (2018).

with pneumatophores as compared to sediments without pneumatophores and that both CH_4 ebullition and fluxes through pneumatophore were higher in the monsoon compared to the premonsoon season (Padhy et al., 2020). Higher wet season compared to dry season diffusive CH_4 fluxes have also been observed from mangrove creek waters, likely driven by freshwater riverine inputs that increase nutrient loadings, increase the supply of labile organic matter and reduce sulfate reduction, all of which can enhance CH_4 production (Ovalle et al., 1990; Sotomayor et al., 1994; Rosentreter et al., 2018c). Methane uptake from the atmosphere into mangrove waters (sink) has not been reported but is possible given that there is evidence of no flux or negative CH_4 fluxes from sediment-air interfaces associated with high salinity and the occurrence of sulfate-reducing bacteria in anoxic sediments that outcompete methanogenic bacteria and archaea (Alongi et al., 2004; Biswas et al., 2007; Poffenbarger et al., 2011; Chen et al., 2014a). However, the large majority of studies reports positive CH_4 fluxes from mangrove waters, that like CO_2 emissions, follow a typical tidal trend with higher fluxes during the low tide and lower fluxes during the high tide (Fig. 6.1) (Call et al., 2015; Jacotot et al., 2018; Rosentreter et al., 2018c). This tidal pattern is related to a mechanism termed "tidal pumping" which describes the tidally mediated exchange of interstitial porewater or groundwater (i.e., high in pCO_2 and CH_4) with surface waters (Ovalle et al., 1990; Stieglitz et al., 2013; Maher et al., 2013), in particular in macro-tidal mangrove ecosystems (Call et al., 2019).

6.2.2 Saltmarsh water-air fluxes

Tidal saltmarshes comprise the upper, vegetated portion of the intertidal zone and are situated at the interface between terrestrial and coastal aquatic ecosystems (Mcowen et al., 2017). As such, they are less inundated or less flooded than mangrove or seagrass habitats. Therefore, water-air gas flux studies in tidal brackish marshes are rare, and regional and global estimates are highly uncertain. The few studies that have examined saltmarsh water-air fluxes include the studies by Ferrón et al. (2007), Tong et al. (2013), Daniel et al. (2013), and Trifunovic et al. (2020). The study by Ferrón et al. (2007) was conducted in a tidal marsh in Spain and estimated CO_2 and CH_4 water-air fluxes using dissolved gas concentrations and empirical k models. High concentrations in the water column resulted in a positive water-air flux of both gases, following a strong tidal pattern. Tong et al. (2013) measured diurnal variation of CO_2 and CH_4 water-air fluxes during spring tides in a saltmarsh in the Shanyutan Wetlands in China. Based on floating chamber incubations, they found CH_4 fluxes were always positive with higher emissions during the daytime compared to the nighttime. In contrast, CO_2 fluxes showed more complex patterns over the tidal cycle and were a source or a sink from the water to the atmosphere. The study by Trifunovic et al. (2020) estimated water-air fluxes of CO_2 and CH_4 from a tidal marsh creek in the Delaware National Estuarine Research Reserve in Dover, also using the floating chamber technique. Similar to the study in Spain, they found the typical tidal trend in mangrove waters with higher fluxes at low tide and lower fluxes at high tide. Furthermore, the study shows that CO_2 water-air fluxes were

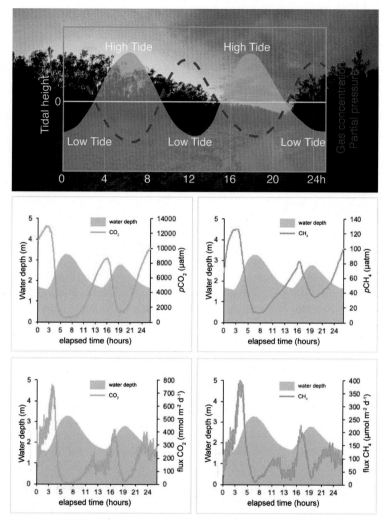

FIG. 6.1

Upper panel: In a semidiurnal mangrove creek, partial pressure or concentration of a gas typically follows a tidal pattern with low concentrations at high tide and high concentrations at low tide driven, e.g., by tidal pumping and water table height. The background photo shows the Burdekin mangrove creek located at the northeastern coast of Queensland in Australia during the low tide at night. Lower panel: Examples of continuous CO_2 and CH_4 partial pressures (µatm) and water-air fluxes (mmol m^{-2} day^{-1}, µmol m^{-2} day^{-1}) over two tidal cycles in the Burdekin mangrove creek during the dry season in 2014. Water-air fluxes were estimated using the gradient flux method and site-specific k_{600} models based on current velocity, wind speed, and water depth, published in Rosentreter et al. (2017). In the dry season, the Burdekin mangrove creek waters were a strong source of CO_2 and CH_4 to the atmosphere. CO_2 and CH_4 fluxes were generally lower in the dry season than in the wet season (not shown here).

CO_2 data are obtained from Rosentreter, J., Maher, D.T. D.V. Erler, R. Murray, Eyre, B.D., 2018. Seasonal and temporal CO_2 dynamics in three tropical mangrove creeks—a revision of global mangrove CO_2 emissions. Geochim. Cosmochim. Acta 222, 729–745. https://doi.org/10.1016/j.gca.2017.11.026 and CH_4 data from Rosentreter, J.A., Maher, D.T., Erler, D.V., Murray, R., Eyre, B.D., 2018. Factors controlling seasonal CO_2 and CH_4 emissions in three tropical mangrove-dominated estuaries in Australia. Estuar. Coast. Shelf Sci. 215, 69–82. https://doi.org/10.1016/j.ecss.2018.10.003. Credit: Judith Rosentreter.

higher than sediment-air CO_2 fluxes, which the authors relate to lateral CO_2 transport from creek sediments to the creek waters via tidal pumping. However, this comparison was not possible for CH_4 because sediment-air fluxes were not measured. Global estimates of water-air CO_2 and CH_4 emissions for saltmarshes presented in Table 6.1 are from Alongi (2020b) that include the studies by Daniel et al. (2013) and Trifunovic et al. (2020) but also fluxes from open-water estuaries and sediment-air interfaces that may not reflect the saltmarsh water flux.

6.2.3 Seagrass water-air fluxes

Compared to mangrove and saltmarsh habitats, little is known about CO_2 and CH_4 - water-air gas exchange over seagrass beds. Submerged in the water, photosynthetic active seagrass plants are assumed to draw CO_2 from the water column during the day, which should result in a negative concentration gradient and thus an influx from the atmosphere to the water (sink) (Frankignoulle, 1988; Gazeau et al., 2005; Duarte et al., 2010; Maher and Eyre, 2012; Tokoro et al., 2014). Nevertheless, there is also evidence of positive water-air CO_2 fluxes from seagrass beds in recent studies (Banerjee et al., 2018; Berg et al., 2019; Burkholz et al., 2019; Van Dam et al., 2019; Berger et al., 2020), suggesting that seagrass beds can be sinks or a sources of CO_2 to the atmosphere. For example, two studied seagrass meadows in Florida Bay were net heterotrophic, indicating a net release of CO_2 from the water to the atmosphere with possible sources from sediment organic matter respiration and carbonate dissolution (Van Dam et al., 2019). Although on average net heterotrophic, the two study sites showed strong diurnal patterns with positive net ecosystem production (NEP) during the day and negative NEP during the night, likely driven by differences in CO_2 water-air gas exchange and contrasting responses to variations in light intensity (Van Dam et al., 2019). Diurnal, tidal, and seasonal changes of CO_2 water-air gas exchange were also found in seagrass meadows in France using the EC technique (Polsenaere et al., 2012). Clearly, more research is needed to better assess the diurnal, seasonal, and annual variability of CO_2 source-sink dynamics in seagrass waters. The global range presented in Table 6.1 summarizes water CO_2 gas exchange over seagrass meadows. The strong diurnal variation of seagrass waters between CO_2 sinks and sources makes seagrass meadows a hotly debated topic with regard to "blue carbon" assessments.

In contrast, seagrass waters are likely a source of CH_4, ranging from 1.9 to 4.9 mg CH_4 m^{-2} day^{-1} (Deborde et al., 2010; Banerjee et al., 2018). In the Chilika lagoon in the Bay of Bengal in India, CO_2 and CH_4 fluxes were compared in the wet and dry season (Banerjee et al., 2018). While CO_2 dynamics were coupled to seasonality, CH_4 fluxes showed no obvious seasonal trends in the seagrass sector of the brackish lagoon. The high CH_4 fluxes from the relatively shallow seagrass waters were likely driven by high carbon mineralization rates in anoxic, carbon-rich sediments, but also linked to seagrass loss in the wet season which may have enhanced CH_4 fluxes. Additionally, river waters entering the Chilika lagoon were highly supersaturated in CH_4, further indicating an allochthonous input. The global CH_4 flux presented in Table 6.1 is based on the two available studies from the Chilika lagoon in India and the Arcachon lagoon in France, therefore should not be used as a predictor of seagrass CH_4 trends on a global scale.

6.2.4 CO_2 versus CH_4: Global warming potentials and switchover times

Coastal wetland waters and sediments exchange CO_2 and CH_4 with the atmosphere, therefore, have the potential to influence global climate. In recent years, CH_4 fluxes have been increasingly expressed as CO_2-equivalent fluxes for the purpose of directly comparing their relative radiative force ($W\,m^{-2}$) and with regard to net warming or cooling effects to one another. The global warming potential (GWP) is the most commonly used metric for quantifying their role in regulating climate change and is defined as "the time-integrated radiative forcing due to a pulse emission of a given component, relative to a pulse emission of an equal mass of CO_2" (Myhre et al., 2013). The recently proposed sustained-flux global warming potential (SGWP) and sustained-flux global cooling potential (SGCP) are two useful alternatives to the GWP and more appropriate when ecosystem GHG fluxes are persistent and not one-time (pulse) events (Neubauer and Megonigal, 2015). The SGWP is defined as the "time-integrated radiative forcing due to sustained emissions of a given component, relative to sustained sequestration of an equal mass of CO_2" and indicates how many kilograms of CO_2 must be sequestered to offset the emission of 1 kg of CH_4. The SGCP is defined as the "time-integrated radiative forcing due to sustained uptake of a given component, relative to sustained sequestration of an equal mass of CO_2" and indicates how many kilograms of CO_2 must be sequestered to have the same cooling effect as the uptake of 1 kg CH_4 (Neubauer and Megonigal, 2015; Neubauer and Verhoeven, 2019). The SGWP is significantly larger than the GWP (by up to 40%), which has substantial implications on the overall interpretation of CO_2-equivalent emissions. For example, an ecosystem that sequestered 40 kg CO_2 in sediments per 1 kg CH_4 emitted to the atmosphere would be considered a carbon sink (net cooling) when using the 100-year GWP (GWP_{100} for $CH_4 = 32$), whereas the same ecosystem would have a small net warming effect when using the 100-year SGWP ($SGWP_{100}$ for $CH_4 = 45$) (Table 6.2) (Neubauer and Megonigal, 2015; Neubauer and Verhoeven, 2019).

The equivalent global flux (efflux or influx) for CH_4 in comparison to CO_2 ($F_{CO_2\ equiv}$) using any global warming potential can be calculated as

$$F_{CO_2\ equiv}\left(Tg\,C\,year^{-1}\right) = F \times (GWP\ or\ SGWP\ or\ SGCP) \times f \qquad (6.1)$$

where F is the flux of CH_4 in gas mass unit ($Tg\,CH_4\,year^{-1}$) multiplied by the GWP, SGWP or SGCP for the appropriate time frames, and f is the conversion factor to $Tg\,C\,year^{-1}$ ($f = 12/44$), the atomic weight of carbon (12 g) divided by the atomic weight of CO_2 (44 g). While the 100-year time horizon is the standard method under the United Nations Framework Convention on Climate Change (UNFCCC), for national climate policies that are based on shorter timescales (National Determined

Table 6.2 Comparison of the global warming potential (GWP) with and without inclusion of climate-carbon feedbacks from Myhre et al. (2013), and the global warming potential, sustained-flux global warming potential (SGWP), sustained-flux global cooling potential (SGCP) proposed by Neubauer and Megonigal (2015), for the 20, 100, and 500-year time horizons.

Global warming potentials						
		Myhre et al. (2013)		Neubauer and Megonigal (2015)		
Gas	Time frame (years)	GWP (no cc fb)	GWP (with cc fb)	GWP	SGWP (Emission)	SGCP (Uptake)
CO_2	any	1	1	1	1	1
CH_4	20	84	86	87	96	96
	100	28	34	32	45	45
	500			11	14	14

Switchover times					
	Time (years)				Study site
Ecosystem	Mean	Median	Min	Max	n
Mangroves	0	0	0	0	1
Saltmarshes	17	17	17	17	1
Tropical peat swamps	26	26	26	26	1
Floodplain	256	256	256	256	1
Freshwater marshes	3117	644	0	20,000	8
Restored freshwater marshes	284	286	266	299	3

Below are listed radiative forcing switchover times in years for selected wetlands proposed by Taillardat et al. (2020).
The switchover time describes the time when a wetland switches over from lifetime warming to lifetime cooling effects and is estimated using the ratio of CO_2 sequestration to CH_4 emission, the magnitude of CH_4 emission, and the wetland age.

Contributions (NDC) under the Paris Agreement, Gallo et al., 2017), the 20-year time horizon may be of interest.

Regardless the choice of global warming potential (GWP, SGWP, SGCP) and time horizon (20, 100, and 500-years), the different time scales of carbon sequestration and CH_4 emissions from coastal wetlands, and the unequal lifetimes of the two gases in the atmosphere, further complicate a direct comparison. It has been suggested recently that the global warming potential over any fixed time period may not representative of an ecosystem's warming or cooling effect because these metrics do not integrate the age of an ecosystem, which is usually much older than the steady state of CH_4 in atmosphere (Neubauer, 2014; Neubauer and Verhoeven, 2019; Taillardat et al., 2020). CH_4 has a much shorter lifetime than CO_2 and reaches a plateau after roughly 50 years (4 atmospheric lifetimes), where CH_4 emissions are balanced by CH_4 oxidation in the atmosphere (Neubauer and Verhoeven, 2019). In contrast, CO_2 never fully reaches the equilibrium with other biotic or abiotic reservoirs such as geological scale weathering of continental rocks. Thus, the wetland warming effect by CH_4 emissions will reach a steady state, while on the other hand the cooling effect by CO_2 sequestration will continue to grow as the age of the wetland increases (see details in Neubauer and Verhoeven, 2019). Accordingly, the term "radiative forcing switchover time" has been introduced, which describes the time when the cumulative lifetime radiative forcing of a wetland switches from a lifetime warming to a lifetime cooling effect (Frolking et al., 2006; Neubauer and Verhoeven, 2019). The median switchover time for coastal wetlands has been estimated at 8.5 ± 8.5 years (mangroves 0 years, saltmarshes 17 years), proposing coastal wetlands that are older than 8.5 years should have a net cooling effect when undisturbed (Table 6.2) (Taillardat et al., 2020).

6.3 Drivers of water-air gas exchange

The diffusive gas exchange of CO_2 and CH_4 between surface waters and the atmosphere depends mainly on the concentration gradient, but also on physical, chemical, and biotic drivers that control gas transfer velocities (see Section 6.4.2). The gas concentration gradient, which is the difference of gas concentration between sediment, water, and air phase, can be controlled by multiple direct and indirect factors as summarized in Fig. 6.2. *Regional-climatic* drivers include weather-driven events, such as strong rainfall that can dilute water concentrations, cyclones, or storm that can increase river discharge or increase wind-driven gas transfer at the water boundary layer. Droughts or long dry periods can decrease flushing times, which impact retention times and general carbon cycling in coastal waters. Sea-level rise induced by rising temperatures and global warming puts increased pressure on the available accommodation space of coastal wetlands (Kirwan and Megonigal, 2013; Rogers et al., 2019) (also see Chapters 7 and 9). Closely linked to climatic drivers are *hydrological* and *geomorphological* drivers such as river discharge, submarine, and coastal groundwater inputs or porewater exchange with surface water (tidal pumping), bed rock geology, and weathering rates (in particular for CO_2). Coastal wetlands are particularly influenced by water table height and tidal regimes such as tidal amplitude (macro, meso, and micro-tidal) and elevation (e.g., low, mid, and high marsh). *Physicochemical* drivers of coastal wetland CO_2 and CH_4 water concentrations include salinity, oxygen, pH, temperature, and redox-potential, all of which greatly influence microbial production and consumption processes in coastal sediments, thus gradients in the water column. For example, methanogenesis occurs at the final stage of organic matter degradation, which leads to the expectation that microbes using terminal electron acceptors (TEAs), such as O_2, NO_3^-, Mn_4^+, Fe_3^+, and SO_4^{2}, will outcompete methanotrophs. Therefore, methanogenesis is usually higher in freshwater sediments, while sulfate reduction dominates in marine sediments (Middelburg et al.,

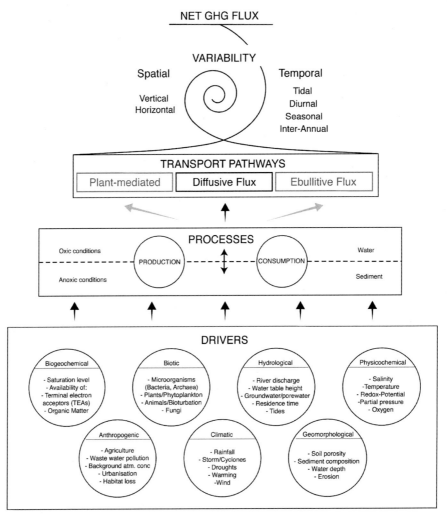

FIG. 6.2

Summary of factors controlling CO_2 and CH_4 fluxes in coastal wetlands. Direct and indirect factors drive production and consumption processes, transport pathways, and spatial and temporal variability. The net flux to the atmosphere is the balance of influx and efflux between sediment-water and water-air interfaces in response to the direct and indirect factors. While possible, there is little or no evidence of ebullitive and plant-mediated fluxes (in *gray*) from mangrove, saltmarshes, and seagrass meadows, and their relative contribution to total fluxes is yet unknown.

From Rosentreter et al. (under review). Credit: Judith Rosentreter.

2002; Kristensen, 2007). The availability of nutrients (NH_4, NO_3^-, NO_2, and PO_4^{3-}) and organic matter supply that fuel microbial decomposition in coastal sediments can be described as *biogeochemical* drivers. In recent years, it has been of increased scientific interest whether organic matter and nutrient inputs to coastal waters are of allochthonous (from outside the system, e.g., upstream rivers) or autochthonous (from within the system, e.g., mangrove plant litter) origin, which is important for the understanding of carbon cycling in coastal wetlands. Microorganisms (bacteria and archaea) are the dominant *biotic* producers of CO_2 and CH_4 in coastal sediments. There is also evidence of fungal production of CH_4 (Lenhart et al., 2012; Schroll et al., 2020), although no studies have investigated the GHG production by fungi in coastal wetlands; therefore, it remains unknown whether this is an important production pathway. As mentioned earlier, plants can act as conduits for gas transport from sediments to the atmosphere. Impacts of bioturbation (crab burrows and pneumatophores) are discussed in detail in Chapter 8. *Anthropogenic* factors have direct or indirect impacts on all above listed drivers, hence water gas concentrations in coastal wetlands. In particular, land-use changes such as agriculture and urbanization have strong impacts by for instance increased nutrient supplies from farmlands via rivers to coastal waters that can shift coastal water from a GHG sink to a GHG source, waste water pollution (sewage discharge into estuaries), reduced flushing by river damming, or habitat loss through the conversion from natural coastal habitats to aquaculture/mariculture farm. Because of the various direct and indirect factors that drive vertical and horizontal concentration gradients, high spatial and temporal (tidal, diurnal, seasonal, and interannual) variability of water-air fluxes is typical in coastal wetland waters.

6.4 Available methods to quantify water-air gas exchange

Currently available methods to measure CO_2 and CH_4 exchange at the water-air interface include the floating chamber method, the gradient flux technique based on gas concentration (or partial pressure) gradients and gas transfer velocity (k), and eddy-covariance. Ebullition traps can be used to estimate the CH_4 ebullition from sediments through the water column to the atmosphere, however, this method measures the ebullitive flux only and cannot be used to estimate the total CH_4 flux to the atmosphere. The different methods described in this chapter have their advantages and disadvantages with regard to cost efficiency, accuracy, reliability, and time-consuming procedure. Choosing the most appropriate method for estimating gas exchange in coastal waters also depends on the spatial and temporal scope of a study.

6.4.1 Floating chamber method

The floating chamber technique is a widely discussed but commonly used method to determine CO_2 and CH_4 gas exchange at the water-air interface. A cylindrical or rectangular box, open at the bottom and equipped with floats, is placed on the water surface during a predefined time during which the gas concentration inside the chamber is measured either by discrete headspace sampling (at least at the start and the end of an incubation) or preferably by continuous measurements via a closed loop connected to an automated gas analyzer (e.g., infra-red gas analyzer, cavity-ring-down spectrometer) (Fig. 6.3A and B) (Leopold et al., 2016a; Rosentreter et al., 2017; Jacotot et al., 2018). The gas flux from each floating chamber incubation is calculated as

$$F = [s(V/RT_{air}A)]t \qquad (6.2)$$

where s is the slope of the linear regression of a gas measured inside the chamber over time (deployments should be conducted between 10 and 20 min depending on the gas) (ppm per second), V is the chamber volume (m^3), R is the universal gas constant, T_{air} is the temperature measured inside the chamber (°K), A is the surface area of the chamber (m^2), and t refers to the conversion factor to µmol (or mmol) per day. The floating chamber incubations result in a linear increase (or decrease) of a gas inside the chamber if molecular diffusion is responsible for the gas concentration gradient. For accurate flux estimates, the linear regression from each incubation should have a high r^2 (>0.9). Chamber incubations in water with very low or undersaturated gas concentrations can result in linear regressions with lower r^2 (<0.9). While they may still be used to estimate gas fluxes, they should be analyzed and interpreted with caution. As a rule of thumb, chamber incubations with linear regression $r^2 > 0.9$ are generally acceptable, whereas incubations resulting in $r^2 < 0.8$ may be excluded. If ebullition (CH_4 bubbles released from sediments) occurs during the deployments, this is indicated by a sudden jump of the CH_4 concentration (nonlinear regression) measured inside the chamber. In this case, the regression slope cannot be used to estimate the diffusive water-air flux. Details on how to measure the ebullitive CH_4 flux rate in a coastal wetland can be found in Section 6.4.2.

The diverse designs of floating chambers found in the literature make them difficult to compare. Floating chambers can differ in geometric shape, height, penetration depth of the chamber walls, surrounding floats, and volume, ranging from simple bucket deployments to well-constructed "flying chambers" (Mazot and Taran, 2009; Xiao et al., 2014; Lorke et al., 2015, 2019; Rosentreter et al., 2017; Huang et al., 2019). The main criticism of the floating chamber technique is that the chamber walls can disturb the water-turbulence regime, thus may overestimate the gas flux (Raymond and Cole, 2001; Kremer et al., 2003; Vachon et al., 2010). However, several studies have found good agreement between the floating chamber and energy dissipation method (Tokoro et al., 2008; Gålfalk et al., 2013) and estimates of k_{600} derived from floating chamber incubations and a deliberate gas tracer release experiment (Rosentreter et al., 2017). Nevertheless, the floating chamber method is restricted to locations with low to moderate winds and water turbulences. For example, the method is not reliable under wavy conditions, during rainfall or very strong currents. To avoid an artificial increase (overestimation) of the gas flux created by the chamber walls penetrating the water column, the deployments should ideally be conducted drifting in opposition to anchored (Lorke et al., 2015).

FIG. 6.3

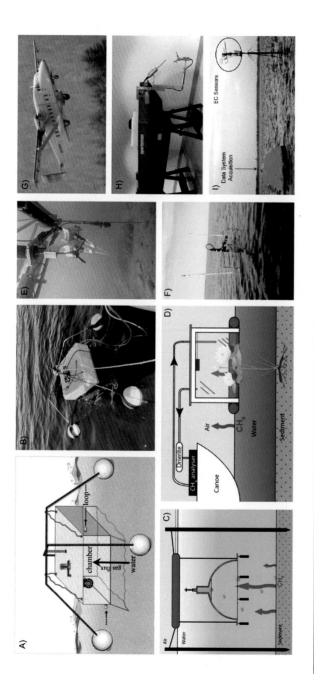

(A) and (B) are an example of a floating "flying" chamber. The chamber is connected through a thin chain to four stainless steel legs with floats that sit on the water surface. Two different types of walls were used in this chamber design: flexible walls at the air-water interface allow for the propagation of water-side turbulence into the chamber and rigid walls for the main chamber to ensure a consistent chamber volume, which is necessary for the flux calculation. This chamber model was built to minimize disturbance of the water turbulence regime by allowing free vertical and horizontal movement of the chamber through loose attachment to the spider-like frame and has been successfully used in the studies by Rosentreter et al. (2017) and Jeffrey et al. (2018). (C) is a schematic diagram of an ebullition dome that can be used to measure the ebullitive CH_4 flux. The design features include a floating ring at the top, sinker weights attached to the dome, and stakes in the sediments ~2 m either side of the dome, designed by Jeffrey et al. (2019a). (D) Shows an example of a closed loop floating chamber that was used by Jeffrey et al. (2019a) to measure diffusive CH_4 fluxes from open water and plant areas. (E)–(I) show examples of the multifunctional applications of Eddy covariance (EC): Aquatic EC has been used to measure primary production in seagrasses by (E) Attard et al. (2019) and (F) Berger et al. (2020). (G) Shows the NASA Carbon Airborne Flux Experiment (CARAFE) Sherpa aircraft quantifying CO_2 and CH_4 surface fluxes from wetlands and waters (Wolfe et al. 2018). (H) shows the "mobile" application of aquatic EC by Berg et al. (2020) measuring gas exchange from a moving platform, and (I) shows the EC station by Polsenaere et al. (2012) that measured CO_2 gas exchange over seagrass beds in an intertidal lagoon in France.

(A and B) Credit: Judith Rosentreter; (C and D) From Jeffrey, L.C., Maher, D.T., Johnston, S.G., Kelaher, B.P., Steven, A., Tait, D.R., 2019. Wetland methane emissions dominated by plant-mediated fluxes: contrasting emissions pathways and seasons within a shallow freshwater subtropical wetland. Limnol. Oceanogr. 64, 1895–1912. https://doi.org/10.1002/lno.11158; (E) From Attard, K.M., Rodil, I.F., Glud, R.N., Berg, P., Norkko, J., Norkko, A., 2019. Seasonal ecosystem metabolism across shallow benthic habitats measured by aquatic eddy covariance. Limnol. Oceanogr. Lett. 4, 79–86. https://doi.org/10.1002/lol2.10107; (F) from Berger, A.C., Berg, P., McGlathery, K.J., Delgard, M.L., 2020. Long-term trends and resilience of seagrass metabolism: a decadal aquatic eddy covariance study. Limnol. Oceanogr. 65, 1423–1438. https://doi.org/10.1002/lno.11397; (G) From Wolfe, G.M., Kawa, S.R., Hanisco, T.F., et al., 2018. The NASA carbon airborne flux experiment (CARAFE): instrumentation and methodology. Atmos. Meas. Tech. 11, 1757–1776. https://doi.org/10.5194/amt-11-1757-2018; (H) From Berg, P., Pace, M.L., Buelo, C.D., 2020. Air–water gas exchange in lakes and reservoirs measured from a moving platform by underwater eddy covariance. Limnol. Oceanogr. Methods 18, 424–436. https://doi.org/10.1002/lom3.10373; and (I) From Polsenaere, P., Lamaud, E., Lafon, V., et al., 2012. Spatial and temporal CO_2 exchanges measured by eddy covariance over a temperate intertidal flat and their relationships to net ecosystem production. Biogeosciences 9, 249–268. https://doi.org/10.5194/bg-9-249-2012.

Considering the relatively small surface of the chamber, this method is particularly useful to identify small-scale and short-time water-air fluxes that are often of interest in tidal coastal wetlands.

6.4.2 Ebullition funnel traps

Gas bubble ebullition describes the direct transport of CH_4 trapped in bubbles from sediments to the atmosphere, bypassing the potential attenuating effects of oxygenated water (methane oxidation) or plant rhizomes (Chanton et al., 1989). The ebullitive CH_4 flux can be measured using ebullition funnel traps (also called ebullition domes), where a conventional funnel is suspended below the water level and the headspace gas is sampled after an incubation time of hours to days (Fig. 6.3C). To minimize the disturbance of the sediment and to avoid creating any artificial CH_4 release from sediments, it is advisable to deploy ebullition funnel traps by boat or canoe. The headspace gas from the ebullition dome can be either collected as a discrete sample with a syringe and analyzed in situ using a portable methane analyzer (Jeffrey et al., 2019a,b) or transported to the lab and analyzed on a gas chromatograph (GC) (Chanton et al., 1989; Purvaja et al., 2004; Padhy et al., 2020), or the headspace can be connected to an automated methane analyzer that autonomously logs both volumetric ebullition rate and methane concentrations (Maher et al., 2019). A detailed description and sampling procedure of an inexpensive Automated Methane Ebullition Sensor (iAMES) can be found in Maher et al. (2019).

From the funnel trap, the ebullition rate (E_b) can be calculated as

$$E_b \left(\text{mmol m}^{-2} \text{day}^{-1} \right) = \left([CH_4] \, CH_{4\text{Vol}} \right) / A \, V_m \, T_d \tag{6.3}$$

where $[CH_4]$ is the % CH_4 concentration in the collected gas, $CH_{4\text{Vol}}$ is the volume (mL) of the gas sample, A is the area (m^2) of the funnel or dome, V_m is the molar volume (L) of CH_4 at the in situ temperature, and T_d is the time (day) of the incubation (Jeffrey et al., 2019a). Further details on how to calculate CH_4 oxidation rates during ebullition transport can be found in Jeffrey et al. (2019a, and references therein).

6.4.3 Gradient flux method

When floating chambers are not practicable or as an addition to the floating chamber incubations, the second most commonly used method to evaluate water-air gas exchange is the gradient flux method (e.g., Kristensen et al., 2008; Aimé et al., 2018; Trifunovic et al., 2020). This method is generally preferred when gas fluxes are estimated over larger scales (spatially, e.g., along an estuary gradient) or longer terms (time series, e.g., over a 24 h tidal cycle) because it is based on a flux computation that requires parameters that can be measured at high resolution and therefore allows to estimate high-resolution gas fluxes. Water-air gas exchange follows a simple model where the flux (F) equals the gradient of the concentration (or partial pressure, µatm) in water (C_w) and air (C_a) multiplied by the gas transfer velocity k (m day^{-1}) and the solubility coefficient K_0 ($\text{mol kg}^{-1} \text{atm}^{-1}$):

$$F\left(\text{mmol}\,\text{m}^{-2}\,\text{day}^{-1}\right) = k\,K_0\left(C_w - C_a\right) \qquad (6.4)$$

While today the concentrations in water and atmosphere can be measured relatively precisely and at high resolution, and the solubility coefficient is depending on temperature and salinity (see in Wanninkhof, 2014), which can also be measured precisely and at high resolution, the largest uncertainty in the flux computation remains in the estimate of k. At the water-air boundary layer, the gas transfer velocity usually normalized to the Schmidt number of 600 (k_{600}) (Wanninkhof, 2014), depends on water-side turbulences (Liss and Merlivat, 1986). Understanding the rates and controls of the gas transfer velocity in aquatic ecosystems has been of great research interest to the scientific community (see reviews by Raymond and Cole, 2001; Hall and Ulseth, 2020; Klaus and Vachon, 2020). For instance, k_{600} can show temporal and spatial variability (Rosentreter et al., 2017; Jeffrey et al., 2018) and can also be gas-specific (Beaulieu et al., 2012; Prairie and del Giorgio, 2013; McGinnis et al., 2015). In coastal wetlands that are located between rivers, estuaries, and shelf environments, site- and gas-specific k_{600} remains largely unexplored. The majority of studies that estimated gas fluxes from coastal wetland waters using the gradient flux technique predicted k_{600} based on empirical models (k parameterizations) that are generally intended for estuarine (Raymond and Cole, 2001; Borges et al., 2004; Ho et al., 2016; Rosentreter et al., 2017) or ocean environments (Liss and Merlivat, 1986; Wanninkhof, 1992, 2014). These k models commonly use wind speed and/or current velocity and water depth to estimate k_{600} at the water boundary layer. By conversion of Eq. (6.2), the floating chamber method can also provide site-specific k_{600} values that can be applied to the flux computation. The one study that compared k_{600} in mangrove creeks, a small mangrove lake, an adjacent estuary channel and bay in the Everglades National Park, found that k_{600} was generally lower in the narrow and wind-protected mangrove creeks compared to the larger estuary channel and bay (Rosentreter et al., 2017). No studies are available that investigate k_{600} in tidal marsh or over seagrass beds. There is also evidence that k_{600} behaves differently for CO_2 and CH_4 at the water-air interface (e.g., Emerson, 1995; Beaulieu et al., 2012). For example, a so called "microbubble flux" component can enhance the flux of CH_4 relative to CO_2 in mangrove dominated estuaries (Rosentreter et al., 2017), and also in lakes (Vagle et al., 2010; Prairie and del Giorgio, 2013; McGinnis et al., 2015). In contrast, k_{600} of CO_2 can be chemically enhanced compared to CH_4 (Hoover and Berkshire, 1969; Wanninkhof and Knox, 1996). CO_2 is a chemically reactive gas that can bypass the rate-limiting molecular diffusion step via two hydration reactions at the water-air boundary layer (Hoover and Berkshire, 1969; Wanninkhof and Knox, 1996). As such, oceanic waters ($pH > 8$) favor CO_2-OH- hydration reactions, which can increase the water-air gas transfer of CO_2 relative to CH_4 (Wanninkhof and Knox, 1996; Kuss and Schneider, 2004). Although the concept of chemical enhancement is known for over four decades and has been applied in ocean models, it remains relatively unexplored in coastal waters (Jeffrey et al., 2018). The active research field around gas transfer velocities in freshwater, coastal, and marine ecosystems is promising, and it is expected that system-specific and gas-specific empirical k_{600} models will be available in the future. This will increase our confidence in water-air gas fluxes from coastal wetland waters estimated by the gradient flux method.

6.4.4 Eddy-covariance

Eddy-covariance (EC), also called turbulent flux method, is a well-established widely used micrometeorological technique and an effective way to quantify large scale net ecosystem carbon exchange (NEE) over long time periods, ranging from hours to years (Baldocchi, 2003, 2014). Traditionally, the EC method has been used primarily to measure the canopy-atmosphere gas flux (Law et al., 2003; Baldocchi et al., 2005; Stoy et al., 2005) or soil/land-atmosphere gas flux (Wolf et al., 2011; Xiao et al., 2011; Schwalm et al., 2012), but it can also be used to measure benthic (sediment-water) gas exchange (Attard et al., 2019; Berg et al., 2019; Berger et al., 2020) or fluxes from water and wetland surfaces (Alberto et al., 2014; Gutiérrez-Loza et al., 2019; Morin, 2019; Benítez-Valenzuela and Sanchez-Mejia, 2020; Berg et al., 2020). Historical information and a practical guide on how to measure and analyze eddy-covariance data can be found in Aubinet et al. (2012), Burba (2013), Baldocchi (2014), and Rebmann et al. (2018) (also see Chapter 4). With regard to coastal wetlands, ecosystem-scale CO_2 fluxes (NEE) using the EC technique have been studied in mangrove forest mainly in China (Chen et al., 2014b; Cui et al., 2018; Liu and Lai, 2019) but also in India (Rodda et al., 2016), the United States (Barr et al., 2010), and New Caledonia (Leopold et al., 2016b), whereas only two studies have used EC to measure canopy-air fluxes of CH_4 over mangrove forests (Ganguly et al., 2008; Liu et al., 2020). Net ecosystem carbon exchange has also been studied over saltmarsh canopy, particularly in the Northern Hemisphere (e.g., Xiao et al., 2013; Artigas et al., 2015; Li et al., 2018; Nahrawi et al., 2020; Trifunovic et al., 2020), while EC measurements are still scarce from saltmarshes in the Southern Hemisphere (Tonti et al., 2018). Few studies have estimated benthic metabolism (sediment-water O_2 and CO_2 fluxes) in seagrass meadows using EC, and "aquatic Eddy-covariance" is generally focused on benthic oxygen fluxes rather than on GHGs fluxes (Attard et al., 2019; Berg et al., 2019; Berger et al., 2020). Atmospheric EC has been used to measure spatial and temporal CO_2 water-air exchanges over seagrass meadows in the Arcachon lagoon in France (Polsenaere et al., 2012), in different climate zones in Japan (Tokoro et al., 2014), and over a shallow coastal lagoon in the Gulf of California (Benítez-Valenzuela and Sanchez-Mejia, 2020). In contrast, CH_4 flux measurements over coastal waters using EC are rare (Gutiérrez-Loza et al., 2019).

While EC is usually used as a land-based application, moving vehicles, airborne, and shipborne installations are increasingly emerging in flux networks (Burba, 2013). For example, the NASA Carbon Airborne Flux Experiment (CARAFE) has used a Sherpa aircraft equipped with EC instrumentation to acquire CO_2 and CH_4 fluxes from forest, cropland, wetlands, and water in the eastern United States (Fig. 6.3G) (Wolfe et al., 2018). Furthermore, EC measurements from moving platforms, for example in lakes and reservoirs, are newly emerging applications of the EC technique (Fig. 6.3H) (Berg et al., 2020). Such "mobile" aquatic EC measurements enable the quantification of high-resolution temporal and spatial gas exchange in aquatic ecosystems, yet to my knowledge have not been applied in coastal wetlands. Clearly, the EC technique is used primarily to study ecosystem-scale exchange (NEE), i.e., canopy-air or soil-air fluxes, while this technique is more challenging to execute in intertidal coastal ecosystems. Whether immobile or mobile, EC is a precise and effective technique adaptable on land, air or water, that can help to better assess long-term and large-scale GHG fluxes from different physical interfaces (sediment-water, sediment-air, water-air), which is needed to better predict coastal wetland GHG trends on regional and global scale.

6.5 Summary and conclusion

Quantifying water-air gas exchange of CO_2 and CH_4 in mangrove, saltmarsh, and seagrass ecosystems is an important component of coastal ocean carbon budgets. This review shows that surface waters of coastal wetland can be a source or a sink of CO_2 and CH_4 to the atmosphere, overall ranging from -3168 to $9726\,mg\,CO_2\,m^{-2}\,day^{-1}$ and -1.5 to $1510\,mg\,CH_4\,m^{-2}\,day^{-1}$, respectively. The high spatial and temporal variability of water-air gas fluxes is driven by various direct and indirect factors including regional-climatic, hydrological, geomorphological, physicochemical, biogeochemical, biotic, and anthropogenic drivers. The relative contribution of the different controlling factors driving surface water concentrations and the relative contribution of the three gas flux pathways (diffusive, ebullitive, and plant-mediated fluxes) to total fluxes, in particular for CH_4, remain uncertain in coastal wetlands and warrant further research efforts. Effective and high-resolution techniques, such as eddy-covariance, but also the combined approach of available methods, have the potential to better assess long-term and large-scale gas exchange in tidal wetland surface waters, which will help to quantify more accurately current and future GHG trends in coastal wetlands. Determining the net warming or net cooling effect of coastal wetlands is important for blue carbon assessments, yet may be more complicated than simply applying the global warming potential (i.e., CH_4 as CO_2-equivalent). An alternative approach to the global warming potential over a fixed time period is to estimate the switchover time of a wetland and by integrating the age of an ecosystem. For example, it has been proposed that the switchover time from lifetime warming to lifetime cooling in coastal wetlands is 8.5 ± 8.5 years. It is crucial to consider the different atmospheric lifetime and equilibrium of CO_2 and CH_4 because warming effects by CH_4 emissions will reach a steady state after roughly 50 years, whereas cooling effects by CO_2 sequestration will continue to grow. Thus, theoretically, any wetland will have a net cooling effect when emissions and sequestration are stable over a certain period of time and the ecosystem is not disturbed, which is relevant for restoration and management policies. As of today, the net radiative forcing effect of coastal blue carbon wetlands is a highly debated but an urgent topic. Until the spatial and temporal variability and controlling processes of CO_2 and CH_4 water-air fluxes can be better resolved, global emissions and global blue carbon offsets in coastal wetlands remain uncertain.

References

Aimé, J., Allenbach, M., Bourgeois, C., et al., 2018. Variability of CO_2 emissions during the rearing cycle of a semi-intensive shrimp farm in a mangrove coastal zone (New Caledonia). Mar. Pollut. Bull. 129, 194–206. https://doi.org/10.1016/j.marpolbul.2018.02.025.

Alberto, M.C.R., Wassmann, R., Buresh, R.J., Quilty, J.R., Correa, T.Q., Sandro, J.M., Centeno, C.A.R., 2014. Measuring methane flux from irrigated rice fields by eddy covariance method using open-path gas analyzer. Field Crop Res. 160, 12–21. https://doi.org/10.1016/j.fcr.2014.02.008.

Al-Haj, A.N., Fulweiler, R.W., 2020a. A synthesis of methane emissions from shallow vegetated coastal ecosystems. Glob. Chang. Biol. 26, 2988–3005. https://doi.org/10.1111/gcb.15046.

Al-Haj, A.N., Fulweiler, R.W., 2020b. Corrigendum. Glob. Chang. Biol. 26, 5342. https://doi.org/10.1111/gcb.15192.

Alongi, D.M., 2014. Carbon cycling and storage in mangrove forests. Ann. Rev. Mar. Sci. 6, 195–219. https://doi.org/10.1146/annurev-marine-010213-135020.

Alongi, D.M., 2020a. Carbon cycling in the world' s mangrove ecosystems revisited : significance of non-steady state diagenesis and subsurface linkages between the forest floor and the coastal ocean. Forests 9, 1–17.

Alongi, D.M., 2020b. Carbon Balance in Salt Marsh and Mangrove Ecosystems : A Global Synthesis Preprint., https://doi.org/10.20944/preprints202009.0236.v1.

Alongi, D.M., Wattayakorn, G., Pfitzner, J., Tirendi, F., Zagorskis, I., Brunskill, G.J., Davidson, A., Clough, B.F., 2001. Organic carbon accumulation and metabolic pathways in sediments of mangrove forests in southern Thailand. Mar. Geol. 179, 85–103. https://doi.org/10.1016/S0025-3227(01)00195-5.

Alongi, D.M., Sasekumar, A., Chong, V.C., Pfitzner, J., Trott, L.A., Tirendi, F., Dixon, P., Brunskill, G.J., 2004. Sediment accumulation and organic material flux in a managed mangrove ecosystem: estimates of land-ocean-atmosphere exchange in peninsular Malaysia. Mar. Geol. 208, 383–402. https://doi.org/10.1016/j.margeo.2004.04.016.

Artigas, F., Shin, J.Y., Hobble, C., Marti-Donati, A., Schäfer, K.V.R., Pechmann, I., 2015. Long term carbon storage potential and CO_2 sink strength of a restored salt marsh in New Jersey. Agric. For. Meteorol. 200, 313–321. https://doi.org/10.1016/j.agrformet.2014.09.012.

Attard, K.M., Rodil, I.F., Glud, R.N., Berg, P., Norkko, J., Norkko, A., 2019. Seasonal ecosystem metabolism across shallow benthic habitats measured by aquatic eddy covariance. Limnol. Oceanogr. Lett. 4, 79–86. https://doi.org/10.1002/lol2.10107.

Aubinet, M., Vesala, T., Papale, D., 2012. Eddy Covariance: A Practical Guide to Measurement and Data Analysis. Springer Science & Business Media.

Bahlmann, E., Weinberg, I., Lavrič, J.V., Eckhardt, T., Michaelis, W., Santos, R., Seifert, R., 2015. Tidal controls on trace gas dynamics in a seagrass meadow of the Ria Formosa lagoon (southern Portugal). Biogeosciences 12, 1683–1696. https://doi.org/10.5194/bg-12-1683-2015.

Baldocchi, D.D., 2003. Assessing the eddy covariance technique for evaluating carbon dioxide exchange rates of ecosystems: past, present and future. Glob. Chang. Biol. 9, 479–492. https://doi.org/10.1046/j.1365-2486.2003.00629.x.

Baldocchi, D., 2014. Measuring fluxes of trace gases and energy between ecosystems and the atmosphere—the state and future of the eddy covariance method. Glob. Chang. Biol. 20, 3600–3609. https://doi.org/10.1111/gcb.12649.

Baldocchi, D.D., Black, T.A., Curtis, P.S., et al., 2005. Predicting the onset of net carbon uptake by deciduous forests with soil temperature and climate data: a synthesis of FLUXNET data. Int. J. Biometeorol. 49, 377–387. https://doi.org/10.1007/s00484-005-0256-4.

Banerjee, K., Paneerselvam, A., Ramachandran, P., Ganguly, D., Singh, G., Ramesh, R., 2018. Seagrass and macrophyte mediated CO_2 and CH_4 dynamics in shallow coastal waters. PLoS One 13. https://doi.org/10.1371/journal.pone.0203922, e0203922.

Barnes, J., Ramesh, R., Purvaja, R., et al., 2006. Tidal dynamics and rainfall control N_2O and CH_4 emissions from a pristine mangrove creek. Geophys. Res. Lett. 33, 4–9. https://doi.org/10.1029/2006GL026829.

Barr, J.G., Engel, V., Fuentes, J.D., Zieman, J.C., O'Halloran, T.L., Smith, T.J., Anderson, G.H., 2010. Controls on mangrove forest-atmosphere carbon dioxide exchanges in western Everglades National Park. Eur. J. Vasc. Endovasc. Surg. 115, G02020. https://doi.org/10.1029/2009JG001186.

Bauer, J.E., Cai, W.J., Raymond, P.A., Bianchi, T.S., Hopkinson, C.S., Regnier, P.A.G.G., 2013. The changing carbon cycle of the coastal ocean. Nature 504, 61–70. https://doi.org/10.1038/nature12857.

Beaulieu, J.J., Shuster, W.D., Rebholz, J.A., 2012. Controls on gas transfer velocities in a large river. Eur. J. Vasc. Endovasc. Surg. 117, 1–13. https://doi.org/10.1029/2011JG001794.

Benítez-Valenzuela, L.I., Sanchez-Mejia, Z.M., 2020. Observations of turbulent heat fluxes variability in a semiarid coastal lagoon (Gulf Of California). Atmosphere (Basel) 11, 15–17. https://doi.org/10.3390/atmos11060626.

Berg, P., Delgard, M.L., Polsenaere, P., McGlathery, K.J., Doney, S.C., Berger, A.C., 2019. Dynamics of benthic metabolism, O_2, and pCO_2 in a temperate seagrass meadow. Limnol. Oceanogr. 64, 2586–2604. https://doi.org/10.1002/lno.11236.

Berg, P., Pace, M.L., Buelo, C.D., 2020. Air–water gas exchange in lakes and reservoirs measured from a moving platform by underwater eddy covariance. Limnol. Oceanogr. Methods 18, 424–436. https://doi.org/10.1002/lom3.10373.

Berger, A.C., Berg, P., McGlathery, K.J., Delgard, M.L., 2020. Long-term trends and resilience of seagrass metabolism: a decadal aquatic eddy covariance study. Limnol. Oceanogr. 65, 1423–1438. https://doi.org/10.1002/lno.11397.

Biswas, H., Mukhopadhyay, S.K., Sen, S., Jana, T.K., 2007. Spatial and temporal patterns of methane dynamics in the tropical mangrove dominated estuary, NE coast of Bay of Bengal, India. J. Mar. Syst. 68, 55–64. https://doi.org/10.1016/j.jmarsys.2006.11.001.

Borges, A.V., Delille, B., Schiettecatte, L.S., Talence, F., Frankignoulle, M., 2004. Gas transfer velocities of CO_2 in three European estuaries (Randers Fjord). Limnol. Oceanogr. 49, 1630–1641.

Bouillon, S., Rao, A.V.V.S., Koedam, N., Dehairs, F., 2003. Sources of organic carbon in mangrove sediments: variability and possible ecological implications. Hydrobiologia 495, 33–39.

Bouillon, S., Dehairs, F., Velimirov, B., Abril, G., Borges, A.V., 2007. Dynamics of organic and inorganic carbon across contiguous mangrove and seagrass systems (Gazi Bay, Kenya). Eur. J. Vasc. Endovasc. Surg. 112, 1–14. https://doi.org/10.1029/2006JG000325.

Bunting, P., Rosenqvist, A., Lucas, R., et al., 2018. The global mangrove watch—a new 2010 global baseline of mangrove extent. Remote Sens. (Basel) 10, 1669. https://doi.org/10.3390/rs10101669.

Burba, G., 2013. Eddy Covariance Method-for Scientific, Industrial, Agricultural, and Regulatory Applications. LI-COR Biosciences (ISBN: 978-0-61576827-4).

Burkholz, C., Garcias-Bonet, N., Duarte, C.M., 2019. Warming enhances carbon dioxide and methane fluxes from Red Sea seagrass (Halophila stipulacea) sediments. Biogeosci. Discuss., 1–20. https://doi.org/10.5194/bg-2019-93.

Call, M., Maher, D.T., Santos, I.R., et al., 2015. Spatial and temporal variability of carbon dioxide and methane fluxes over semi-diurnal and spring–neap–spring timescales in a mangrove creek. Geochim. Cosmochim. Acta 150, 211–225. https://doi.org/10.1016/j.gca.2014.11.023.

Call, M., Santos, I.R., Dittmar, T., de Rezende, C.E., Asp, N.E., Maher, D.T., 2019. High pore-water derived CO_2 and CH_4 emissions from a macro-tidal mangrove creek in the Amazon region. Geochim. Cosmochim. Acta 247, 106–120. https://doi.org/10.1016/j.gca.2018.12.029.

Cameron, C., Hutley, L.B., Friess, D.A., 2019a. Estimating the full greenhouse gas emissions offset potential and profile between rehabilitating and established mangroves. Sci. Total Environ. 665, 419–431. https://doi.org/10.1016/j.scitotenv.2019.02.104.

Cameron, C., Hutley, L.B., Friess, D.A., Munksgaard, N.C., 2019b. Hydroperiod, soil moisture and bioturbation are critical drivers of greenhouse gas fluxes and vary as a function of land use change in mangroves of Sulawesi, Indonesia. Sci. Total Environ. 654, 365–377. https://doi.org/10.1016/j.scitotenv.2018.11.092.

Chanton, J.P., Martens, C.S., Kelley, C.A., 1989. Gas transport from methane-saturated, tidal freshwater and wetland sediments. Limnol. Oceanogr. 34, 807–819. https://doi.org/10.4319/lo.1989.34.5.0807.

Chen, G.C., Ulumuddin, Y.I., Pramudji, S., et al., 2014a. Rich soil carbon and nitrogen but low atmospheric greenhouse gas fluxes from North Sulawesi mangrove swamps in Indonesia. Sci. Total Environ. 487, 91–96. https://doi.org/10.1016/j.scitotenv.2014.03.140.

Chen, H., Lu, W., Yan, G., Yang, S., Lin, G., 2014b. Typhoons exert significant but differential impacts on net ecosystem carbon exchange of subtropical mangrove forests in China. Biogeosciences 11, 5323–5333. https://doi.org/10.5194/bg-11-5323-2014.

Cui, X., Liang, J., Lu, W., et al., 2018. Stronger ecosystem carbon sequestration potential of mangrove wetlands with respect to terrestrial forests in subtropical China. Agric. For. Meteorol. 249, 71–80. https://doi.org/10.1016/j.agrformet.2017.11.019.

Daniel, I., DeGrandpre, M., Farías, L., 2013. Greenhouse gas emissions from the Tubul-Raqui estuary (central Chile 36°S). Estuar. Coast. Shelf Sci. 134, 31–44. https://doi.org/10.1016/j.ecss.2013.09.019.

Deborde, J., Anschutz, P., Guérin, F., et al., 2010. Methane sources, sinks and fluxes in a temperate tidal lagoon: the Arcachon lagoon (SW France). Estuar. Coast. Shelf Sci. 89, 256–266. https://doi.org/10.1016/j.ecss.2010.07.013.

Duarte, C.M., Marbà, N., Gacia, E., Fourqurean, J.W., Beggins, J., Barrón, C., Apostolaki, E.T., 2010. Seagrass community metabolism: assessing the carbon sink capacity of seagrass meadows. Global Biogeochem. Cycles 24. https://doi.org/10.1029/2010GB003793.

Duarte, C.M., Losada, I.J., Hendriks, I.E., Mazarrasa, I., Marbà, N., 2013. The role of coastal plant communities for climate change mitigation and adaptation. Nat. Clim. Chang. 3, 961–968. https://doi.org/10.1038/nclimate1970.

Emerson, S., 1995. Enhanced transport of carbon dioxide during gas exchange. In: Jähne, B., Monahan, E.C. (Eds.), Air-Water Gas Transfer. AEON Verlad & Studio, pp. 23–36.

Emery, H.E., Fulweiler, R.W., 2014. Spartina alterniflora and invasive Phragmites australis stands have similar greenhouse gas emissions in a New England marsh. Aquat. Bot. 116, 83–92. https://doi.org/10.1016/j.aquabot.2014.01.010.

Ferrón, S., Ortega, T., Gómez-Parra, A., Forja, J.M., 2007. Seasonal study of dissolved CH_4, CO_2 and N_2O in a shallow tidal system of the bay of Cádiz (SW Spain). J. Mar. Syst. 66, 244–257. https://doi.org/10.1016/j.jmarsys.2006.03.021.

Frankignoulle, M., 1988. Field measurements of air-sea CO_2 exchange. Limnol. Oceanogr. 33, 313–322.

Friedlingstein, P., Jones, M.W., O'Sullivan, M., et al., 2019. Global carbon budget 2019. Earth Syst. Sci. Data 11, 1783–1838. https://doi.org/10.5194/essd-11-1783-2019.

Frolking, S., Roulet, N., Fuglestvedt, J., 2006. How northern peatlands influence the Earth's radiative budget: sustained methane emission versus sustained carbon sequestration. Eur. J. Vasc. Endovasc. Surg. 111, 1–10. https://doi.org/10.1029/2005JG000091.

Fuss, S., Lamb, W.F., Callaghan, M.W., et al., 2018. Negative emissions—part 2: costs, potentials and side effects. Environ. Res. Lett. 13. https://doi.org/10.1088/1748-9326/aabf9f.

Gålfalk, M., Bastviken, D., Fredriksson, S., Arneborg, L., 2013. Determination of the piston velocity for water-air interfaces using flux chambers, acoustic Doppler velocimetry, and IR imaging of the water surface. Eur. J. Vasc. Endovasc. Surg. 118, 770–782. https://doi.org/10.1002/jgrg.20064.

Gallo, N.D., Victor, D.G., Levin, L.A., 2017. Ocean commitments under the Paris Agreement. Nat. Clim. Chang. 7, 833–838. https://doi.org/10.1038/nclimate3422.

Ganguly, D., Dey, M., Mandal, S.K., De, T.K., Jana, T.K., 2008. Energy dynamics and its implication to biosphere-atmosphere exchange of CO_2, H_2O and CH_4 in a tropical mangrove forest canopy. Atmos. Environ. 42, 4172–4184. https://doi.org/10.1016/j.atmosenv.2008.01.022.

Garcias-Bonet, N., Duarte, C.M., 2017. Methane production by seagrass ecosystems in the Red Sea. Front. Mar. Sci. 4, 340. https://doi.org/10.3389/fmars.2017.00340.

Gazeau, F., Duarte, C.M., Gattuso, J.-P., et al., 2005. Whole-system metabolism and CO_2 fluxes in a Mediterranean Bay dominated by seagrass beds (Palma Bay, NW Mediterranean). Biogeosci. Discuss. 1, 755–802. https://doi.org/10.5194/bgd-1-755-2004.

Gutiérrez-Loza, L., Wallin, M.B., Sahlée, E., Nilsson, E., Bange, H.W., Kock, A., Rutgersson, A., 2019. Measurement of air-sea methane fluxes in the Baltic Sea using the eddy covariance method. Front. Earth Sci. 7, 1–13. https://doi.org/10.3389/feart.2019.00093.

Hall, R.O., Ulseth, A.J., 2020. Gas exchange in streams and rivers. WIREs Water 7, 1–18. https://doi.org/10.1002/wat2.1391.

Hansen, J., Kharecha, P., Sato, M., et al., 2013. Assessing "dangerous climate change": required reduction of carbon emissions to protect young people, future generations and nature. PLoS One 8. https://doi.org/10.1371/journal.pone.0081648.

He, Y., Guan, W., Xue, D., et al., 2019. Comparison of methane emissions among invasive and native mangrove species in Dongzhaigang, Hainan Island. Sci. Total Environ. 697. https://doi.org/10.1016/j.scitotenv.2019.133945.

Ho, D.T., Coffineau, N., Hickman, B., Chow, N., Koffman, T., Schlosser, P., 2016. Influence of current velocity and wind speed on air-water gas exchange in a mangrove estuary. Geophys. Res. Lett. 43, 3813–3821. https://doi.org/10.1002/2016GL068727.

Höglund-Isaksson, L., 2012. Global anthropogenic methane emissions 2005-2030: technical mitigation potentials and costs. Atmos. Chem. Phys. 12, 9079–9096. https://doi.org/10.5194/acp-12-9079-2012.

Hoover, T.E., Berkshire, D.C., 1969. Effects of hydration on carbon dioxide exchange across an air-water interface. J. Geophys. Res. 74, 456–464. https://doi.org/10.1029/JB074i002p00456.

Howard, J., Sutton-Grier, A., Herr, D., Kleypas, J., Landis, E., Mcleod, E., Pidgeon, E., Simpson, S., 2017. Clarifying the role of coastal and marine systems in climate mitigation. Front. Ecol. Environ. 15, 42–50. https://doi.org/10.1002/fee.1451.

Huang, J., Luo, M., Liu, Y., Zhang, Y., Tan, J., 2019. Effects of tidal scenarios on the methane emission dynamics in the subtropical tidal marshes of the Min River estuary in Southeast China. Int. J. Environ. Res. Public Health 16, 2790. https://doi.org/10.3390/ijerph16152790.

Huertas, I.E., de la Paz, M., Perez, F.F., Navarro, G., Flecha, S., 2019. Methane emissions from the Salt Marshes of Doñana wetlands: Spatio-temporal variability and controlling factors. Front. Ecol. Evol. 7. https://doi.org/10.3389/fevo.2019.00032.

Jacotot, A., Marchand, C., Allenbach, M., 2018. Tidal variability of CO_2 and CH_4 emissions from the water column within a Rhizophora mangrove forest (New Caledonia). Sci. Total Environ. 631–632, 334–340. https://doi.org/10.1016/j.scitotenv.2018.03.006.

Jeffrey, L.C., Maher, D.T., Santos, I.R., Call, M., Reading, M.J., Holloway, C., Tait, D.R., 2018. The spatial and temporal drivers of pCO_2, pCH_4 and gas transfer velocity within a subtropical estuary. Estuar. Coast. Shelf Sci. 208, 83–95. https://doi.org/10.1016/j.ecss.2018.04.022.

Jeffrey, L.C., Maher, D.T., Johnston, S.G., Kelaher, B.P., Steven, A., Tait, D.R., 2019a. Wetland methane emissions dominated by plant-mediated fluxes: contrasting emissions pathways and seasons within a shallow freshwater subtropical wetland. Limnol. Oceanogr. 64, 1895–1912. https://doi.org/10.1002/lno.11158.

Jeffrey, L.C., Maher, D.T., Johnston, S.G., Maguire, K., Steven, A.D.L., Tait, D.R., 2019b. Rhizosphere to the atmosphere: contrasting methane pathways, fluxes, and geochemical drivers across the terrestrial-aquatic wetland boundary. Biogeosciences 16, 1799–1815. https://doi.org/10.5194/bg-16-1799-2019.

Kirwan, M.L., Megonigal, J.P., 2013. Tidal wetland stability in the face of human impacts and sea-level rise. Nature 504, 53–60. https://doi.org/10.1038/nature12856.

Klaus, M., Vachon, D., 2020. Challenges of predicting gas transfer velocity from wind measurements over global lakes. Aquat. Sci. 82, 53. https://doi.org/10.1007/s00027-020-00729-9.

Kremer, J.N., Nixon, S.W., Buckley, B., Roques, P., 2003. Technical note: conditions for using the floating chamber method to estimate air-water gas exchange. Estuaries 26, 985–990. https://doi.org/10.1007/BF02803357.

Kristensen, E., 2007. Carbon balance in mangrove sediments: the driving processes and their controls. In: Greenhouse Gas and Carbon Balances in Mangrove Coastal Ecosystem. Maruzen Publishing, pp. 61–78.

Kristensen, E., Alongi, D.M., 2006. Control by fiddler crabs (Uca vocans) and plant roots (Avicennia marina) on carbon, iron, and sulfur biogeochemistry in mangrove sediment. Limnol. Oceanogr. 51, 1557–1571. https://doi.org/10.4319/lo.2006.51.4.1557.

Kristensen, E., King, G.M., Holmer, M., Banta, G.T., Jensen, M.H., Hansen, K., Bussarawit, N., 1994. Sulfate reduction, acetate turnover and carbon metabolism in sediments of the Ao Nam Bor Mangrove, Phuket, Thailand. Mar. Ecol. Prog. Ser. 109, 245–256.

Kristensen, E., Andersen, F., Holmboe, N., Holmer, M., Thongtham, N., 2000. Carbon and nitrogen mineralization in sediments of the Bangrong mangrove area, Phuket, Thailand. Aquat. Microb. Ecol. 22, 199–213. https://doi.org/10.3354/ame022199.

Kristensen, E., Flindt, M.R., Ulomi, S., Borges, A.V., Abril, G., Bouillon, S., 2008. Emission of CO_2 and CH_4 to the atmosphere by sediments and open waters in two Tanzanian mangrove forests. Mar. Ecol. Prog. Ser. 370, 53–67. https://doi.org/10.3354/meps07642.

Krithika, K., Purvaja, R., Ramesh, R., 2008. Fluxes of methane and nitrous oxide from an Indian mangrove. Curr. Sci. 94, 218–224.

Kuss, J., Schneider, B., 2004. Chemical enhancement of the CO_2 gas exchange at a smooth seawater surface. Mar. Chem. 91, 165–174. https://doi.org/10.1016/j.marchem.2004.06.007.

Law, B.E., Bakwin, P.S., Baldocchi, D.D., et al., 2003. The AmeriFlux network paradigm for measuring and understanding the role of the terrestrial biosphere in global climate change. Bioscience.

Le Quéré, C., Andrew, R.M., Canadell, J.G., et al., 2016. Global carbon budget 2016. Earth Syst. Sci. Data 8, 605–649. https://doi.org/10.5194/essd-8-605-2016.

Lenhart, K., Bunge, M., Ratering, S., et al., 2012. Evidence for methane production by saprotrophic fungi. Nat. Commun. 3, 1046. https://doi.org/10.1038/ncomms2049.

Leopold, A., Marchand, C., Deborde, J., Chaduteau, C., Allenbach, M., 2013. Influence of mangrove zonation on CO_2 fluxes at the sediment–air interface (New Caledonia). Geoderma 202–203, 62–70. https://doi.org/10.1016/j.geoderma.2013.03.008.

Leopold, A., Marchand, C., Deborde, J., Allenbach, M., 2016a. Water biogeochemistry of a mangrove-dominated estuary under a semi-arid climate (New Caledonia). Estuar. Coasts. https://doi.org/10.1007/s12237-016-0179-9.

Leopold, A., Marchand, C., Renchon, A., Deborde, J., Quiniou, T., Allenbach, M., 2016b. Net ecosystem CO_2 exchange in the "Coeur de Voh" mangrove, New Caledonia: effects of water stress on mangrove productivity in a semi-arid climate. Agric. For. Meteorol. 223, 217–232. https://doi.org/10.1016/j.agrformet.2016.04.006.

Li, H., Dai, S., Ouyang, Z., et al., 2018. Multi-scale temporal variation of methane flux and its controls in a subtropical tidal salt marsh in eastern China. Biogeochemistry 137, 163–179. https://doi.org/10.1007/s10533-017-0413-y.

Linto, N., Barnes, J., Ramachandran, R., Divia, J., Ramachandran, P., Upstill-Goddard, R.C., 2014. Carbon dioxide and methane emissions from mangrove-associated waters of the Andaman Islands, Bay of Bengal. Estuar. Coasts 37, 381–398. https://doi.org/10.1007/s12237-013-9674-4.

Liss, P.S., Merlivat, L., 1986. Air-sea gas exchange rates: introduction and synthesis. In: The Role of Air-Sea Exchange in Geochemical Cycling. Springer, Netherlands, pp. 113–127.

Liu, J., Lai, D.Y.F., 2019. Subtropical mangrove wetland is a stronger carbon dioxide sink in the dry than wet seasons. Agric. For. Meteorol. 278. https://doi.org/10.1016/j.agrformet.2019.107644, 107644.

Liu, J., Zhou, Y., Valach, A., et al., 2020. Methane emissions reduce the radiative cooling effect of a subtropical estuarine mangrove wetland by half. Glob. Chang. Biol., 1–19. https://doi.org/10.1111/gcb.15247.

Livesley, S.J., Andrusiak, S.M., 2012. Temperate mangrove and salt marsh sediments are a small methane and nitrous oxide source but important carbon store. Estuar. Coast. Shelf Sci. 97, 19–27. https://doi.org/10.1016/j.ecss.2011.11.002.

Lorke, A., Bodmer, P., Noss, C., et al., 2015. Technical note: drifting vs. anchored flux chambers for measuring greenhouse gas emissions from running waters. Biogeosci. Discuss. 12, 14619–14645. https://doi.org/10.5194/bgd-12-14619-2015.

Lorke, A., Bodmer, P., Koca, K., Noss, C., 2019. Hydrodynamic Control of Gas-Exchange Velocity in Small Streams. EarthArXiv, pp. 1–23, https://doi.org/10.31223/osf.io/8u6vc.

Lovelock, C.E., Duarte, C.M., 2019. Dimensions of blue carbon and emerging perspectives. Biol. Lett. 15, 1–5. https://doi.org/10.1098/rsbl.2018.0781.

Lovelock, C.E., Ruess, R.W., Feller, I.C., 2006. Fine root respiration in the mangrove Rhizophora mangle over variation in forest stature and nutrient availability. Tree Physiol. 26, 1601–1606. https://doi.org/10.1093/treephys/26.12.1601.

Luijendijk, E., Gleeson, T., Moosdorf, N., 2020. Fresh groundwater discharge insignificant for the world's oceans but important for coastal ecosystems. Nat. Commun. 11. https://doi.org/10.1038/s41467-020-15064-8.

Macreadie, P.I., Anton, A., Raven, J.A., et al., 2019. The future of blue carbon science. Nat. Commun. 10, 1–13. https://doi.org/10.1038/s41467-019-11693-w.

Magenheimer, J.F., Moore, T.R., Chmura, G.L., Daoust, R.J., 1996. Methane and carbon dioxide flux from a macrotidal salt marsh, bay of Fundy, New Brunswick. Estuaries 19, 139. https://doi.org/10.2307/1352658.

Maher, D.T., Eyre, B.D., 2012. Carbon budgets for three autotrophic Australian estuaries: implications for global estimates of the coastal air-water CO_2 flux. Global Biogeochem. Cycles 26, GB1032. https://doi.org/10.1029/2011GB004075.

Maher, D.T., Santos, I.R., Golsby-Smith, L., Gleeson, J., Eyre, B.D., 2013. Groundwater-derived dissolved inorganic and organic carbon exports from a mangrove tidal creek: the missing mangrove carbon sink? Limnol. Oceanogr. 58, 475–488. https://doi.org/10.4319/lo.2013.58.2.0475.

Maher, D.T., Drexl, M., Tait, D.R., Johnston, S.G., Jeffrey, L.C., 2019. IAMES: an inexpensive, automated methane ebullition sensor. Environ. Sci. Technol. 53, 6420–6426. https://doi.org/10.1021/acs.est.9b01881.

Martin, R.M., Moseman-Valtierra, S., 2015. Greenhouse gas fluxes vary between Phragmites Australis and native vegetation zones in coastal wetlands along a salinity gradient. Wetlands 35, 1021–1031. https://doi.org/10.1007/s13157-015-0690-y.

Mazot, A., Taran, Y., 2009. CO_2 flux from the volcanic lake of El Chichón (Mexico). Geofis. Int. 48, 73–83.

McGinnis, D.F., Kirillin, G., Tang, K.W., Flury, S., Bodmer, P., Engelhardt, C., Casper, P., Grossart, H., 2015. Enhancing surface methane fluxes from an oligotrophic lake: exploring the microbubble hypothesis. Environ. Sci. Technol. 49, 873–880. https://doi.org/10.1021/es503385d.

McKenzie, L.J., Nordlund, L.M., Jones, B.L., Cullen-Unsworth, L.C., Roelfsema, C., Unsworth, R.K.F., 2020. The global distribution of seagrass meadows. Environ. Res. Lett. 15. https://doi.org/10.1088/1748-9326/ab7d06, 074041.

Mcleod, E., Chmura, G.L., Bouillon, S., et al., 2011. A blueprint for blue carbon: toward an improved understanding of the role of vegetated coastal habitats in sequestering CO_2. Front. Ecol. Environ. 9, 552–560. https://doi.org/10.1890/110004.

Mcowen, C., Weatherdon, L., Bochove, J.-W., et al., 2017. A global map of saltmarshes. Biodivers. Data J. 5. https://doi.org/10.3897/BDJ.5.e11764, e11764.

Middelburg, J.J., Nieuwenhuize, J., Iversen, N., Høgh, N., Wilde, H.D.E., Helder, W., Seifert, R., Christof, O., 2002. Methane distribution in European tidal estuaries. Biogeochemistry 59, 95–119. https://doi.org/10.1023/A:1015515130419.

Minx, J.C., Lamb, W.F., Callaghan, M.W., et al., 2018. Negative emissions—part 1: research landscape and synthesis. Environ. Res. Lett. 13. https://doi.org/10.1088/1748-9326/aabf9b.

Morin, T.H., 2019. Advances in the Eddy covariance approach to CH_4 monitoring over two and a half decades. Eur. J. Vasc. Endovasc. Surg. 124, 453–460. https://doi.org/10.1029/2018JG004796.

Myhre, G., Shindell, D., Bréon, F.-M., et al., 2013. IPCC AR5 (2013) Chapter 8: anthropogenic and natural radiative forcing. In: Climate Change 2013: The Physical Science Basis. Contribution of Working Group I to the Fifth Assessment Report of the Intergovernmental Panel on Climate Change, pp. 659–740.

Nahrawi, H., Leclerc, M.Y., Pennings, S., Zhang, G., Singh, N., Pahari, R., 2020. Impact of tidal inundation on the net ecosystem exchange in daytime conditions in a salt marsh. Agric. For. Meteorol. 294. https://doi.org/10.1016/j.agrformet.2020.108133, 108133.

Nellemann, C., Corcoran, E., Duarte, C.M., Valdes, L., De Young, C., Fonseca, L., Grimsditch, G., 2009. Blue Carbon. A Rapid Response Assessment. United Nations Environment Programme, GRID-Arendal.

Neubauer, S.C., 2014. On the challenges of modeling the net radiative forcing of wetlands: reconsidering Mitsch et al. 2013. Landsc. Ecol. 29, 571–577. https://doi.org/10.1007/s10980-014-9986-1.

Neubauer, S.C., Megonigal, J.P., 2015. Moving beyond global warming potentials to quantify the climatic role of ecosystems. Ecosystems 18, 1000–1013. https://doi.org/10.1007/s10021-015-9879-4.

Neubauer, S.C., Verhoeven, J.T.A., 2019. In: An, S., Verhoeven, J.T.A. (Eds.), Wetland Effects on Global Climate: Mechanisms, Impacts, and Management Recommendations BT—Wetlands: Ecosystem Services, Restoration and Wise Use. Springer International Publishing, pp. 39–62.

Nirmal Rajkumar, A., Barnes, J., Ramesh, R., Purvaja, R., Upstill-Goddard, R.C., 2008. Methane and nitrous oxide fluxes in the polluted Adyar River and estuary, SE India. Mar. Pollut. Bull. 56, 2043–2051. https://doi.org/10.1016/j.marpolbul.2008.08.005.

Nóbrega, G.N., Ferreira, T.O., Siqueira Neto, M., Queiroz, H.M., Artur, A.G., Mendonça, E.D.S., Silva, E.D.O., Otero, X.L., 2016. Edaphic factors controlling summer (rainy season) greenhouse gas emissions (CO_2 and CH_4) from semiarid mangrove soils (NE-Brazil). Sci. Total Environ. 542, 685–693. https://doi.org/10.1016/j.scitotenv.2015.10.108.

Olsson, L., Ye, S., Yu, X., Wei, M., Krauss, K.W., Brix, H., 2015. Factors Influencing CO_2 and CH_4 Emissions from Coastal Wetlands in the Liaohe Delta, Northeast China.

Oreska, M.P.J., McGlathery, K.J., Aoki, L.R., Berger, A.C., Berg, P., Mullins, L., 2020. The greenhouse gas offset potential from seagrass restoration. Sci. Rep. 10, 7325. https://doi.org/10.1038/s41598-020-64094-1.

Ovalle, A.R.C., Rezende, C.E., Lacerda, L.D., Silva, C.A.R., 1990. Factors affecting the hydrochemistry of a mangrove tidal creek, Sepetiba Bay, Brazil. Estuar. Coast. Shelf Sci. 31, 639–650. https://doi.org/10.1016/0272-7714(90)90017-L.

Padhy, S.R., Bhattacharyya, P., Dash, P.K., Reddy, C.S., Chakraborty, A., Pathak, H., 2020. Seasonal fluctuation in three mode of greenhouse gases emission in relation to soil labile carbon pools in degraded mangrove, Sundarban, India. Sci. Total Environ. 705. https://doi.org/10.1016/j.scitotenv.2019.135909, 135909.

Poffenbarger, H.J., Needelman, B.A., Megonigal, J.P., 2011. Salinity influence on methane emissions from tidal marshes. Wetlands 31, 831–842. https://doi.org/10.1007/s13157-011-0197-0.

Polsenaere, P., Lamaud, E., Lafon, V., et al., 2012. Spatial and temporal CO_2 exchanges measured by eddy covariance over a temperate intertidal flat and their relationships to net ecosystem production. Biogeosciences 9, 249–268. https://doi.org/10.5194/bg-9-249-2012.

Prairie, Y., del Giorgio, P., 2013. A new pathway of freshwater methane emissions and the putative importance of microbubbles. Inland Waters 3, 311–320. https://doi.org/10.5268/IW-3.3.542.

Purvaja, R., Ramesh, R., Frenzel, P., 2004. Plant-mediated methane emission from an Indian mangrove. Glob. Chang. Biol. 10, 1825–1834. https://doi.org/10.1111/j.1365-2486.2004.00834.x.

Raymond, P.A., Cole, J.J., 2001. Gas exchange in rivers and estuaries: choosing a gas transfer velocity. Estuaries 24, 312. https://doi.org/10.2307/1352954.

Rebmann, C., Aubinet, M., Schmid, H., et al., 2018. ICOS eddy covariance flux-station site setup: a review. Intensiv. Agric. 32, 471–494. https://doi.org/10.1515/intag-2017-0044.

Rodda, S.R., Thumaty, K.C., Jha, C.S., Dadhwal, V.K., 2016. Seasonal variations of carbon dioxide, water vapor and energy fluxes in tropical Indian mangroves. Forests 7, 1–18. https://doi.org/10.3390/f7020035.

Rogers, K., Kelleway, J.J., Saintilan, N., et al., 2019. Wetland carbon storage controlled by millennial-scale variation in relative sea-level rise. Nature 567, 91–95. https://doi.org/10.1038/s41586-019-0951-7.

Rosentreter, J.A., Williamson, P., 2020. Concerns and uncertainties relating to methane emissions synthesis for vegetated coastal ecosystems. Glob. Chang. Biol. 26, 5351–5352. https://doi.org/10.1111/gcb.15201.

Rosentreter, J.A., Maher, D.T., Ho, D.T., Call, M., Barr, J.G., Eyre, B.D., 2017. Spatial and temporal variability of CO_2 and CH_4 gas transfer velocities and quantification of the CH_4 microbubble flux in mangrove dominated estuaries. Limnol. Oceanogr. 62, 561–578. https://doi.org/10.1002/lno.10444.

Rosentreter, J., Maher, D.T., Erler, D.V., Murray, R., Eyre, B.D., 2018a. Seasonal and temporal CO_2 dynamics in three tropical mangrove creeks—a revision of global mangrove CO_2 emissions. Geochim. Cosmochim. Acta 222, 729–745. https://doi.org/10.1016/j.gca.2017.11.026.

Rosentreter, J.A., Maher, D.T., Erler, D.V., Murray, R., Eyre, B.D., 2018b. Factors controlling seasonal CO_2 and CH_4 emissions in three tropical mangrove-dominated estuaries in Australia. Estuar. Coast. Shelf Sci. 215, 69–82. https://doi.org/10.1016/j.ecss.2018.10.003.

Rosentreter, J.A., Maher, D.T., Erler, D.V., Murray, R.H., Eyre, B.D., 2018c. Methane emissions partially offset "blue carbon" burial in mangroves. Sci. Adv. 4, eaao4985. https://doi.org/10.1126/sciadv.aao4985.

Saderne, V., Geraldi, N.R., Macreadie, P.I., et al., 2019. Role of carbonate burial in blue carbon budgets. Nat. Commun. 10, 1106. https://doi.org/10.1038/s41467-019-08842-6.

Schroll, M., Keppler, F., Greule, M., Eckhardt, C., Zorn, H., Lenhart, K., 2020. The stable carbon isotope signature of methane produced by saprotrophic fungi. Biogeosciences 17, 3891–3901. https://doi.org/10.5194/bg-17-3891-2020.

Schwalm, C.R., Williams, C.A., Schaefer, K., et al., 2012. Reduction in carbon uptake during turn of the century drought in western North America. Nat. Geosci. 5, 551–556. https://doi.org/10.1038/ngeo1529.

Sea, M.A., Garcias-Bonet, N., Saderne, V., Duarte, C.M., 2018. Carbon dioxide and methane fluxes at the air–sea interface of Red Sea mangroves. Biogeosciences 15, 5365–5375. https://doi.org/10.5194/bg-15-5365-2018.

Sotomayor, D., Corredor, J.E., Morell, J.M., 1994. Methane flux from mangrove sediments along the southwestern coast of Puerto Rico. Estuaries 17, 140–147. https://doi.org/10.2307/1352563.

Stieglitz, T.C., Clark, J.F., Hancock, G.J., 2013. The mangrove pump: the tidal flushing of animal burrows in a tropical mangrove forest determined from radionuclide budgets. Geochim. Cosmochim. Acta 102, 12–22. https://doi.org/10.1016/j.gca.2012.10.033.

Stoy, P.C., Katul, G.G., Siqueira, M.B.S., Juang, J.Y., McCarthy, H.R., Kim, H.S., Oishi, A.C., Oren, R., 2005. Variability in net ecosystem exchange from hourly to inter-annual time scales at adjacent pine and hardwood forests: a wavelet analysis. Tree Physiol. 25, 887–902. https://doi.org/10.1093/treephys/25.7.887.

Taillardat, P., Thompson, B.S., Garneau, M., Trottier, K., Friess, D.A., 2020. Climate Change Mitigation Potential of Wetlands and the Cost-Effectiveness of their Restoration: Wetlands for Climate Change Mitigation. Royal Society Publishing.

Tokoro, T., Kayanne, H., Watanabe, A., Nadaoka, K., Tamura, H., Nozaki, K., Kato, K., Negishi, A., 2008. High gas-transfer velocity in coastal regions with high energy-dissipation rates. J. Geophys. Res. 113, C11006. https://doi.org/10.1029/2007JC004528.

Tokoro, T., Hosokawa, S., Miyoshi, E., Tada, K., Watanabe, K., Montani, S., Kayanne, H., Kuwae, T., 2014. Net uptake of atmospheric CO_2 by coastal submerged aquatic vegetation. Glob. Chang. Biol. 20, 1873–1884. https://doi.org/10.1111/gcb.12543.

Tong, C., Huang, J.F., Hu, Z.Q., Jin, Y.F., 2013. Diurnal variations of carbon dioxide, methane, and nitrous oxide vertical fluxes in a subtropical estuarine marsh on neap and spring tide days. Estuar. Coasts 36, 633–642. https://doi.org/10.1007/s12237-013-9596-1.

Tonti, N.E., Gassmann, M.I., Pérez, C.F., 2018. First results of energy and mass exchange in a salt marsh on southeastern South America. Agric. For. Meteorol. 263, 59–68. https://doi.org/10.1016/j.agrformet.2018.08.001.

Trifunovic, B., Vázquez-Lule, A., Capooci, M., Seyfferth, A.L., Moffat, C., Vargas, R., 2020. Carbon dioxide and methane emissions from temperate salt marsh tidal creek. Eur. J. Vasc. Endovasc. Surg. 125, 1–16. https://doi.org/10.1029/2019jg005558.

Vachon, D., Prairie, Y.T., Cole, J.J., 2010. The relationship between near-surface turbulence and gas transfer velocity in freshwater systems and its implications for floating chamber measurements of gas exchange. Limnol. Oceanogr. 55, 1723–1732. https://doi.org/10.4319/lo.2010.55.4.1723.

Vagle, S., McNeil, C., Steiner, N., 2010. Upper Ocean bubble measurements from the NE Pacific and estimates of their role in air-sea gas transfer of the weakly soluble gases nitrogen and oxygen. J. Geophys. Res. 115, C12054. https://doi.org/10.1029/2009JC005990.

Van Dam, B.R., Lopes, C., Osburn, C.L., Fourqurean, J.W., 2019. Net heterotrophy and carbonate dissolution in two subtropical seagrass meadows. Biogeosciences 16, 4411–4428. https://doi.org/10.5194/bg-16-4411-2019.

Wanninkhof, R., 1992. Relationship between wind speed and gas exchange over the ocean. J. Geophys. Res. 97, 7373–7382. https://doi.org/10.1029/92JC00188.

Wanninkhof, R., 2014. Relationship between wind speed and gas exchange over the ocean revisited. Limnol. Oceanogr. Methods 12, 351–362. https://doi.org/10.4319/lom.2014.12.351.

Wanninkhof, R., Knox, M., 1996. Chemical enhancement of CO_2 exchange in natural waters. Limnol. Oceanogr. 41, 689–697. https://doi.org/10.4319/lo.1996.41.4.0689.

Wilson, B.J., Mortazavi, B., Kiene, R.P., 2015. Spatial and temporal variability in carbon dioxide and methane exchange at three coastal marshes along a salinity gradient in a northern Gulf of Mexico estuary. Biogeochemistry 123, 329–347. https://doi.org/10.1007/s10533-015-0085-4.

Wolf, S., Eugster, W., Potvin, C., Buchmann, N., 2011. Strong seasonal variations in net ecosystem CO_2 exchange of a tropical pasture and afforestation in Panama. Agric. For. Meteorol. 151, 1139–1151. https://doi.org/10.1016/j.agrformet.2011.04.002.

Wolfe, G.M., Kawa, S.R., Hanisco, T.F., et al., 2018. The NASA carbon airborne flux experiment (CARAFE): instrumentation and methodology. Atmos. Meas. Tech. 11, 1757–1776. https://doi.org/10.5194/amt-11-1757-2018.

Wood, N., Roelich, K., 2019. Tensions, capabilities, and justice in climate change mitigation of fossil fuels. Energy Res. Soc. Sci. 52, 114–122. https://doi.org/10.1016/j.erss.2019.02.014.

Xiao, J., Zhuang, Q., Law, B.E., et al., 2011. Assessing net ecosystem carbon exchange of U.S. terrestrial ecosystems by integrating eddy covariance flux measurements and satellite observations. Agric. For. Meteorol. 151, 60–69. https://doi.org/10.1016/j.agrformet.2010.09.002.

Xiao, J., Sun, G., Chen, J., et al., 2013. Carbon fluxes, evapotranspiration, and water use efficiency of terrestrial ecosystems in China. Agric. For. Meteorol. 182–183, 76–90. https://doi.org/10.1016/j.agrformet.2013.08.007.

Xiao, S., Yang, H., Liu, D., et al., 2014. Gas transfer velocities of methane and carbon dioxide in a subtropical shallow pond. Tellus B Chem. Phys. Meteorol. 66. https://doi.org/10.3402/tellusb.v66.23795.

Zeebe, R.E., Ridgwell, A., Zachos, J.C., 2016. Anthropogenic carbon release rate unprecedented during the past 66 million years. Nat. Geosci. 9, 325–329. https://doi.org/10.1038/ngeo2681.

Zheng, X., Guo, J., Song, W., Feng, J., Lin, G., 2018. Methane emission from mangrove wetland soils is marginal but can be stimulated significantly by anthropogenic activities. Forests 9, 738. https://doi.org/10.3390/f9120738.

Impact of climate change and related disturbances on CO$_2$ and CH$_4$ cycling in coastal wetlands

Cyril Marchand[a], Xiaoguang Ouyang[b,c,d], Faming Wang[e], and Audrey Leopold[f]

[a]*University of New Caledonia, ISEA, Noumea, New Caledonia,*
[b]*Southern Marine Science and Engineering Guangdong Laboratory (Guangzhou), Guangzhou, China,*
[c]*Guangdong Provincial Key Laboratory of Water Quality Improvement and Ecological Restoration for Watersheds, School of Ecology, Environment and Resources, Guangdong University of Technology, Guangzhou, China,*
[d]*Simon F.S. Li Marine Science Laboratory, School of Life Sciences, The Chinese University of Hong Kong, Hong Kong Special Administrative Region, China,*
[e]*Xiaoliang Research Station for Tropical Coastal Ecosystems, Key Laboratory of Vegetation Restoration and Management of Degraded Ecosystems, and the CAS engineering Laboratory for Ecological Restoration of Island and Coastal Ecosystems, South China Botanical Garden, Chinese Academy of Sciences, Guangzhou, China,*
[f]*Institut Agronomique néo-Calédonien, SolVeg, Noumea, New Caledonia*

7.1 Introduction

Coastal wetlands are vegetated ecosystems that develop at the interface between oceanic and terrestrial ecosystems, with abilities to thrive in stressful environments (e.g., anoxia, waterlogged soils, and saline conditions). They can develop in a variety of climatic regions (tropical, subtropical, arid, semiarid, temperate, and boreal) and eco-geomorphic settings (e.g., deltas, lagoons, estuaries, and oceanic islands). Among these wetlands, mangroves develop along coastlines between 30°N and 40°S, with the majority occurring between 20°N and 20°S (Giri et al., 2011). Despite limited extent (135,873 km^2 in 2016, Worthington et al., 2020), mangroves are considered key ecosystems in carbon cycling along those coastlines. They are characterized by high primary productivity, equivalent to that of tropical humid evergreen forests, being among the most productive plants in the sea (Alongi, 2014). Bouillon et al. (2008) estimated a global mangrove production rate of $218 \pm 72 \, \mathrm{Tg \, C \, year^{-1}}$, and Alongi (2020a,b,c) proposed a net primary production average of $18.4 \, \mathrm{Mg \, C \, ha^{-1} \, year^{-1}}$. Additionally, mangroves are characterized by high organic carbon burial rates; Breithaupt et al. (2012) estimated a value at 163 (+40; −31) $\mathrm{g \, OC \, m^{-2} \, year^{-1}}$,

Carbon Mineralization in Coastal Wetlands. https://doi.org/10.1016/B978-0-12-819220-7.00010-8

while Wang et al. (2021) reported a higher value ($194 \pm 15\,\mathrm{g\,OC\,m^{-2}\,year^{-1}}$) based on a global data set. Despite being organic-rich, mineralization rates and greenhouse gas (GHG) emissions from mangrove soils are limited, mainly due to the prevailing waterlogging and anoxic conditions (Lovelock, 2008; Kristensen et al., 2008). Salt-marshes are also vegetated coastal wetlands, but their distribution extends to higher latitudes, even in the Arctic. Compared to the woody plants in mangroves, halophytes of saltmarshes mainly consist of shrubby and herbaceous plants. Local net primary production of saltmarshes reaches $17.6\,\mathrm{Mg\,C\,ha\,year^{-1}}$ (Duarte et al., 2013). Like mangroves, they also have high soil carbon burial rates. Ouyang and Lee (2014) estimated the global average carbon accumulation rate in saltmarshes at $244.7\,\mathrm{g\,C\,m^{-2}\,year^{-1}}$, while Wang et al. (2021) reported, more recently, a much lower value ($168 \pm 7\,\mathrm{g\,OC\,m^{-2}\,year^{-1}}$), likely due to the expanded data set. However, saltmarshes sequester proportionally more (24%) net primary production (NPP) than mangroves (12%) (Alongi, 2020b). The biogeochemical conditions in saltmarshes are similar to mangroves, which hamper carbon mineralization and thus GHG emissions.

Climate change is affecting and will strongly affect coastal wetlands. Changes in atmospheric CO_2, air and sea temperatures, precipitation, and sea-level will influence their spatial distributions as well as their structure and function (McKee et al., 2012; Osland et al., 2016; Feller et al., 2017). Air and water temperatures determine distributional limits of coastal wetlands. For instance, low temperatures at high latitudes impede mangrove reproduction (Duke et al., 1998). Aridity and regional changes in rainfall patterns will also have an influence on coastal wetland development. Increasing temperature and atmospheric CO_2 concentration will modify plant growth notably through changes in photosynthetic and respiration rates (Alongi, 2008). Restricted to the intertidal zone, coastal wetlands are not static entities but continuously subjected to disturbance due to sea-level fluctuations (Alongi, 2015; Woodroffe, 1981, 1992). Even if some mangroves will be able to keep pace with current fast sea-level rise (Woodroffe et al., 2016), most of them will have to move landward (Lovelock et al., 2015). Mangrove encroachment on saltmarshes will result in increase in the carbon storage capacity of coastal wetlands (Saintilan and Rogers, 2015; Osland et al., 2016). The analysis of stratigraphic sequences in mangrove peat deposits indicate that mangroves gradually moved inland, while the fringing seaward stands died back as sea level rose during the Holocene (Woodroffe, 1981; Plaziat, 1995; Alongi, 2015). In addition, current temperature increase may promote mangrove poleward migration (Cavanaugh et al., 2014), even if they are not consistently extending their latitudinal range across the globe due to local constraints, including local anthropogenic pressures and impacts, oceanographic, hydrologic, and topographic conditions (Hickey et al., 2017).

Climate change not only influences coastal wetland distribution and functioning but also affects C cycling in and GHG emissions from these ecosystems (Simard et al., 2019; Spivak et al., 2019). GHG fluxes between coastal wetlands and the atmosphere can be separated into three main categories: (i) ecosystem plant photosynthesis and respiration; (ii) soil organic carbon mineralization by microorganisms in different redox conditions that can produce both CO_2 and CH_4; and (iii) OC mineralization in tidal creeks and waterways that are main conduits of C export to the coastal ocean. Mainly, these fluxes are studied through eddy-covariance systems or incubation chambers. GHG emissions from mangrove stems (Jeffrey et al., 2019; Vinh et al., 2019) and coarse woody debris (Troxler et al., 2015) have also been recently evaluated. Ecosystem photosynthesis and respiration are influenced by rise in air and water temperatures as well as the evolution of water pressure deficit and changes in precipitation that modify soil salinity (Barr et al., 2010; Leopold et al., 2016; Liu and Lai, 2019). GHG emissions from coastal wetland soils, without considering root respiration, depend on a sequence of processes: (i) ecosystem productivity; (ii) carbon burial in their soils; (iii) carbon mineralization efficiency; and (iv) diffusion to the atmosphere (Kristensen et al., 2017). All these processes vary with numerous factors that can be influenced by climate change. Coastal wetland productivity notably depends on latitude, season, and the position of the stand in the intertidal zone, mainly because these parameters control temperature, rainfall, and nutrient inputs that have direct influence on plant photosynthesis and growth. Local organic carbon burial is strongly influenced by sedimentation rates and the tidal regime, the latter parameter controlling tidal flushing, length of inundation, and pore-water seepage. Soil organic carbon burial in coastal wetlands is controlled by latitude and tidal range and varies among geographic regions (Ouyang and Lee, 2014). For mangroves, roughly 50% of litterfall is exported to the adjacent coastal zone, with particulate organic carbon export accounting for approximately ~11% of total terrestrial carbon input to the ocean (Dittmar et al., 2006; Alongi, 2014). Temperature and soil water content control OC mineralization in coastal wetland soils. In addition, mineralization will depend on the response of bacterial communities and benthic photosynthetic microorganisms to these changes (Lovelock, 2008; Leopold et al., 2015). Eventually, GHG emissions at the soil surface do not represent total GHG production in soils. Indeed, tidal advection and pore-water seepage can transport up to 70% of total soil respiration to adjacent waters in the form of inorganic carbon, which can be further emitted to the atmosphere (Maher et al., 2013). Mangroves inhabit only 0.3% of the global coastal ocean area but contribute to 55% of air-sea exchange (Alongi, 2020c). Sea-level rise will modify this process.

The objective of this chapter is to assess the latest data and analyses on the influence of climate change on coastal wetland productivity, organic carbon burial, CO_2 and CH_4 production, and emission. We reviewed in situ studies that were interested in the variability of CO_2 and CH_4 emissions from coastal wetlands notably with temperature and soil water content, both at the ecosystem scale using eddy-covariance data and from the various interfaces using incubation chambers. In addition, we reviewed the few studies that were performed in atmosphere-controlled greenhouses or open-top chambers to assess the role of climate change on coastal wetland productivity.

7.2 Influence of climate change on coastal wetland productivity

Ecosystem productivity is a key driver of organic matter (OM) accumulation in coastal wetland soils, and the largest component of CO_2 exchange between ecosystems and the atmosphere is canopy respiration (Alongi, 2009). Until recently,

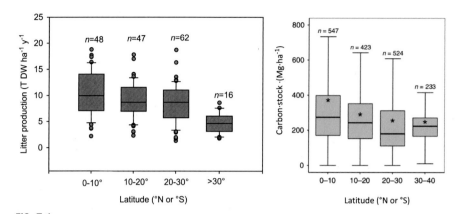

FIG. 7.1

Boxplot compilation of mangrove leaf litter fall and carbon stock data, grouped for different latitudinal zones.

From Bouillon, S., Borges, A.V., Castañeda-Moya, E., Diele, K., Dittmar, T., Duke, N.C., Kristensen, E., Lee, S.Y., Marchand, C., Middelburg, J.J., Rivera-Monroy, V.H., Smith III, T.J., Twilley, R.R., 2008. Mangrove production and carbon sinks: a revision of global budget estimates. Global Biogeochem. Cycles 22, GB 2013. and Ouyang, X., Lee, S.Y., 2020. Improved estimates on global carbon stock and carbon pools in tidal wetlands Nat. Commun. 11, 317. https:/doi.org/10.1038/ s41467-019-14120-2.

mangrove productivity has been assessed using annual litterfall as a proxy. Litterfall shows a latitudinal gradient, being significantly higher at latitudes lower than 10° and significantly lower in the latitudes higher than 30° (Twilley et al., 1992; Bouillon et al., 2008), consistent with the change in sediment organic carbon stock in global mangroves (Ouyang and Lee, 2020) (Fig. 7.1). This variability reflects the influence of climate (temperature, rainfall, and solar energy) on mangrove productivity (Lugo and Patterson-Zucca, 1977; Twilley et al., 1992; Sanders et al., 2016). Saltmarsh productivity also demonstrates a latitudinal gradient that is consistent with solar energy inputs at a 0.2%–0.35% net conversion efficiency (Turner, 1976). In contrast to the litterfall method, saltmarsh (*Spartina alterniflora*) productivity was assessed at a national scale (United States) using data on annual aboveground productivity (Kirwan et al., 2009). The productivity shows a $27\,\mathrm{g\,m^{-2}\,year^{-1}}$ increase with an increase of mean annual temperature by 1°C, and a significant latitudinal gradient partly attributable to plant phenology, i.e., the length of growing seasons. Additionally, precipitation increases coastal wetlands plant productivity and growth by providing freshwater and nutrients (Ewel et al., 1998). Therefore, increasing precipitation in tropical coastal regions due to climate change (Tan et al., 2015) should also increase the C accumulation capacity in some of the world's largest mangrove systems in the Indo-Pacific and tropical South Americas regions. Eventually, rises in sea-level and atmospheric CO_2 will affect mangrove productivity through changes in photosynthetic rates and water-use efficiency.

7.2.1 **Temperature**

Temperature increase will result in enhanced halophyte growth rates and respiration in coastal wetlands until a threshold is reached and then productivity is expected to decrease (Couto et al., 2014). On a leaf scale, assimilation and transpiration rates are highest at leaf temperatures ranging from 25°C to 30°C (Ball, 1988). Cheeseman et al. (1997) also indicated that most mangrove species show a sharp decline of photosynthetic rate at temperatures higher than 33°C. High temperatures result in stomatal conductance reduction; in addition, temperatures higher than 40°C may inhibit photosynthesis (Andrews et al., 1984; Gilman et al., 2008; Reef et al., 2016). However, temperature for optimum photosynthesis seems to be species-specific. In their review, Gilman et al. (2008) indicated that the optimum lies between 28°C and 32°C; however, lower values have been reported, e.g., 24.5°C for *Avicennia germinans* (Reef et al., 2016), 26.8°C for *Avicennia marina* (Fig. 7.2; Leopold et al., 2016). Temperature responses are complex, depend on a series of parameters, notably solar radiation, and vapor pressure deficit (VPD). In tropical climates, Liu and Lai (2019) observed that temperature exerts a positive influence on ecosystem respiration (Re) and a negative influence on gross primary production (GPP) of mangrove forests. As a result, temperature is a dominant control of the temporal variations of net ecosystem exchange (NEE) during the wet seasons, when freshwater availability is not a limiting factor. Temperature influence is weaker during dry seasons, the lack of fresh water exerting a stronger influence (see Section 7.2.2 of this chapter). Temperature also influences evaporation processes and thus pore-water salinity. In tropical coastal wetlands with low salinity (5–15), temperature-related salinity increases may result

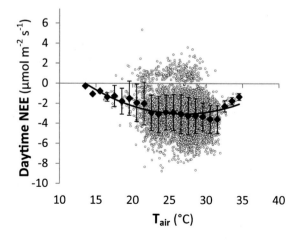

FIG. 7.2

Half-hourly daytime NEE response to air temperature.

From Leopold, A., Marchand, C., Renchon, A., Deborde, J., Quiniou, T., Allenbach, M., 2016. Net ecosystem CO₂ exchange of "Le Coeur de Voh" mangrove: evidences of water stress controls on mangrove productivity under semi-arid climate (New Caledonia). Agric. For. Meteorol. 223, 217–232.

in greater availability of ions supporting photosynthesis and thus greater ecosystem productivity (Liu and Lai, 2019). Conversely, in arid climate, temperature increase usually results in salinity that limits ecosystem productivity (Leopold et al., 2016).

Consequently, in tropical areas with abundant water supply, temperature is likely to be more important than moisture-related parameters in controlling coastal wetland productivity. Conversely, in arid and semiarid climates, water availability from tides and rainfall appeared to be more critical factors (Leopold et al., 2016; Liu and Lai, 2019).

7.2.2 Precipitation

On a global scale, precipitation strongly affects coastal wetland net primary productivity, and the alteration of their patterns by climate change are expected to modify C cycling in these systems (Osland et al., 2017, 2018; Wang et al., 2020). Increased precipitation will enhance nutrient inputs, length of immersion of estuarine systems, and decrease pore-water salinity of coastal wetland soils, possibly leading to higher tree growth (Sanders et al., 2016). Conversely, coastal wetlands developing in arid or semiarid climates that are expected to be subject to decrease in rainfall will probably have a lower productivity, notably due to higher salt-stress (Ward et al., 2016). Studies on seasonal variations of net ecosystem productivity (NEP) and Re may help to assess the influence of different precipitation regime on C cycling. On a global scale, there is a positive relationship between NEP and precipitation (Leopold et al., 2016; Liu and Lai, 2019) (Fig. 7.3). However, depending on climate, different responses have been observed at local scales. In tropical mangroves, positive relationships between NEE and rainfall have been reported (Barr et al., 2010; Liu and Lai, 2019). Conversely, in semiarid climate, where water availability is a limiting factor, a strong negative relationship between monthly NEP and rainfall was observed (Leopold et al., 2016). During the rainy season, both gross ecosystem production (GEP) and Re were enhanced (Leopold et al., 2016), but high Re should be responsible for the negative relationship between precipitation and NEP. This positive relationship between rainfall and Re would result in enhanced tree respiration because of growth stimulation and resumption related to freshwater inputs from rain. Indeed, during the rainy season, both maintenance and growth respirations occur; while in the dry season, only maintenance respiration occurs.

Consequently, precipitation patterns altered by climate change will not only have an influence on ecosystem gross productivity but also on ecosystem respiration, subsequently modifying greenhouse gas emissions as well as the net balance of the carbon storage in coastal wetlands.

7.2.3 Sea-level rise

Sea-level rise will certainly affect coastal wetland photosynthesis and thus productivity, notably by modifying nutrient availability, and inundation frequency that partly controls physico-chemical properties like pore-water salinity and soil redox conditions (Morris et al., 2013).

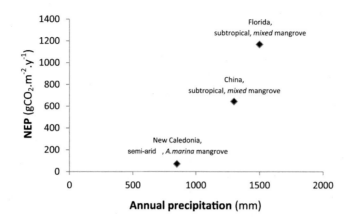

FIG. 7.3

Influence of precipitation on mangrove net ecosystem productivity (gCO_2 m^{-2} year^{-1}) measured from eddy-covariance studies.

From Leopold, A., Marchand, C., Renchon, A., Deborde, J., Quiniou, T., Allenbach, M., 2016. Net ecosystem CO_2 exchange of "Le Coeur de Voh" mangrove: evidences of water stress controls on mangrove productivity under semi-arid climate (New Caledonia). Agric. For. Meteorol. 223, 217–232. With data from China (Chen, G.C., Ulumuddin, Y.I., Pramudji, S., Chen, S.Y., Chen, B., Ye, Y., Ou, D.Y., Ma, Z.Y., Huang, H., Wang, J. K., 2014. Rich soil carbon and nitrogen but low atmospheric greenhouse gas fluxes from North Sulawesi mangrove swamps in Indonesia. Sci. Total Environ. 487, 91–96, Chen, H., Lu, W., Yan, G., Yang, S., Lin, G., 2014. Typhoons exert significant but differential impacts on net ecosystem carbon exchange of subtropical mangrove forests in China. Biogeosciences 11. https:/doi.org/10.5194/bg-11-5323-2014), Florida (Barr, J.G., Engel, V., Fuentes, J.D., Zieman, J.C., O'Halloran, T.L., Smith, T.J., Anderson, G.H., 2010. Controls on mangrove forest-atmosphere carbon dioxide exchanges in western Everglades National Park J. Geophys. Res. Biogeo. 115: G02020. https:/doi.org/10.1029/2009JG001186) and New Caledonia (Leopold, A., Marchand, C., Renchon, A, Deborde, J., Quiniou, T, Allenbach, M., 2016. Net ecosystem CO_2 exchange of "Le Coeur de Voh" mangrove: evidences of water stress controls on mangrove productivity under semi-arid climate (New Caledonia). Agric. For. Meteorol. 223, 217–232).

Coastal wetlands are dynamic ecosystems in intertidal zones, which are continually evolving because of natural or human-induced sea-level variations (Alongi, 2011). Ancestral mangroves appeared 65 million years ago and then moved with shorelines (Duke, 1992). Coastal wetlands always had to adapt and migrated with sea-level variations and changes in rates of accretion, subsidence, or uplift (Schuerch et al., 2018). Woodroffe (2002) described six potential patterns of mangrove evolution with sea-level rise, including the conditions in which mangroves can keep up, and those that lead to migration. During the Holocene, the analysis of coastal sediments indicates that mangroves either moved seaward as sea-level dropped, inducing their disappearance on the landward side of the system (Jacotot et al., 2018a), or moved inland as sea-level rose while the fringing seaward stands died back (Kim et al., 2005; Di Nitto et al., 2014; Saintilan et al., 2020). Both types of moves led to colonization of new areas and the decline of some stands, modifying ecosystems productivity.

Some CO_2 eddy-covariance studies measured the influence of tidal immersion on NEE and NEP. Leopold et al. (2016) observed that water availability either from tides or rainfall was a key factor driving dwarf mangrove productivity in semiarid climate. When the mangrove was submerged at high tide, rate of tree photosynthesis was enhanced and GEP was increased. Li et al. (2014) reported higher maximum rate of photosynthesis and lower value of light point compensation during spring tides and suggested that more carbon uptake occurs under submerged conditions. Conversely, the absence of flooding for a few days rapidly decreased the NEE in semiarid mangroves, indicating that the trees experienced water-stress. Similarly, longer inundation increased the gross ecosystem exchange (GEE) and NEE of saltmarshes on the west coast of the Atlantic (Morris et al., 2013; Wang et al., 2020).

Salinity is also one of the main parameters driving coastal wetland functioning, distribution, and productivity, regulating photosynthesis and stomatal conductance (Ball and Farquhar, 1984; Krauss and Ball, 2013). Liu and Lai (2019) suggested that tidal inundation influences mangrove net productivity, through its influence on water salinity. High pore-water salinity induces decreases in specific leaf area (SLA), stomatal conductance, and chlorophyll content, resulting in lower leaf photosynthesis (Naidoo, 2010; Naidoo et al., 2011). In semiarid environments, the upper intertidal zone is colonized by dwarf stands because of the high pore-water salinity (Bourgeois et al., 2019). Barr et al. (2010) observed that daily light use efficiency (LUE) was reduced by 46% at surface water salinity higher than 34 compared to salinity lower than 17. They concluded that the long-term carbon balance of mangroves largely depends on the factors controlling salinity levels and are evolving with climate change. However, some studies showed opposite results. Liu and Lai (2019) observed that salinity had a positive effect on both LUE and GPP with water salinity ranging from 1 to 17. Conversely, at salinity lower than 15, some studies in mangrove (Cui et al., 2015) and saltmarsh (Knox et al., 2018) did not observe any correlation between salinity and LUE. It is highly possible that the photosynthetic rate of coastal wetland plants increases with salinity until an optimal value is reached (Krauss and Ball, 2013; Liu and Lai, 2019). For some mangrove species, this limit lies between 15 and 25 (Hwang and Chen, 2001; Chen and Ye, 2014). Consequently, salinity modification resulting from temperature increase, rainfall pattern evolution, and sea-level rise will influence coastal wetland productivity. Depending on the location of the ecosystem in the intertidal zone, pore-water salinity will decrease with sea-level rise, e.g., in the upper zone, due to more frequent tidal flooding (Bourgeois et al., 2019) possibly resulting in an increase of the ecosystem productivity.

Additionally, coastal wetland species have different tolerance to flooding and salinity (Baldwin and Mendelssohn, 1998; Buffington et al., 2020), suggesting possible whole-forest changes in community composition in relation to sea-level rise. In mesocosm experiments, He et al. (2007) found that tolerance to flooding differed among four mangrove species (from most to least tolerant): *Avicennia marina > Aegiceras corniculatum > Rhizophora stylosa > Bruguiera gymnorhiza*. In the same way, responses of biomass productivity of tidal marsh to salinity and flooding vary between species. Moreover, the combined effects of these two factors could be greater than their individual effects (Buffington et al., 2020). In greenhouses simulating climate change, Jacotot et al. (2018b) observed that enhanced length of tidal immersion resulted in a decrease of 5% and 3% in photosynthesis (Fig. 7.4) and 23% and 2% in water-use efficiency (WUE) for *Avicennia marina* and *R. stylosa*, respectively.

These decreases were also observed on mature mangrove trees (Youssef and Saenger, 1996; Naidoo et al., 1997; Chen et al., 2005) and can result from the fact that tidal flooding may induce a rapid deficit in oxygen in the soil due to root and microorganism respiration (Naidoo et al., 1997). In anaerobic conditions, the

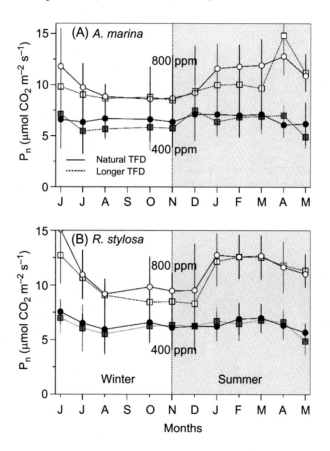

FIG. 7.4

Monthly variation ($\pm$SD) of leaf net photosynthetic rates (P_n, μmol m^{-2} s^{-1}) for (A) *Avicennia marina*; and (b) *Rhizophora stylosa*. Colored markers: ambient CO_2 concentrations ($\sim$400 ppm), *white* markers: elevated CO_2 concentrations (800 ppm). *TFD*, tidal flooding duration. Shaded area represents the warm season.

From Jacotot, A., Marchand, C., Gensous, S., Allenbach, M., 2018. Effects of elevated atmospheric CO_2 and increased tidal flooding on leaf gas-exchange parameters of two common mangrove species: Avicennia marina and Rhizophora stylosa. Photosynth. Res. https:/doi.org/10.1007/s11120-018-0570-4.

production of abscisic acid (ABA) increased (Ellison and Farnsworth, 1997; Pezeshki et al., 1997), and RuBisCO enzyme activity is inhibited (Chen et al., 2005), which can result in limited photosynthesis. However, decrease of photosynthesis with increasing tidal immersion may not always result in lower biomass but depends on the resulting modification of pore-water salinity as well as nutrient inputs. In mesocosms, longer tidal flooding increased the growth of *Avicennia marina*, whereas it reduced the growth of *R. stylosa* in comparison to the control treatment, which was set to the length of tidal immersion of these two species measured in the field (Jacotot et al., 2019a) (Fig. 7.5). They attributed this result to the

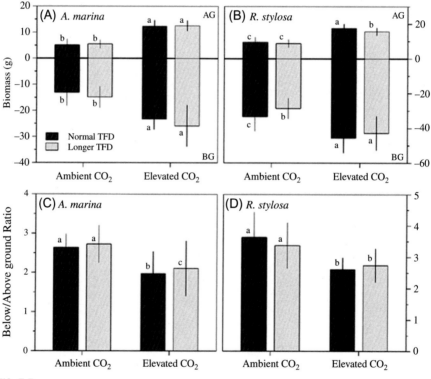

FIG. 7.5

Final above-ground (AG) and below-ground (BG) biomass (g) after 1 year of experiment and below/above ground biomass ratios of (A and C) *Avicennia marina* and (B and D) *Rhizophora stylosa* grown in ambient CO_2 concentrations (~400 ppm) or elevated CO_2 concentrations (800 ppm) and under normal (*dark-gray* bars) and increased tidal flooding duration (*gray* bars). Values are means ± SD ($n=30$). Different letters indicate significant differences after ANOVA analysis ($P<.05$).

From Jacotot, A., Marchand, C., Allenbach, M., 2019. Increase in growth and alteration of C:N ratios of Avicennia marina and Rhizophora stylosa subject to elevated CO_2 concentrations and longer tidal flooding duration. Front. Ecol. Evol.

specific zonation, with *Rhizophora* spp. developing on the seaside of the mangrove while *Avicennia marina* develops at higher elevation. *Rhizophora* spp. already had the optimum conditions for growth, while *Avicennia marina* occurs at the limit of its ability, in zones that have low water inputs and high pore-water salinity. Increasing length of tidal inundation reduces the water stress, possibly inducing more nutrient inputs and lower salinity, resulting in biomass increase. Lu et al. (2013) observed that inundation period exerted significant effects on biomass accumulation, photosynthetic rate, leaf electron transportation, and water-use efficiency. In addition, they showed that canopy immersion of *Avicennia marina* seedlings has a great negative effect on early growth and may limit their development and colonization of new areas. In tanks simulating sea-level rise, Ellison and Farnsworth (1997) observed that *R. mangle* seedlings initially grew faster than controls, but at the sapling stage, growth was slowed. After 2.5 year, controls were 10%–20% larger than those in the experimental tanks.

Consequently, coastal wetlands vary greatly in how they respond to a sea-level rise, and the effect of tidal flooding will be species-specific and will depend on the climate and the position of the stand in the intertidal zone, and whether the current position is optimal. Sea-level rise will influence coastal wetland productivity by modifying the length of immersion, pore-water salinity, the soil redox conditions, and the nutrient inputs.

7.2.4 Atmospheric CO_2 concentration

Higher atmospheric CO_2 concentration will probably lead to increased coastal ecosystem growth, notably because of the modification of some leaf-gas exchange parameters (Reef et al., 2016). Farnsworth et al. (1996) observed that *R. mangle* exhibited increased growth and biomass, as well as increased branching activity and earlier maturation under elevated CO_2 concentrations. However, they also measured photosynthetic acclimation after long-term exposure to elevated CO_2 concentrations, as well as a slight decrease of dark respiration, which will possibly limit the positive influence of elevated CO_2 concentrations on coastal wetland growth over the long term. Plant responses will probably be species-specific (Snedaker and Araujo, 1998). Recently, Jacotot et al. (2018b) showed that mangroves response to elevated CO_2 concentrations was dependent on the species. They notably showed that with CO_2 concentration at 800 ppm, photosynthesis was enhanced by more than 37% for *Avicennia marina* and by more than 45% for *R. stylosa* compared to ambient atmospheric concentrations (Fig. 7.4).

Elevated CO_2 concentrations will also result in decreases in stomatal density and transpiration rates, and as a result will lead to increase in water-use efficiency (Reef et al., 2015; Jacotot et al., 2018b). Better water-use efficiency might be a key for mangroves to resist drought episodes (Dai, 2013). Contrary to other studies (McKee and Rooth, 2008; Reef and Lovelock, 2014), Jacotot et al. (2019a) measured SLA increases under elevated CO_2, which can lead to higher carbon acquisition by photosynthesis but also higher light interception. Stimulation of photosynthesis under elevated CO_2 will result in increased growth rates (Jacotot et al., 2018b). However, earlier studies (Ball and Munns, 1992; Ball et al., 1997) showed that elevated CO_2 concentrations did not increase mangrove growth rates at high salinity conditions. Consequently, plant responses to elevated CO_2 will probably be confounded by variations in salinity and nutrient inputs (Alongi, 2018). Concerning tidal marshes, root growth under elevated CO_2 (twice ambient CO_2 concentrations) in open top chambers led to an 83% increase in root dry mass in comparison with roots from ambient-growth C3 plants (*Scirpus* spp.), whereas C4 plants (*Spartina patens*) show little response to elevated CO_2 (Curtis et al., 1990). Parallel to this finding, the photosynthetic capacity of *Scirpus olneyi* was enhanced by 31% when exposed to elevated CO_2 (ambient air +340 ppm) for four growing seasons (Arp and Drake, 1991). These results reflect that photosynthesis of some saltmarsh halophytes (C3 photosynthetic pathway) responds to elevated levels of atmospheric CO_2. In essence, C4 species show a CO_2 concentration mechanism at the site of assimilation and are usually less responsive to elevated CO_2 in the range of 200–600 ppm, while C3 species increase net photosynthetic rates over the same CO_2 concentration range due to the lack of this CO_2-concentrating mechanism (McKee et al., 2012). Overall, the productivity of mangroves may respond more significantly to elevated CO_2 than that of saltmarshes since almost all mangroves photosynthesize under C3 pathways, while most of saltmarsh species photosynthesize under C4 pathways.

Consequently, coastal wetland productivity responses to increased CO_2 concentrations will probably depend on the biogeographic settings and the position of the stand in the intertidal zone (both influencing pore-water salinity and nutrient inputs), as well as photosynthetic pathways.

7.2.5 Cyclonic events

Tropical cyclones, also called hurricanes or typhoons, are meteorological phenomena characterized by high and multidirectional wind speed, heavy rainfall inducing pulses in freshwater inputs. They are accompanied by high seas and storm surge that can strongly modify coastal areas and geochemical functioning (i.e., sediment deposition, erosion and hydrologic pattern). Tree vegetated coastal wetlands, especially near-shore and tall mangroves, are particularly damaged by wind, while saltmarshes and herbaceous coastal wetlands are impacted by storm surge, notably wave action, sediment burial, and saltwater intrusion.

Sippo et al. (2018) showed that tropical cyclones are responsible for the greatest area of mangrove mortality, equivalent to 45% of the reported global mangrove mortality area from events over six decades. Among damages, mass foliation, tree uprooting, and mortality directly decrease the mangrove aboveground productivity (Barr et al., 2012; Chen et al., 2014b; Danielson et al., 2017). Highlighting the link between the global distribution of canopy height and the tropical cyclone relative frequency, Simard et al. (2019) recently proposed that cyclone landfall frequency would limit the growth of mangrove trees. However, the magnitude of cyclone impact on productivity, between and inside mangrove forests, would depend on the intensity of exposure (i.e., distance from the eye path, side of the eye track) (Zhang et al., 2019), as well as on site-specific attributes (mangrove structure and species, geochemical and hydrologic functioning, and human-induced degradation) over the precyclone period. It will also depend on the recovery time between two cyclones (Aung et al., 2013; Danielson et al., 2017; Walcker et al., 2019; Castañeda-Moya et al., 2020; Krauss and Osland, 2020; Taillie et al., 2020). Cyclone Wilma (2005) significantly impacted the mangrove litterfall net primary productivity inducing a stand-specific mangrove defoliation up to $4.72 \, MgC \, ha^{-1} \, year^{-1}$, i.e., two times the prestorm defoliation average, while in the upper canopy, trees were 100% defoliated (Barr et al., 2012; Danielson et al., 2017). However, the low cumulative tree mortality and long-term recovery from defoliation (after ~ 10 years) highlighted hurricane resistance and ecosystem recovery (Rivera-Monroy et al., 2019). In the Philippines, after Super Typhoon Haiyan, Long et al. (2016) studied normalized difference vegetation index (NDVI) values of mangroves and showed that despite an initial damage caused to 3.5% of the Philippines' total mangrove area, the majority of damaged mangrove recovered within 18 months in the absence of further typhoons or other perturbations. Mangrove forests may be strongly damaged by cyclones; however, numerous studies highlighted fast recovery after damages by storms, thanks to specific ecophysiological traits (e.g., resource use-efficiency, resprouting) (Alongi, 2008; Lugo, 2008; Long et al., 2016; Danielson et al., 2017; Imbert, 2018). However, the trajectory of cyclone response may differ from the initial state, notably because of modification in community structure, ecosystem functioning, or ecosystem loss, and can subsequently affect productivity (Lugo, 2008; Smith III et al., 2009). In Louisiana, saltmarsh NDVI values and length of growing season were significantly impacted by Hurricanes Katrina (2005), Gustave (2008), and Isaac (2012) (Mo et al., 2020), suggesting an important impact on productivity. Nevertheless, general recovery of these marshes was observed in the following years, highlighting high recovery capacity for mangrove forests (Paling et al., 2008) dependent on storm surge and cyclone intensity, stress tolerance of species, and geomorphologic specificity of site (e.g., Hill et al., 2020).

Cyclones can affect net ecosystem productivity of coastal wetlands, both carbon uptake and carbon release. In China, Chen et al. (2014b) showed significant but different impacts of typhoons on NEE of different mangroves. Daily total NEE values decrease by 26%–50% following some typhoons but significantly increased (43%–131%) following some others. If decrease of NEE was directly related to GPP decrease because of defoliation, NEE increase could be related to a higher decrease of Re because of soil waterlogging. In the Everglades National Park (Florida, United States), heating of the sediment surface after hurricane Wilma, resulting from the increase of solar irradiance penetration through the damaged canopy, was partially responsible for higher CO_2 efflux from soil and subsequently for higher ecosystem respiration (Barr et al., 2012). In this wetland system, in 2009, i.e., 4 years after hurricane Wilma, the net carbon assimilation rates remained approximately $250 \, gC \, m^{-2} \, year^{-1}$ lower compared to the average annual values determined for the period 2004–05.

During and after disturbance, litter input strongly increases at the soil surface, and its decomposition increases soil and ecosystem respiration (Ostertag et al., 2003; Barr et al., 2012; Ouyang et al., 2020). Litter input can also be associated with large sediment deposition. After hurricane Wilma, vertical accretion was 8–17 times greater than the annual accretion rate averaged over the last 50 years (Castañeda-Moya et al., 2010). Tweel and Turner (2012) estimated sediment deposition on coastal wetlands in Louisiana of 68, 48, and 21 million metric tons from hurricanes Katrina, Rita, and Gustave, respectively. Sediment deposition and decomposition of supply of OM increase nutrient inputs, providing notably N and P to species growing in limited-nutrient ecosystem (Lovelock et al., 2014; Ouyang et al., 2020; Xu et al., 2004). Castañeda-Moya et al. (2020) showed that P-deposition related to hurricane Irma (2017) contributed up to 98% of the soil nutrient pool, suggesting fertilization effects over longer periods on net ecosystem productivity. However, for some coastal wetlands, these inputs of nutrients could also drive negative feedbacks. In Florida, Feller et al. (2015) showed that nutrient enrichment increased the productivity of scrub mangroves, but this benefit was offset by a decrease in their resistance and resilience to hurricane damages. In another case, in Louisiana, Mo et al. (2020) showed that intermediate and brackish wetlands were more impacted by cyclones than saline marsh dominated shoreline because nutrient enrichment could decrease belowground biomass and increase root decomposition (e.g., Bulseco et al., 2019), leading to a weaker soil profile more sensible to erosion, especially in more inland saltmarshes.

NEP may also be affected by pore-water salinity variations following cyclone. Indeed, soil pore-water salinity represents a major factor controlling the structure and productivity of coastal vegetated wetlands (Twilley and Rivera-Monroy, 2009, see also Section 7.2.3 of this chapter). Promoting sediment elevation in some cases and increasing pore-water salinity, cyclone could favor the development of saltmarshes and saltflats instead of mangroves (Paling et al., 2008). In the same way, according to impact of storm surge inland, cyclones may permit saltmarsh development through mortality of freshwater forests (Middleton, 2016). Conversely, heavy cyclone-induced precipitation causes freshwater pulses, which can increase competition for space with freshwater vegetation. Consequently, through their ability of deeply modified plant communities, cyclones may indirectly influence ecosystem productivity as carbon uptake and burial rate are different between marshes and mangrove forests.

Climate change is expected to modify cyclone intensity and frequency at a global scale. While their frequency could decrease, their intensity and associated precipitation and storm surge will increase (Collins et al., 2019). Indeed, future cyclones will move more slowly, particularly in the mid latitudes, increasing the magnitude of damages (Zhang et al., 2020). Increasing cyclone intensity will lead to higher damages on coastal wetlands in cases of cyclone landfall. Even if fertilization effects are expected to occur following high intensity cyclones, costal wetland productivity may be particularly reduced if tree uprooting and shoreline erosion increase. Ecosystem structure and recovery rate and time could be impacted, likely modifying coastal wetland productivity and capacity for storing carbon. In addition, the increased intensity of cyclones associated with sea-level rise could particularly affect the productivity of coastal wetlands through their cumulative effects. Indeed, in some cases, tree mortality rates have been correlated with declines in peat surface elevation (Barr et al., 2012), suggesting that impacts induced by intense cyclones could decrease the capacity of coastal wetlands to cope with sea-level rise. While, in some cases, if storm-induced sedimentation is sufficient, stability of coastal marshes may be enhanced, helping to cope with sea-level rise (Baustian and Mendelssohn, 2015). However, as suggested by McKee and Cherry (2009), this true positive feedback will also depend on characteristics of sediment deposit (i.e., sediment texture, resistance to compaction). Eventually, by increasing poleward movement of cyclones (Altman et al., 2018), climate change may impair mangrove poleward migration.

7.3 Influence of climate change on C burial in coastal wetlands

Coastal wetland primary production can be consumed by detritivores, exported to adjacent systems through tidal flushing and pore-water seepage, or buried within the soil (Bouillon et al., 2008; Maher et al., 2013). In mangroves, carbon burial rates have been studied since the early 1990s (Twilley et al., 1992; Jennerjahn and Ittekkot, 2002; Chmura et al., 2003; Duarte et al., 2005; Bouillon et al., 2008; Alongi, 2009; Breithaupt et al., 2012). Recently, Alongi (2020a,b,c) estimated the mean global burial rate at $1.62\,Mg\,C\,ha^{-1}\,year^{-1}$. In their budget, Breithaupt et al. (2012) concluded that burial represented 10%–15% of estimated annual mangrove production. However, burial can be highly variable, depending notably on the rate of soil accretion and thus on the tidal regime and the elevation of the zone considered (Adame et al., 2010). Sea-level variations have thus a great impact on burial rates in coastal wetlands (Wang et al., 2019b). In some places, mangroves keep pace with sea-level rise through vertical surface elevation change (Alongi, 2008), but in other places, they do not like in the Indo-Pacific area (Lovelock et al., 2015). On a long time scale, to maintain their physiological requirements coastal wetlands have to migrate with the intertidal zone and to colonize new areas (Schuerch et al., 2018), which has an influence on carbon burial rates. Recently, Jacotot et al. (2018a) suggested that a period of sea-level stability that lasted ~3000 years resulted in high organic carbon accumulation in mangrove soils, whereas during a period when the sea level was continuously decreasing, soil C stocks were lower. Current sea-level rise is inducing a landward migration of coastal wetlands that will influence C burial in their soils, modifying C stocks and thus GHG production and emission. Coastal wetlands may directly influence soil accretion processes through the development of their root system, which can also fluctuates with climate change (Krauss et al., 2014). In controlled atmosphere greenhouses, Jacotot et al. (2019a) demonstrated that elevated CO_2 concentrations resulted in an increase in the below-ground biomass of the two mangrove species studied (Fig. 7.5). They suggested that the capacity of mangrove ecosystems to face sea-level rise could be enhanced, as higher root density will increase soil volume and soil organic matter content, two important factors contributing not only to vertical accretion but also to carbon burial rates.

   A continuous migration of coastal wetlands resulting from sea-level rise will limit the development of coastal wetlands in a given area, and thus the organic matter accumulation in their soils. Along with sea-level rise, climate change factors, like rainfall and increasing temperature, greatly shape the subsidence, root development, sediment inputs from watersheds, and eventually influence burial rates in coastal ecosystems (Woodroffe et al., 2016). Sanders et al. (2016) showed that a model using temperature, tidal range, latitude, and rainfall explained almost 90% of the variability in the studied coastal ecosystem carbon stocks, which showed a much stronger relationship with rainfall than any other factor. They concluded that rainfall will increase ecosystem productivity, notably due to reduced salt-stress, and lower rates of organic matter decay processes in waterlogged soil, but we also suggest that enhanced rainfall induces increased erosion in watersheds and thus sedimentation in the coastal areas, which results in higher C burial rates (Fig. 7.6). They hypothesized that global mangrove forest carbon stocks could increase by almost 10% by the beginning of the next century because of increased rainfall in the tropical regions. This hypothesis is consistent with recent reports on how precipitation controls canopy height, aboveground biomass C, and function of mangrove forests (Osland et al., 2017; Simard et al., 2019). A recent study found that temperature and mean annual precipitation explained 57% of the variation in the C burial of mangrove forests (Wang et al., 2021). A positive relationship between mangrove C burial rate and temperature was described in recent studies (Osland et al., 2018; Wang et al., 2019b).

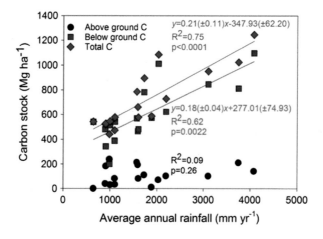

FIG. 7.6

Above-ground and below-ground carbon stocks in mangroves versus annual rainfall.
Precipitation data taken from NOAA GPCP. From Sanders, C.J., Maher, D.T., Tait, D.R., Williams, D., Holloway, C., Sippo, J.Z., Santos, I.R., 2016. Are global mangrove carbon stocks driven by rainfall? J. Geophys. Res. G: Biogeosci. https://doi.org/10.1002/2016JG003510.

However, if precipitation is an important driver of soil C burial in warmer climate mangrove soils, it is not the case for tidal marshes developing in cooler regions.

Consequently, sea-level rise will probably enhanced tidal flushing and pore-water seepage, limiting carbon burial. Conversely, higher carbon burial rates in coastal wetlands can result from i) increased productivity related to increased temperature (except in arid and semiarid climates) and atmospheric CO_2 concentration and ii) increased precipitation that may enhance sedimentation rate in the coastal zone.

7.4 Influence of climate change on GHG production in coastal wetlands

Despite their organic-rich soils, low production rates of greenhouses gases (GHG), carbon dioxide (CO_2), or methane (CH_4) usually characterize coastal wetlands (Kristensen et al., 2017). In the soil, GHG production through microbial processes is driven by many parameters including oxygen availability, soil and water temperatures, soil water content, grain size, porosity, redox potential (Eh), salinity, pH, and OM quality (Kristensen et al., 2017; Ouyang et al., 2017). In coastal wetlands, oxic conditions are usually restricted to the upper mm of the soils and near root systems or crab burrows that can translocate oxygen into the soil (Kristensen and Alongi, 2006). Usually, the bulk soil remains largely suboxic or anoxic (Marchand et al., 2006). Under these conditions, organic molecules can be oxidized completely to CO_2 by a wide variety of anaerobic microorganisms using electron acceptors in the following sequence: Mn^{4+}, NO_3^-, Fe^{3+}, and SO_4^{2-}. Methane production can occur in coastal wetland soils by either CO_2 reduction and acetate fermentation or conversion of formate, methanol, methylamines, and CO to CH_4. Methanogenesis can occur in anoxic soils when all other electron acceptors have been exhausted. However, recent studies showed that high number of methanogenic archaea can be present and highly active in coastal wetland soils and that sulfate-reducing and methanogen bacteria can coexist (Lyimo et al., 2002, 2009; Chauhan et al., 2015). Consequently, even if GHG production is limited, it is highly variable in relation to soil carbon stocks, tidal immersion frequency, position along the intertidal zone, development of microphytobenthos at the soil surface, presence of crab burrows or pneumatophores, and also seasons (radiation, temperature, and rainfall) (Lovelock, 2008; Lovelock et al., 2014; Chen et al., 2012, 2014a,b; Leopold et al., 2013, 2015; Wang et al., 2019a; Jacotot et al., 2019b).

7.4.1 Temperature influence on GHG production

Temperature increase will induce faster rates of OM decomposition, and thus GHG production. Increased temperature is a positive driver of carbon mineralization in mangrove soils (Lewis et al., 2014). Temperature, bacterial production, and microbial biomass are positively correlated (Alongi, 1988). In summer, enhanced CO_2 and CH_4 emissions from coastal wetland soils to the atmosphere were attributed to

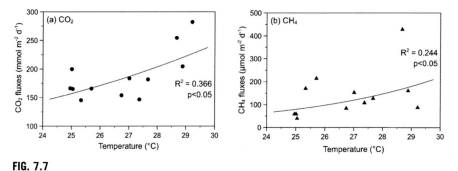

FIG. 7.7

Influence of temperature on CO_2 and CH_4 emissions from *Rhizophora* mangrove soil.

From Jacotot, A., Marchand, C., Allenbach, M., 2019. Biofilm and temperature controls on greenhouse gas (CO$_2$ and CH$_4$) emissions from a mangrove soil (New Caledonia). Sci. Total Environ.

temperature increases (Allen et al., 2007; Chen et al., 2012; Chanda et al., 2014; Chauhan et al., 2015; Jacotot et al., 2019b; Wang et al., 2020) (Fig. 7.7).

Methane emissions from saltmarsh soils were observed to be correlated with both air and soil temperatures under transparent/opaque chambers and clipping conditions, reflecting that methanogenic bacteria are better adapted to warm soil conditions than methanotrophic bacteria (Yin et al., 2015). Several authors measured Q_{10} values ranging between 2 and 3 for mangroves (Lovelock, 2008; Leopold et al., 2015; Ha et al., 2018), similar to those measured for terrestrial ecosystems (Reich and Schlesinger, 1992). Lovelock (2008) observed that soil respiration did not increase with temperature over the whole temperature range but showed a decline beyond 26°C in scrub mangrove forests. At the water-air interface, CO_2 production in mangroves was also observed to increase significantly with increasing water temperature (Chambers et al., 2014). However, most of the studies did not observe any optimal threshold temperature for emission rates (Chanda et al., 2013). In addition, the highest CO_2 emissions from mangrove soils, reaching up to $500\,mmol\,CO_2\,m^{-2}\,day^{-1}$, were attributed to the high temperatures prevailing during summer at low latitude (Vinh et al., 2019). Nevertheless, most of these in situ studies quantified GHG fluxes and not GHG production, thus temperature influence may be difficult to identify, notably due to confounding effects. Other factors influencing the emissions, e.g., soil water content, microphytobenthos, may be responsible for these variations (detailed in Section 7.5 of this chapter).

7.4.2 Influence of precipitation and sea-level rise

Evolution of rainfall patterns and sea-level rise may influence litter and OM decomposition as well as GHG production within coastal wetland soils, notably influencing the soil water content, oxygen supply, and renewal of electron acceptors in soils (Davidson and Janssens, 2006; Sanders et al., 2016; Ouyang et al., 2017; Spivak et al., 2019; Vinh et al., 2020). Coastal wetlands are usually low energy environments that trap fine sediments, which result in soils with low permeability. With increased precipitation and increased length of tidal immersion, the whole soil column would be waterlogged without water renewable, leading to more anoxic conditions (Clark et al., 1998; Marchand et al., 2004), sulfate reduction would become the dominant pathway resulting in low mineralization rates (Kristensen et al., 2008). During the rainy season, it was observed that CO_2 fluxes decreased as the redox potential did in coastal wetlands (Chanda et al., 2013). Chambers et al. (2014) observed that mangrove soil respiration decreased with increased length of inundation due to the limited renewal of electron acceptors compared to period of soil exposure. Lewis et al. (2014) observed that soil CO_2 efflux declined by 65% as soil moisture increased from 75% to 85%, concluding that soil saturation and inundation suppressed short-term C mineralization from near-surface soils in coastal wetlands. However, if the soil is permeable, the residence time of the water in the soil will be short, the renewal of electron acceptors will be high, oxic to suboxic conditions will prevail, and OM mineralization may be enhanced. Conversely, in areas that will be subject to a drier climate, the soil water table would be lower, oxygen would penetrate deeper into the soil through crab burrows and cracks, allowing oxic to suboxic oxidation of organic carbon and enhanced GHG production (Agusto et al., 2020). However, microbial metabolism is stimulated by ideal moisture, soil water content lower than 10% limits the normal metabolic activity in the soil, and droughts were reported to limit OM mineralization and thus GHG production (Janssens et al., 2003).

7.4.3 Influence of CO_2 concentration

Atmospheric CO_2 concentration increase can induce an increase of C:N ratios of coastal wetland tissues, as observed for mangrove seedlings (Jacotot et al., 2019b). Decay constants are most of the time negatively correlated to C:N ratios (see Fig. 2.3 in Chapter 2). Consequently, climate change may result in more refractory organic matter, which will lead to lower decomposition rates and GHG production. In addition, increased CO_2 concentrations are inducing ocean acidification, which may influence microbial activity in coastal wetland soils. For instance, methanogens are pH sensitive and most of them grow over a relatively narrow pH range of about 6–8, with an optimum pH of 7.7 (Chang and Yang, 2003). Consequently, lower pH might inhibit methanogen activity and thus GHG production. At the ecosystem level, the pattern of ecosystem respiration responding to elevated atmospheric CO_2 is patchy. Elevated atmospheric CO_2 (ambient + 340 ppmv) stimulated ecosystem emission of CO_2 and CH_4 for C3 plants in a brackish marsh (Marsh et al., 2005).

7.5 Influence of climate change on GHG diffusion in coastal wetlands

Diffusion of GHG from coastal wetland soils to the atmosphere depends on several drivers, including soil water content, bioturbation, and the development of microphytobenthos at the soil surface. Sea-level rise and increase in rainfall patterns will result in higher frequency of immersion of coastal wetlands and thus higher soil water content. GHG emissions may decrease when soils are water saturated (Livesley and Andrusiak, 2012), notably because molecular diffusion is slower for fluids than for gas. Very few studies measured GHG emission at high tide in coastal wetlands. Recently, Jacotot et al. (2018c) showed that tide characteristics (flow/ebb, water column thickness, neap/spring) drove GHG emissions at the water-air interface in mangroves. They notably demonstrated that increased residence time of the water over the forest floor, as well as higher exchange surface between the soil and the water column, resulted in higher CO_2 and CH_4 emissions. Additionally, pore-water seepage, which is a sink of coastal wetland primary productivity (Borges et al., 2003; Bouillon et al., 2007; Maher et al., 2013; Stieglitz et al., 2013), may be increased due to modified residence time of water in soils and greater area submerged. As a result, more GHG would be exported toward adjacent tidal creeks and further seawards, and then emitted to the atmosphere or fueling trophic chains (David et al., 2018a,b). However, for coastal wetlands that have to migrate landward, they will colonize new areas at higher elevation and less frequently submerged. In these new settlement areas, soil water content will be lower, the unsaturated zone deeper, and GHG diffusion faster, as observed in the higher intertidal zone in semiarid climate (Leopold et al., 2013).

GHG diffusion also depends on the response of benthic photosynthetic microorganisms to climate change. For example, biofilm development can reduce GHG diffusion from coastal wetland soils (Lovelock, 2008; Leopold et al., 2013, 2015; Chen et al., 2014a,b; Bulmer et al., 2015; Jacotot et al., 2019b). On the one hand, biofilm could form a protective barrier, decreasing soil surface permeability; on the other hand, the CO_2 produced within the soil may be used by the benthic microorganisms. Soil water content influenced the growth, biomass, and metabolic activity of the biofilm (Coelho et al., 2009; Molnar et al., 2014). Leopold et al. (2015) measured an increasing gradient of differences between light and dark CO_2 fluxes from the landside to the seaside of a mangrove. They attributed this variability to different activity of the biofilm. Along the intertidal zone, there is an increasing gradient in the level of hydration and a decreasing one in the length of soil desiccation. Additionally, temperature increase, sea-level rise, and different rainfall pattern can modify these gradients, and thus biofilm development, influencing GHG diffusion.

Opposite to the biofilm, which reduces the emission of CO_2, crab burrows and aerial roots like pneumatophores, acting as a conduit and allowing the ascent of CO_2 to the atmosphere, induce higher GHG emissions (Kristensen et al., 2008; Troxler et al., 2015; Padhy et al., 2020). Bioturbation increases the renewal of electron acceptor at depth, favoring OC mineralization. In addition, it was recently demonstrated that total below-ground air-sediment surface area per m^2 is strongly increased by crab burrows (Agusto et al., 2020). Kristensen (2008) found that the contribution of 100 *Sonneratia alba* pneumatophores m^{-2} is about 170 mmol CO_2 day^{-1}, and 100 *Avicennia marina* pneumatophores m^{-2} is roughly 60 mmol CO_2 day^{-1}, while 100 *Uca* spp. burrows m^{-2} may add 90 mmol CO_2 day^{-1} to the basic rate measured for bare sediments. Recently, Padhy et al. (2020) observed that the gaseous fluxes through pneumatophores were higher than ebullition. With increasing sea level, some coastal wetland soils will be more water saturated and as a result may become more anoxic and sulfidic. McKee et al. (1988) established a negative correlation between pneumatophore density and sulfide concentrations. There are strong reciprocal effects between soil characteristics and tree roots (McKee, 1993). In fact, to thrive in anoxic environments, some wetland species, notably *Avicennia*, can develop their root systems and increase the number of pneumatophores, which can facilitate GHG diffusion to the atmosphere. Additionally, in greenhouses experiments, Jacotot et al. (2019a) observed that with elevated atmospheric CO_2 concentrations and longer length of tidal immersion, the root biomass of two mangrove species was enhanced (Fig. 7.5). Consequently, soil water content, microphytobenthos, and root structure will be impacted by climate change, influencing GHG diffusion. Depending on the wetland species, its position in the intertidal zone, diffusion will be either reduced or enhanced.

7.6 Synthesis and research directions

Climate change is not only influencing coastal wetland spatial distribution, structure, and functioning but also GHG cycling in these ecosystems. Over the past three decades, many case studies have significantly increased our knowledge on greenhouse gas cycling in coastal wetlands and on the importance of climate-related parameters on carbon stocks and fluxes in these ecosystems, notably from soils (Fig. 7.8).

Coastal wetland productivity notably depends on temperature, rainfall, and nutrient inputs that have direct influence on photosynthesis and growth. Increasing temperature and atmospheric CO_2 concentration will modify plant growth notably through changes in photosynthetic and respiration rates. However, coastal wetland productivity responses to increase of these parameters will probably depend on the biogeographic settings and the position of the stand in the intertidal zone, both influencing pore-water salinity and nutrient input. Aridity and regional changes in rainfall patterns will also modify coastal wetland development. Evolution of rainfall pattern will not only have an influence on ecosystem gross productivity but also on ecosystem respiration, modifying subsequently greenhouse gas emission as well as the net C budget of coastal wetlands. In tropical areas with abundant water supply, temperature is likely to be more important than moisture-related parameters in controlling coastal wetlands productivity. Conversely, in regions with more arid climates, water availability from tides and rainfall appeared to be the main driving factor. Sea-level rise will influence coastal wetland productivity by modifying the length of immersion, pore-water salinity, redox conditions of the soil, and nutrient input. However, coastal wetlands vary greatly in how they respond to a sea-level rise. It can, thus, be expected that the effect of tidal flooding will be species-specific and will depend on the prevalent climate and the position of the stand along the intertidal zone (optimal or not). Eventually, increasing cyclone intensity, both wind speed and

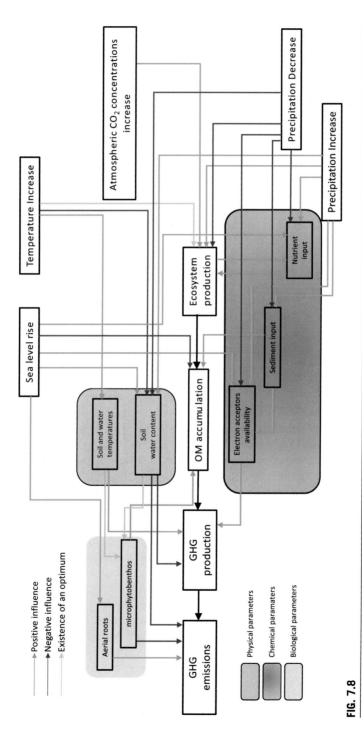

FIG. 7.8

Flow diagram of GHG cycling in coastal wetlands, from ecosystem production to organic carbon (OC) accumulation, greenhouse gas (GHG) production, and soil GHG emission. Important controlling factors are indicated, and marked with *green* arrows if they have a positive effect, *red* arrows if they have a negative effect, and *yellow* arrows if they have an intermediate optimum.

Modified from Kristensen, E, Connolly, R., Ferreira, T.O., Marchand, C., Otero, X.L., Rivera-Monroy, V.H., 2017. Biogeochemical cycles; global approaches and perspectives. In: Mangrove Ecosystems: A Global Biogeographic Perspective Structure, Function and Services. Springer.

storm surge, will lead to higher damages on coastal wetlands, notably their productivity may be reduced if tree uprooting, and shoreline erosion increase.

Climate change will also affect soil organic carbon burial in coastal wetlands, the latter being controlled by the productivity of the ecosystem, varying with latitudes, tidal range, and geomorphologic settings. Sea-level rise will probably enhance tidal flushing and pore-water seepage, limiting carbon burial. Conversely, increased productivity related to rising temperature and atmospheric CO_2 concentration as well as increased precipitation that may enhance sedimentation rates in the coastal zones will result in higher carbon burial rates in these ecosystems.

OC mineralization in coastal wetland soils notably depends on temperature and moisture that control microbial metabolism, and on the rate of electron acceptor renewal varying with tidal immersion and precipitation, both factors evolving with climate change. With increased precipitation and increased length of tidal immersion, mineralization may be reduced or enhanced depending on soil permeability. In areas that will be subject to a drier climate, the soil water table would be lower, allowing oxic to suboxic oxidation of organic carbon and enhanced GHG production until a threshold is reached. In arid areas, microbial activity is usually low, resulting in low GHG production and emissions. In fact, the effects of climate change on soil respiration in coastal wetlands will also depend on the answers of microorganism communities to these changes, notably soil water content and pH. Microbial metabolism is stimulated by ideal moisture levels; soil water content lower than 10% reduces metabolic activity in the soil and droughts limit OM mineralization. In addition, methanogens are pH sensitive and most of them grow over a relatively narrow pH range. Lower pH might inhibit microorganisms' activity and thus GHG production. Eventually, climate change may result in more refractory organic matter, which will lead to lower decomposition rates and GHG production.

Sea-level rise will result in coastal wetland soils becoming more water-saturated (until ecosystem migration), limiting GHG emissions, notably because the molecular diffusion is slower for fluids than for gases. GHG diffusion also depends on the response of the microphytobenthos to climate change. Its development can reduce GHG diffusion from coastal wetlands soils. Conversely, crab burrows and aerial roots can act as a conduit allowing the ascent GHG to the atmosphere. To thrive in anoxic environments, which can result from sea-level rise, some wetland species, notably *Avicennia*, can develop their root systems and increase the number of pneumatophores, inducing higher GHG emissions.

Eventually, coastal wetlands are not static entities, even if some mangroves will be able to keep pace with current sea-level rise, most of them will have to move landward. In addition, current temperature increase may promote mangrove poleward migration, increasing the ability of coastal wetlands to fix and store GHG. Mangrove encroachment into saltmarshes will increase the carbon storage capacity of coastal wetlands. However, given the complexity of the interactions between variables and the multiple potential responses of the ecosystems facing multistressors (climate change and increasing anthropogenic pressure, e.g., eutrophication, fragmentation), limited predictions concerning coastal wetland evolution were proposed.

Eddy-covariance studies, in situ GHG flux measurements at the various interfaces, and atmosphere-controlled greenhouse experiments helped us to refine C cycling. We now have a better understanding of the role of temperature, precipitation, atmospheric concentrations, or cyclonic events on coastal wetland productivity, carbon burial, GHG production, and emissions, but we still lack a complete understanding of the combined effects of those factors as well as on the reciprocal effects between those factors and coastal wetlands. We suggest more studies on longer time scales, not only to assess the seasonal effects on GHG cycling but also to precisely quantify the influence of climate change on GHG cycling. True in situ experiments should be implemented in these ecosystems over decades. Eventually, considering coastal wetland landward migration and mangrove poleward migration, studies in recently colonized areas would be highly relevant.

Acknowledgment

D. Alongi is gratefully acknowledged for his help in manuscript improvement.

References

Adame, M.F., Neil, D., Wright, S.F., Lovelock, C.E., 2010. Sedimentation within and among mangrove forests along a gradient of geomorphological settings. Estuar. Coast. Shelf Sci. 86, 21–30.

Agusto, L.E., Thibodeau, B., Tang, J., Wang, F., Cannicci, S., 2020. Fiddling With The Carbon Budget: Fiddler Crab Burrowing Activity Increases Wetland's Carbon Flux. Earth and Space Science Open Archive, pp. 1–27, https://doi.org/10.1002/essoar.10503504.1.

Allen, D.E., Dalal, R.C., Rennenberg, H., Meyer, R.L., Reeves, S., Schmidt, S., 2007. Spatial and temporal variation of nitrous oxide and methane flux between subtropical mangrove sediments and the atmosphere. Soil Biol. Biochem. 39, 622–631. https://doi.org/10.1016/j.soilbio.2006.09.013.

Alongi, D.M., 1988. Bacterial productivity and microbial biomass in tropical mangrove sediments. Microb. Ecol. 15, 59–79. https://doi.org/10.1007/BF02012952.

Alongi, D.M., 2008. Mangrove forests: resilience, protection from tsunamis, and responses to global climate change. Estuar. Coast. Shelf Sci. 76, 1–13. https://doi.org/10.1016/j.ecss.2007.08.024.

Alongi, D.M., 2009. The Energetics of Mangrove Forests. Springer, Breinigsville.

Alongi, D.M., 2011. Carbon payments for mangrove conservation: ecosystem constraints and uncertainties of sequestration potential. Environ. Sci. Pollut. Res. 14, 462–470.

Alongi, D.M., 2014. Carbon cycling and storage in mangrove forests. Ann. Rev. Mar. Sci. 6, 195–219.

Alongi, D.M., 2015. The impact of climate change on mangrove forests. Curr. Clim. Change Rep. 1, 30–39. https://doi.org/10.1007/s40641-015-0002-x.

Alongi, D.M., 2018. Impact of global change on nutrient dynamics in mangrove forests. Forests 9, 596.

Alongi, D.M., 2020a. Global significance of mangrove blue carbon in climate change mitigation. Science 2 (3), 67. https://doi.org/10.3390/sci2030067.

Alongi, D.M., 2020b. Carbon balance in salt marsh and mangrove ecosystems: a global synthesis. J. Mar. Sci. Eng. 8, 767. https://doi.org/10.3390/jmse8100767.

Alongi, D.M., 2020c. Carbon cycling in the world's mangrove ecosystems revisited: significance of non-steady state diagenesis and subsurface linkages between the forest floor and the coastal ocean. Forests 11, 977. https://doi.org/10.3390/f11090977.

Altman, J., Ukhvatkina, O.N., Omelko, A.M., Macek, M., Plener, T., Pejcha, V., Cerny, T., Petrik, P., Srutek, M., Song, J.S., Zhmerenetsky, A.A., Vozmishcheva, A.S., Krestov, P.V., Petrenko, T.Y., Treydte, K., Dolezal, J., 2018. Poleward migration of the destructive effects of tropical cyclones during the 20th century. Proc. Natl. Acad. Sci. U. S. A. 115 (45), 11543–11548. https://doi.org/10.1073/pnas.1808979115.

Andrews, T.J., Clough, B.F., Muller, G.J., 1984. Photosynthetic gas exchange properties and carbon isotope ratios of some mangroves in North Queensland. In: Teas, H.J. (Ed.), Physiology and Management of Mangroves. Tasks for Vegetation Science, vol. 9. Dr. W. Junk, The Hague, pp. 15–23.

Arp, W.J., Drake, B.G.J.P., 1991. Increased photosynthetic capacity of Scirpus olneyi after 4 years of exposure to elevated CO_2. Plant Cell Environ. 14 (9), 1003–1006.

Aung, T.T., Mochida, Y., Than, M.M., 2013. Prediction of recovery pathways of cyclone-disturbed mangroves in the mega delta of Myanmar. For. Ecol. Manage. 293, 103–113. https://doi.org/10.1016/j.foreco.2012.12.034.

Baldwin, A., Mendelssohn, I., 1998. Effects of salinity and water level on coastal marshes: an experimental test of disturbance as a catalyst for vegetation change. Aquat. Bot. 61, 255–268. https://doi.org/10.1016/S0304-3770(98)00073-4.

Ball, M.C., 1988. Ecophysiology of mangroves. Trees 2, 129–142.

Ball, M.C., Farquhar, G.G., 1984. Photosynthesis and stomatal responses of two mangrove species, Aegiceras corniculatum and Avicennia marina, to long term salinity and humidity conditions. Plant Physiol. 74, 1–6.

Ball, M.C., Munns, R., 1992. Plant responses to salinity under elevated atmospheric concentrations of CO_2. Aust. J. Bot. 40 (4–5), 515–525. https://doi.org/10.1071/BT9920515.

Ball, M.C., Cochrane, M.J., Rawson, H.M., 1997. Growth and water use of the mangroves Rhizophora apiculata and Rhizophora. stylosa in response to salinity and humidity under ambient and elevated concentrations of atmospheric CO2. Plant Cell Environ. 20, 1158–1166. https://doi.org/10.1046/j.1365-3040.1997.d01-144.x.

Barr, J.G., Engel, V., Fuentes, J.D., Zieman, J.C., O'Halloran, T.L., Smith, T.J., Anderson, G.H., 2010. Controls on mangrove forest-atmosphere carbon dioxide exchanges in western Everglades National Park. J. Geophys. Res. Biogeo. 115, G02020. https://doi.org/10.1029/2009JG001186.

Barr, J., Engel, V., Smith III, T.J., Fuentes, J., 2012. Hurricane disturbance and recovery of energy balance, CO_2 fluxes and canopy structure in a mangrove forest of the Florida Everglades. Agric. For. Meteorol. 153, 54–66. https://doi.org/10.1016/j.agrformet.2011.07.022.

Baustian, J., Mendelssohn, I., 2015. Hurricane-induced sedimentation improves marsh resilience and vegetation vigor under high rates of relative sea-level rise. Wetlands 35, 1–8. https://doi.org/10.1007/s13157-015-0670-2.

Borges, A.V., Djenidi, S., Lacroix, G., Théate, J., Delille, B., Frankignoulle, M., 2003. Atmospheric CO_2 flux from mangrove surrounding waters. Geophys. Res. Lett. 30, 1158.

Bouillon, S., Middelburg, J.J., Dehairs, F., Borges, A.V., Abril, G., Flindt, M.R., Ulomi, S., Kristensen, E., 2007. Importance of intertidal sediment processes and porewater exchange on the water column biogeochemistry in a pristine mangrove creek (Ras Dege, Tanzania). Biogeosciences 4, 311–322.

Bouillon, S., Borges, A.V., Castañeda-Moya, E., Diele, K., Dittmar, T., Duke, N.C., Kristensen, E., Lee, S.Y., Marchand, C., Middelburg, J.J., Rivera-Monroy, V.H., Smith III, T.J., Twilley, R.R., 2008. Mangrove production and carbon sinks: a revision of global budget estimates. Global Biogeochem. Cycles 22. GB 2013.

Bourgeois, C., Alfaro, A.C., Leopold, A., Andreoli, R., Bisson, E., Desnues, A., Duprey, J.-L., Marchand, C., 2019. Sedimentary and elemental dynamics as a function of the elevation profile in a semi-arid mangrove toposequence. Catena 173, 289–301.

Breithaupt, J.L., Smoak, J.M., Smith, T.J., Sanders, C.J., Hoare, A., 2012. Organic carbon burial rates in mangrove sediments: strengthening the global budget: mangrove organic carbon burial rates. Global Biogeochem. Cycles, 26. https://doi.org/10.1029/2012GB004375.

Buffington, K.J., Goodman, A.C., Freeman, C.M., Thorne, K.M., 2020. Testing the interactive effects of flooding and salinity on tidal marsh plant productivity. Aquat. Bot. 164. https://doi.org/10.1016/j.aquabot.2020.103231, 103231.

Bulmer, R.H., Lundquist, C.J., Schwendenmann, L., 2015. Sediment properties and CO_2 efflux from intact and cleared temperate mangrove forests. Biogeosciences 12, 6169–6180. https://doi.org/10.5194/bg-12-6169-2015.

Bulseco, A., Giblin, A., Tucker, J., Murphy, A., Sanderman, J., Hiller, K., Bowen, J., 2019. Nitrate addition stimulates microbial decomposition of organic matter in salt marsh sediments. Glob. Chang. Biol., 25. https://doi.org/10.1111/gcb.14726.

Castañeda-Moya, E., Twilley, R.R., Rivera-Monroy, V.H., Zhang, K., Davis, S.E., Ross, M., 2010. Sediment and nutrient deposition associated with hurricane Wilma in mangroves of the Florida coastal everglades. Estuaries Coast 33, 45–58. https://doi.org/10.1007/s12237-009-9242-0.

Castañeda-Moya, E., Rivera-Monroy, V., Chambers, R., Zhao, X., Lamb-Wotton, L., Gorsky, A., Gaiser, E., Troxler, T., Kominoski, J., Hiatt, M., 2020. Hurricanes fertilize mangrove forests in the Gulf of Mexico (Florida Everglades, USA). Proc. Natl. Acad. Sci. U. S. A. 117, 201908597. https://doi.org/10.1073/pnas.1908597117.

Cavanaugh, K.C., Kellner, J.R., Forde, A.J., Gruner, D.S., Parker, J.D., Rodriguez, W., Feller, I.C., 2014. Poleward expansion of mangroves is a threshold response to decreased frequency of extreme cold events. Proc. Natl. Acad. Sci. U. S. A. 111 (2), 723–727. https://doi.org/10.1073/pnas.1315800111.

Chambers, L.G., Davis, S.E., Troxler, T., Boyer, J.N., Downey-Wall, A., Scinto, L.J., 2014. Biogeochemical effects of simulated sea-level rise on carbon loss in an Everglades mangrove peat soil. Hydrobiologia 726 (1), 195–211. https://doi.org/10.1007/s10750-013-1764-6.

Chanda, A., Akhand, A., Manna, S., Dutta, S., Hazra, S., Das, I., Dadhwal, V.K., 2013. Characterizing spatial and seasonal variability of carbon dioxide and water vapour fluxes above a tropical mixed mangrove forest canopy, India. J. Earth Syst. Sci. 122, 503–513.

Chanda, A., Akhand, A., Manna, S., Dutta, S., Das, I., Hazra, S., Rao, K.H., Dadhwal, V.K., 2014. Measuring daytime CO_2 fluxes from the inter-tidal mangrove soils of Indian Sundarbans. Environ. Earth Sci. 72, 417–427. https://doi.org/10.1007/s12665-013-2962-2.

Chang, T.-C., Yang, S.-S., 2003. Methane emissions from wetlands in Thailand. Atmos. Environ. 37, 4551–4558.

Chauhan, R., Datta, A., Ramanathan, A.L., Adhya, T.K., 2015. Factors influencing spatio-temporal variation of methane and nitrous oxide emission from a tropical mangrove of eastern coast of India. Atmos. Environ. 107, 95–106.

Cheeseman, J.M., Herendeen, L.B., Cheeseman, A.T., Clough, B.F., 1997. Photosynthesis and photoprotection in mangroves under field conditions. Plant Cell Environ. 20, 579–588. https://doi.org/10.1111/j.1365-3040.1997.00096 .x.

Chen, Y., Ye, Y., 2014. Effects of salinity and nutrient addition on mangrove Excoecaria agallocha. PLoS One 9 (4). https://doi.org/10.1371/journal.pone.0093337, e93337.

Chen, L., Wang, W., Lin, P., 2005. Photosynthetic and physiological responses of Kandelia candel L. Druce seedlings to duration of tidal immersion in artificial seawater. Environ. Exp. Bot. 54, 256–266. https://doi.org/10.1016/j.envex pbot.2004.09.004.

Chen, G.C., Tam, N.F., Ye, Y., 2012. Spatial and seasonal variations of atmospheric N_2O and CO_2 fluxes from a subtropical mangrove swamp and their relationships with soil characteristics. Soil Biol. Biochem. 48, 175–181.

Chen, G.C., Ulumuddin, Y.I., Pramudji, S., Chen, S.Y., Chen, B., Ye, Y., Ou, D.Y., Ma, Z.Y., Huang, H., Wang, J.K., 2014a. Rich soil carbon and nitrogen but low atmospheric greenhouse gas fluxes from North Sulawesi mangrove swamps in Indonesia. Sci. Total Environ. 487, 91–96.

Chen, H., Lu, W., Yan, G., Yang, S., Lin, G., 2014b. Typhoons exert significant but differential impacts on net ecosystem carbon exchange of subtropical mangrove forests in China. Biogeosciences, 11. https://doi.org/10.5194/bg-11-5323-2014.

Chmura, G.L., Anisfeld, S.C., Cahoon, D.R., Lynch, J.C., 2003. Global carbon sequestration in tidal, saline wetland soils. Global Biogeochem. Cycles 17, 1111.

Clark, M.W., McConchie, D., Lewis, D.W., Saenger, P., 1998. Redox stratification and heavy metal partitioning in Avicennia-dominated mangrove sediments: a geochemical model. Chem. Geol. 149, 147–171.

Coelho, H., Vieira, S., Serôdio, J., 2009. Effects on desiccation on the photosynthetic activity of intertidal microphytobenthos biofilm as studied by optical method. J. Exp. Mar. Biol. Ecol. 381, 98–104.

Collins, M., Sutherland, M., Bouwer, L., Cheong, S.-M., Frölicher, T., Jacot Des Combes, H., Koll Roxy, M., Losada, I., McInnes, K., Ratter, B., Rivera-Arriaga, E., Susanto, R.D., Swingedouw, D., Tibig, L., 2019. Extremes, abrupt changes and managing risk. In: Pörtner, H.-O., Roberts, D.C., Masson-Delmotte, V., Zhai, P., Tignor, M., Poloczanska, E., Weyer, N.M. (Eds.), IPCC Special Report on the Ocean and Cryosphere in a Changing Climate.

Couto, T., Martins, H., Duarte, B., Cacador, I., Marques, J.C., 2014. Modelling the effects of global temperature increase on the growth of salt marsh plants. Appl. Ecol. Environ. Res. 12, 753–764.

Cui, M., Ma, A., Qi, H., Zhuang, X., Zhuang, G., 2015. Anaerobic oxidation of methane: an "active" microbial process. Microbiologyopen 4, 1–11.

Curtis, P.S., Balduman, L.M., Drake, B.G., Whigham, D.F., 1990. Elevated atmospheric CO_2 effects on belowground processes in C_3 and C_4 estuarine marsh communities. Ecology 71 (5), 2001–2006. https://doi.org/10.2307/1937608.

Dai, A., 2013. Increasing drought under global warming in observations and models. Nat. Clim. Chang. 3, 52–58. https://doi.org/10.1038/nclimate1633.

Danielson, T., Rivera-Monroy, V., Castañeda-Moya, E., Briceño, H., Travieso, R., Marx, B., Gaiser, E., Farfan, L., 2017. Assessment of Everglades mangrove forest resilience: implications for above-ground net primary productivity and carbon dynamics. For. Ecol. Manage. 404, 115–125. https://doi.org/10.1016/j.foreco.2017.08.009.

David, F., Marchand, C., Taillardat, P., Nho, N.T., Meziane, T., 2018a. Nutritional composition of suspended particulate matter in a tropical mangrove creek during a tidal cycle (Can Gio, Vietnam). Estuar. Coast. Shelf Sci. 200, 126–130.

David, F., Meziane, T., Nhu-Trang, T.T., Truong, V.V., Nho, N.T., Taillardat, P., Marchand, C., 2018b. Carbon biogeochemistry and CO_2 emissions in a human impacted and mangrove dominated tropical estuary (Can Gio, Vietnam). Biogeochemistry. https://doi.org/10.1007/s10533-018-0444-z.

Davidson, E.A., Janssens, I.A., 2006. Temperature sensitivity of soil carbon decomposition and feedbacks to climate change. Nature 440, 165–173.

Di Nitto, D., Neukermans, G., Koedam, N., Defever, H., Pattyn, F., Kairo, J.G., Dahdouh-Guebas, F., 2014. Mangroves facing climate change : landward migration potential in response to projected scenarios of sea-level rise. Biogeosciences 11 (3), 857–871. https://doi.org/10.5194/bg-11-857-2014.

Dittmar, T., Hertkorn, N., Kattner, G., Lara, R.J., 2006. Mangroves, a major source of dissolved organic carbon to the oceans. Global Biogeochem. Cycles 20. https://doi.org/10.1029/2005GB002570.

Duarte, C.M., Middelburg, J.J., Caraco, N., 2005. Major role of marine vegetation on the oceanic carbon cycle. Biogeosciences 2, 1–8.

Duarte, C.M., Losada, I.J., Hendriks, I.E., Mazarrasa, I., Marbà, N., 2013. The role of coastal plant communities for climate change mitigation and adaptation. Nat. Clim. Chang. 3 (11), 961–968.

Duke, N.C., 1992. Mangrove floristics and biogeography. In: Robertson, A.I., Alongi, D.M. (Eds.), Tropical Mangrove Ecosystems. American Geophysical Union, Washington, DC.

Duke, N.C., Ball, M.C., Ellison, J.C., 1998. Factors influencing biodiversity and distributional gradients in mangroves. Glob. Ecol. Biogeogr. 7, 27–47.

Ellison, A.M., Farnsworth, E.J., 1997. Simulated sea-level change alters anatomy, physiology, growth, and reproduction of red mangrove (Rhizophora mangle L.). Oecologia 112, 435–446.

Ewel, K.C., Bourgeois, J.A., Cole, T.G., Zheng, S., 1998. Variation in environmental characteristics and vegetation in high-rainfall mangrove forests, Kosrae, Micronesia. Glob. Ecol. Biogeogr. Lett. 7, 49–56.

Farnsworth, E.J., Ellison, A.M., Gong, W.K., 1996. Elevated CO_2 alters anatomy, physiology, growth, and reproduction of red mangrove (Rhizophora mangle L.). Oecologia 108, 599–609. https://doi.org/10.1007/BF00329032.

Feller, I., Dangremond, E., Devlin, D., Lovelock, C., Proffitt, E., Rodriguez, W., 2015. Nutrient enrichment intensifies hurricane impact in scrub mangrove ecosystems in the Indian River Lagoon, FL USA. Ecology 96, 2960–2972. https://doi.org/10.1890/14-1853.1.

Feller, I.C., Friess, D.A., Krauss, K.W., Lewis III, R.R., et al., 2017. The state of the world's mangroves in the 21st century under climate change. Hydrobiologia 803, 1–12. https://doi.org/10.1007/s10750-017-3331-z.

Gilman, E.L., Ellison, J., Duke, N.C., Field, C., 2008. Threats to mangroves from climate change and adaptation options: a review. Aquat. Bot. 89, 237–250. https://doi.org/10.1016/j.aquabot.2007.12.009.

Giri, C., Ochieng, E., Tieszen, L.L., Zhu, Z., Singh, A., Loveland, T., Masek, J., Duke, N., 2011. Status and distribution of mangrove forests of the world using earth observation satellite data. Glob. Ecol. Biogeogr. 20, 154–159.

Ha, T.H., Marchand, C., Aimé, J., Thi Kim Cuc, N., 2018. Seasonal variability of CO_2 emissions from mature planted mangroves (Northern Viet Nam). Estuar. Coast. Shelf Sci. 213, 28–39.

He, B., Lai, T., Fan, H., Wang, W., Zheng, H., 2007. Comparison of flooding-tolerance in four mangrove species in a diurnal tidal zone in the Beibu Gulf. Estuar. Coast. Shelf Sci. 74, 254.

Hickey, S.M., Phinn, S.R., Callow, N.J., Van Niel, K.P., Duarte, C.M., et al., 2017. Is climate change shifting the poleward limit of mangroves? Estuaries Coast 40, 1215–1226. https://doi.org/10.1007/s12237-017-0211-8.

Hill, J.M., Petraitis, P.S., Heck Jr., K.L., 2020. Submergence, nutrient enrichment, and tropical storm impacts on Spartina alterniflora in the microtidal northern Gulf of Mexico. Mar. Ecol. Prog. Ser. 644, 33–45.

Hwang, Y.H., Chen, S.C., 2001. Effects of ammonium, phosphate, and salinity on growth, gas exchange characteristics and ionic contents of seedlings of mangrove Kandelia candel (L.) Druce. Bot. Bull. Acad. Sin. 42, 131–139.

Imbert, D., 2018. Hurricane disturbance and forest dynamics in east Caribbean mangroves. Ecosphere 9. https://doi.org/10.1002/ecs2.2231, e02231.

Jacotot, A., Marchand, C., Rosenheim, B.E., Domack, E., Allenbach, M., 2018a. Mangrove soil carbon stocks along an elevation gradient: influence of the late Holocene marine regression (New Caledonia). Mar. Geol. 404, 60670.

Jacotot, A., Marchand, C., Gensous, S., Allenbach, M., 2018b. Effects of elevated atmospheric CO_2 and increased tidal flooding on leaf gas-exchange parameters of two common mangrove species: Avicennia marina and Rhizophora stylosa. Photosynth. Res. https://doi.org/10.1007/s11120-018-0570-4.

Jacotot, A., Marchand, C., Allenbach, M., 2018c. Tidal variability of CO_2 and CH_4 emissions from the water column within a Rhizophora mangrove forest (New Caledonia). Sci. Total Environ. 631–632, 334–340.

Jacotot, A., Marchand, C., Allenbach, M., 2019a. Increase in growth and alteration of C:N ratios of Avicennia marina and Rhizophora stylosa subject to elevated CO_2 concentrations and longer tidal flooding duration. Front. Ecol. Evol. 7. https://doi.org/10.3389/fevo.2019.00098.

Jacotot, A., Marchand, C., Allenbach, M., 2019b. Biofilm and temperature controls on greenhouse gas (CO_2 and CH_4) emissions from a mangrove soil (New Caledonia). Sci. Total Environ. 650, 1019–1028. https://doi.org/10.1016/j.scitotenv.2018.09.093 (Elsevier).

Janssens, I.A., Freibauer, A., Ciais, P., Smith, P., Nabuurs, G.J., Folberth, G., Schlamadinger, B., Hutjes, R.W.A., Ceulemans, R., Schulze, E.D., Valentini, R., Dolman, A.J., 2003. Europe's terrestrial Biosphere Absorbs 7 to 12 % of European anthropogenic CO_2 emissions. Science 300, 1438–1541.

Jeffrey, L.C., Reithmaier, G., Sippo, J.Z., Johnston, S.G., Tait, D.R., Harada, Y., Maher, D.T., 2019. Are methane emissions from mangrove stems a cryptic carbon loss pathway? Insights from a catastrophic forest mortality. New Phytol. 224 (1), 146–154.

Jennerjahn, T.C., Ittekkot, V., 2002. Relevance of mangroves for the production and deposition of organic matter along tropical continental margins. Naturwissenschaften 89, 23–30.

Kim, J.-H., Dupont, L., Behling, H., Versteegh, G.J.M., 2005. Impacts of rapid sea-level rise on mangrove deposit erosion: application of taraxerol and Rhizophora records. J. Quat. Sci. 20, 221–225.

Kirwan, M.L., Guntenspergen, G.R., Morris, J.T., 2009. Latitudinal trends in Spartina alterniflora productivity and the response of coastal marshes to global change. Glob. Chang. Biol. 15 (8), 1982–1989.

Knox, S.H., Windham-Myers, L., Anderson, F., Sturtevant, C., Bergamaschi, B., 2018. Direct and indirect effects of tides on ecosystem-scale CO_2 exchange in a brackish tidal marsh in northern California. J. Geophys. Res. Biogeo. 123, 787–806.

Krauss, K.W., Ball, M.C., 2013. On the halophytic nature of mangroves. Trees 27, 7–11. https://doi.org/10.1007/s00468-012-0767-7.

Krauss, K.W., Osland, M.J., 2020. Tropical cyclones and the organization of mangrove forests: a review. Ann. Bot. 125, 213–234. https://doi.org/10.1093/aob/mcz161.

Krauss, K.W., McKee, K.L., Lovelock, C.E., Cahoon, D.R., Saintilan, N., Reef, R., Chen, L., et al., 2014. How mangrove forests adjust to rising sea-level. New Phytol. 202, 19–34. https://doi.org/10.1111/nph.12605.

Kristensen, E., 2008. Mangrove crabs as ecosystem engineers, with emphasis on sediment processes. J. Sea Res. 59, 30–43.

Kristensen, E., Alongi, D.M., 2006. Control by fiddler crabs (Uca vocans) and plant roots (Avicennia marina) on carbon, iron and sulfur biogeochemistry in mangrove sediment. Limnol. Oceanogr. 51, 1557–1571.

Kristensen, E., Bouillon, S., Dittmar, T., Marchand, C., 2008. Organic carbon dynamics in mangrove ecosystem, a review. Aquat. Bot. 89, 201–219.

Kristensen, E., Connolly, R., Ferreira, T.O., Marchand, C., Otero, X.L., Rivera-Monroy, V.H., 2017. Biogeochemical cycles; global approaches and perspectives. In: Mangrove Ecosystems: A Global Biogeographic Perspective Structure, Function and Services. Springer.

Leopold, A., Marchand, C., Deborde, J., Chaduteau, C., Allenbach, M., 2013. Influence of mangrove zonation on CO_2 fluxes at sediment-air interface (New-Caledonia). Geoderma 202–203, 62–70.

Leopold, A., Marchand, C., Deborde, J., Allenbach, M., 2015. Temporal variability of CO_2 fluxes at the sediment-air interface in mangroves (New Caledonia). Sci. Total Environ. 502, 617–626.

Leopold, A., Marchand, C., Renchon, A., Deborde, J., Quiniou, T., Allenbach, M., 2016. Net ecosystem CO_2 exchange of "Le Coeur de Voh" mangrove: evidences of water stress controls on mangrove productivity under semi-arid climate (New Caledonia). Agric. For. Meteorol. 223, 217–232.

Lewis, D.B., Brown, J.A., Jimenez, K.L., 2014. Effects of flooding and warming on soil organic matter mineralization in Avicennia germinans mangrove forests and Juncus roemerianus salt marshes. Estuar. Coast. Shelf Sci. 139, 11–19.

Li, Q., Lu, W., Chen, H., Luo, Y., Lin, G., 2014. Differential responses of net ecosystem exchange of carbon dioxide to light and temperature between spring and neap tides in subtropical mangrove forests. Sci. World J. https://doi.org/10.1155/2014/943697.

Liu, J., Lai, D.Y.F., 2019. Subtropical mangrove wetland is a stronger carbon dioxide sink in the dry than wet seasons. Agric. For. Meteorol. 278, 107644. ISSN 0168-1923 https://doi.org/10.1016/j.agrformet.2019.107644.

Livesley, S.J., Andrusiak, S.M., 2012. Temperate mangrove and salt marsh sediments are a small methane and nitrous oxide source but important carbon store. Estuar. Coast. Shelf Sci. 97, 19–27.

Long, J., Giri, C., Primavera, J., Trivedi, M., 2016. Damage and recovery assessment of the Philippines' mangroves following Super Typhoon Haiyan. Mar. Pollut. Bull. 109, 734–743. https://doi.org/10.1016/j.marpolbul.2016.06.080.

Lovelock, C.E., 2008. Soil respiration and belowground carbon allocation in mangrove forests. Ecosystems 11, 342–354.

Lovelock, C.E., Feller, I.C., Reef, R., Ruess, R.W., 2014. Variable effects of nutrient enrichment on soil respiration in mangrove forests. Plant and Soil 379, 135–148.

Lovelock, C.E., Cahoon, D.R., Friess, D.A., Guntenspergen, G.R., Krauss, K.W., Reef, R., Rogers, K., Saunders, M.L., Sidik, F., Swales, A., Saintilan, N., Thuyen, L.X., Triet, T., 2015. The vulnerability of Indo-Pacific mangrove forests to sea-level rise. Nature 526, 559–563.

Lu, W.Z., Chen, L.Z., Wang, W.Q., Tam, N.F.Y., Lin, G.H., 2013. Effects of sea-level rise on mangrove Avicennia population growth, colonization and establishment: evidence from a field survey and greenhouse manipulation experiment. Acta Oecol. 49, 83–91.

Lugo, A., 2008. Visible and invisible effects of hurricanes on forest ecosystems: an international review. Austral Ecol. 33. https://doi.org/10.1111/j.1442-9993.2008.01894.x.

Lugo, A.E., Patterson-Zucca, C., 1977. The impact of low temperature stress on mangrove structure and growth. Trop. Ecol. 18, 149–161.

Lyimo, T.J., Pol, A., Op den Camp, H.J.M., 2002. Sulfate reduction and methanogenesis in sediments of Mtoni mangrove forest, Tanzania. Ambio 31, 614–616.

Lyimo, T.J., Pol, A., Jetten, M.S.M., Op den Camp, H.J.M., 2009. Diversity of methanogenic archaea in a mangrove sediment and isolation of a new Methanococcoides strain. FEMS Microbiol. Lett. 291, 247–253.

Maher, D.T., Santos, I.R., Golsby-Smith, L., Gleeson, J., Eyre, B.D., 2013. Groundwater-derived dissolved inorganic and organic carbon exports from a mangrove tidal creek: the missing mangrove carbon sink? Limnol. Oceanogr. 58, 475–488.

Marchand, C., Baltzer, F., Lallier-Vergès, E., Albéric, P., 2004. Pore-water chemistry in mangrove sediments: relationship with species composition and developmental stages. (French Guiana). Mar. Geol. 208, 361–381.

Marchand, C., Lallier-Vergès, E., Baltzer, F., Alberic, P., Cossa, D., Baillif, P., 2006. Heavy metals distribution in mangrove sediments (French Guiana). Mar. Chem. 98, 1–17.

Marsh, A.S., Rasse, D.P., Drake, B., Megonigal, P., 2005. Effect of elevated CO_2 on carbon pools and fluxes in a brackish marsh. Estuaries Coasts 28, 694–704. https://doi.org/10.1007/BF02732908.

McKee, K.L., 1993. Soil physicochemical patterns and mangrove species distribution—reciprocal effects? J. Ecol. 81, 477–487.

McKee, K.L., Cherry, J., 2009. Hurricane Katrina sediment slowed elevation loss in subsiding brackish marshes of the Mississippi river delta. Wetlands 29, 2–15. https://doi.org/10.1672/08-32.1.

McKee, K.L., Rooth, J.E., 2008. Where temperate meets tropical: multi-factorial effects of elevated CO_2, nitrogen enrichment, and competition on a mangrove-salt marsh community. Glob. Chang. Biol. 14, 971–984.

McKee, K.L., Mendelssohn, I.A., Hester, M.W., 1988. Reexamination of pore water sulphide concentrations and redox potentials near the aerial roots of Rhizophora mangle and Avicennia germinans. Am. J. Bot. 75, 1352–1359.

McKee, K., Rogers, K., Saintilan, N., 2012. Response of salt marsh and mangrove wetlands to changes in atmospheric CO_2, climate, and sea-level. In: Global Change and the Function and Distribution of Wetlands. Springer, pp. 63–96.

Middleton, B.A., 2016. Differences in impacts of Hurricane Sandy on freshwater swamps on the Delmarva Peninsula, Mid-Atlantic Coast, USA. Ecol. Eng. 87, 62–70. https://doi.org/10.1016/j.ecoleng.2015.11.035.

Mo, Y., Kearney, M.S., Turner, R.E., 2020. The resilience of coastal marshes to hurricanes: the potential impact of excess nutrients. Environ. Int. 138. https://doi.org/10.1016/j.envint.2019.105409, 105409.

Molnar, N., Marchand, C., Deborde, J., Della Patrona, L., Meziane, T., 2014. Seasonal pattern of the biogeochemical properties of mangrove sediments receiving shrimp farm effluents (New Caledonia). J. Aquac. Res. Dev. 5 (5).

Morris, J.T., Sundberg, K., Hopkinson, C.S., 2013. Salt marsh primary production and its responses to relative sea-level and nutrients in estuaries at Plum Island, Massachusetts, and North Inlet, South Carolina, USA. Oceanography 26, 78–84.

Naidoo, G., 2010. Ecophysiological differences between fringe and dwarf Avicennia marina mangroves. Trees 24, 667–673.

Naidoo, G., Rogall, H., von Willert, D.J., 1997. Gas exchange responses of a mangrove species, Avicennia marina, to waterlogged and drained conditions. Hydrobiologia 352, 39–47.

Naidoo, G., Hiralal, O., Naidoo, Y., 2011. Hypersalinity effects on leaf ultra structure and physiology in the mangrove Avicennia marina. Flora 206, 814–820.

Osland, M.J., Enwright, N.M., Day, R.H., Gabler, C.A., Stagg, C.L., Grace, J.B., 2016. Beyond just sea-level rise: considering macroclimatic drivers within coastal wetland vulnerability assessments to climate change. Glob. Chang. Biol. 22, 1–11.

Osland, M., Feher, L.C., Griffith, K.T., Cavanaugh, K.C., Enwright, N.M., Day, R.H., Stagg, C.L., Krauss, K.W., Howard, R.J., Grace, J.B., Rogers, K., 2017. Climatic controls on the global distribution, abundance, and species richness of mangrove forests. Ecol. Monogr. 87 (2), 341–359.

Osland, M.J., Gabler, C.A., Grace, J.B., et al., 2018. Climate and plant controls on soil organic matter in coastal wetlands. Glob. Chang. Biol. 24, 5361–5379. https://doi.org/10.1111/gcb.14376.

Ostertag, R., Scatena, F., Silver, W., 2003. Forest floor decomposition following hurricane litter inputs in several puerto rican forests. Ecosystems 6, 261–273. https://doi.org/10.1007/PL00021512.

Ouyang, X., Lee, S.Y., 2014. Updated estimates of carbon accumulation rates in coastal marsh sediments. Biogeosciences 11, 5057–5071. https://doi.org/10.5194/bg-11-5057-2014.

Ouyang, X., Lee, S.Y., 2020. Improved estimates on global carbon stock and carbon pools in tidal wetlands. Nat. Commun. 11, 317. https://doi.org/10.1038/s41467-019-14120-2.

Ouyang, X., Lee, S.Y., Connolly, R.M., 2017. The role of root decomposition in global mangrove and saltmarsh carbon budgets. Earth Sci. Rev. 166, 53–63. https://doi.org/10.1016/j.compchemeng.2016.09.009.

Ouyang, X., Guo, F., Lee, S., 2020. The impact of super-typhoon Mangkhut on sediment nutrient density and fluxes in a mangrove forest in Hong Kong. Sci. Total Environ. https://doi.org/10.1016/j.scitotenv.2020.142637.

Padhy, R., Bhattacharyya, P., Dash, P.K., Reddy, C.S., Chakraborty, A., Pathak, H., 2020. Seasonal fluctuation in three mode of greenhouse gases emission in relation to soil labile carbon pools in degraded mangrove, Sundarban, India. Sci. Total Environ. 705. https://doi.org/10.1016/j.scitotenv.2019.135909, 135909.

Paling, E.I., Kobryn, H., Humphreys, G., 2008. Assessing the extent of mangrove change caused by Cyclone Vance in the eastern Exmouth Gulf, northwestern Australia. Estuar. Coast. Shelf Sci. 77, 603–613. https://doi.org/10.1016/j.ecss.2007.10.019.

Pezeshki, S.R., DeLaune, R.D., Meeder, J.F., 1997. Carbon assimilation and biomass in Avicennia germinans and Rhizophora mangle seedlings in response to soil redox conditions. Environ. Exp. Bot. 37, 161–171.

Plaziat, J.-C., 1995. Modern and fossil mangroves and mangals: their climatic and biogeographic variability. In: Bosence, D.W.J., Allison, P.A. (Eds.), Marine Palaeoenvironmental Analysis from Fossils, pp. 73–96. Geological Society Special Publications No. 83.

Reef, R., Lovelock, C.E., 2014. Historical analysis of mangrove leaf traits throughout the 19th and 20th centuries reveals differential responses to increases in atmospheric CO_2. Glob. Ecol. Biogeogr. 23, 1209–1214. https://doi.org/10.1111/geb.12211.

Reef, R., Winter, K., Morales, J., Adame, F., Reef, D.L., Lovelock, C.E., et al., 2015. The effect of atmospheric carbon dioxide concentrations on the performance of the mangrove Avicennia germinans over a range of salinities. Physiol. Plant. 154, 358–368. https://doi.org/10.1111/ppl.12289.

Reef, R., Slot, M., Motro, U., Motro, M., Motro, Y., Adame, M.F., Garcia, M., Aranda, J., Lovelock, C.E., Winter, K., et al., 2016. The effects of CO_2 and nutrient fertilisation on the growth and temperature response of the mangrove Avicennia germinans. Photosynth. Res. 129, 159–170. https://doi.org/10.1007/s1112 0-016-0278-2.

Reich, J.W., Schlesinger, W.H., 1992. The global carbon dioxide flux in soil respiration and its relationship to climate. Tellus 44B, 81–99.

Rivera-Monroy, V.H., Danielson, T.M., Castañeda-Moya, E., Marx, B.D., Travieso, R., Zhao, X., Gaiser, E.E., Farfan, L.M., 2019. Long-term demography and stem productivity of Everglades mangrove forests (Florida, USA): resistance to hurricane disturbance. For. Ecol. Manage. 440, 79–91. https://doi.org/10.1016/j.foreco.2019.02.036.

Saintilan, N., Rogers, K., 2015. Woody plant encroachment of grasslands: a comparison of terrestrial and wetland settings. New Phytol. 205, 1062–1070.

Saintilan, N., Khan, N.S., Ashe, E., Kelleway, J.J., Rogers, K., Woodroffe, C.D., Horton, B.P., 2020. Thresholds of mangrove survival under rapid sea-level rise. Science 368, 1118–1121.

Sanders, C.J., Maher, D.T., Tait, D.R., Williams, D., Holloway, C., Sippo, J.Z., Santos, I.R., 2016. Are global mangrove carbon stocks driven by rainfall? J. Geophys. Res. G: Biogeosci. https://doi.org/10.1002/2016JG003510.

Schuerch, M., Spencer, T., Temmerman, S., Kirwan, M.L., Wolff, C., Lincke, D., McOwen, C. J., Pickering, M.D., Reef, R., Vafeidis, A.T., Hinkel, J., Nicholls, R.J., Brown, S., 2018. Future response of global coastal wetlands to sea-level rise. Nature 561, 231–234.

Simard, M., Fatoyinbo, L., Smetanka, C., Rivera-Monroy, V.H., Castañeda-Moya, E., Thomas, N., Van der Stocken, T., 2019. Mangrove canopy height globally related to precipitation, temperature and cyclone frequency. Nat. Geosci. 12, 40–45.

Sippo, J.Z., Lovelock, C.E., Santos, I.R., Sanders, C.J., Maher, D.T., 2018. Mangrove mortality in a changing climate: an overview. Estuar. Coast. Shelf Sci. 215, 241–249. https://doi.org/10.1016/j.ecss.2018.10.011.

Smith III, T.J., Anderson, G., Balentine, K., Tiling, G., Ward, G., Whelan, K., 2009. Cumulative impacts of hurricanes on Florida mangrove ecosystems: sediment deposition, storm surges and vegetation. Wetlands 29, 24–34. https://doi.org/10.1672/08-40.1.

Snedaker, S.C., Araujo, R.J., 1998. Stomatal conductance and gas exchange in four species of Caribbean mangroves exposed to ambient and increased CO_2. Mar. Freshw. Res. 49, 325–327.

Spivak, A.C., Sanderman, J., Bowen, J.L., Canuel, E.A., Hopkinson, C.S., 2019. Global-change controls on soil-carbon accumulation and loss in coastal vegetated ecosystems. Nat. Geosci. 12, 685–692.

Stieglitz, T.C., Clark, J.F., Hancock, G.J., 2013. The mangrove pump: the tidal flushing of animal burrows in a tropical mangrove forest determined from radionuclide budgets. Geochim. Cosmochim. Acta 102, 12–22.

Taillie, P.J., Roman-Cuesta, R., Lagomasino, D., Cifuentes-Jara, M., Fatoyinbo, T., Ott, L.E., Poulter, B., 2020. Widespread mangrove damage resulting from the 2017 Atlantic mega hurricane season. Environ. Res. Lett. 15. https://doi.org/10.1088/1748-9326/ab82cf, 064010.

Tan, J., Jakob, C., Rossow, W.B., Tselioudis; G., 2015. Increases in tropical rainfall driven by changes in frequency of organized deep convection. Nature 519, 451–454.

Troxler, T.G., Barr, J.G., Fuentes, J.D., Engel, V., Anderson, G., Sanchez, C., Lagomasino, D., Price, R., Davis, S.E., 2015. Component-specific dynamics of riverine mangrove CO_2 efflux in the Florida coastal Everglades. Agric. For. Meteorol. 213, 273–282.

Turner, R., 1976. Geographic variations in salt marsh macrophyte production: a review. Contrib. Mar. Sci. 20, 47–68.

Tweel, A., Turner, R., 2012. Landscape-scale analysis of wetland sediment deposition from four tropical cyclone events. PLoS One 7. https://doi.org/10.1371/journal.pone.0050528, e50528.

Twilley, R.R., Rivera-Monroy, V.H., 2009. Ecogeomorphic models of nutrient biogeochemistry for mangrove wetlands. In: Perillo, G.M.E., et al. (Eds.), Coastal Wetlands: An Integrated Ecosystem Approach. Elsevier, New York, pp. 641–683.

Twilley, R.R., Chen, R.H., Hargis, T., 1992. Carbon sinks in mangrove forests and their implications to the carbon budget of tropical coastal ecosystems. Water Air Soil Pollut. 64, 265–288.

Vinh, T.V., Allenbach, M., Aimée, J., Marchand, C., 2019. Seasonal variability of CO_2 fluxes at different interfaces and vertical CO_2 concentrations profiles in a Rhizophora mangrove stand (Can Gio, Viet Nam). Atmos. Environ. https://doi.org/10.1016/j.atmosenv.2018.12.049.

Vinh, T.V., Allenbach, M., Linh, T.V.K., Marchand, C., 2020. Changes in leaf litter quality during its decomposition in a tropical planted mangrove forest (Can Gio, Vietnam). Frontiers in Environmental. Science 13|. https://doi.org/10.3389/fenvs.2020.00010.

Walcker, R., Laplanche, Herteman, M., Lambs, L., Fromard, F., 2019. Damages caused by hurricane Irma in the human-degraded mangroves of Saint Martin (Caribbean). Sci. Rep. 9. https://doi.org/10.1038/s41598-019-55393-3.

Wang, F., Kroeger, K.D., Gonneea, M.E., Pohlman, J.W., Tang, J., 2019a. Water salinity and inundation control soil carbon decomposition during salt marsh restoration: an incubation experiment. Ecol. Evol. 9, 1911–1921.

Wang, F., Lu, X., Sanders, C.J., Tang, J., 2019b. Tidal wetland resilience to Sea-level rise increases their carbon sequestration capacity in United States. Nat. Commun. 10, 5434.

Wang, F., Eagle, M., Kroeger, K.D., Spivak, A.C., Tang, J., 2020. Plant biomass and rates of carbon dioxide uptake are enhanced by successful restoration of tidal connectivity in salt marshes. Sci. Total Environ., 141566.

Wang, F., Sanders, C.J., Santos, I.R., Tang, J., Schuerch, M., Kirwan, M.L., Kopp, R.E., Zhu, K., Li, X., Yuan, J., Liu, W., Li, Z., 2021. Global blue carbon accumulation in tidal wetlands increases with climate change. Natl. Sci. Rev. https://doi.org/10.1093/nsr/nwaa1296.

Ward, R.D., Friess, D.A., Day, R.H., MacKenzie, R.A., 2016. Impacts of climate change on mangrove ecosystems: a region by region overview. Ecosyst. Health Sustain. 2 (4). https://doi.org/10.1002/ehs2.1211, e01211.

Woodroffe, C.D., 1981. Mangrove swamp stratigraphy and Holocene transgression, Grand Cayman Island, West Indies. Mar. Geol. 41, 271–294.

Woodroffe, C., 1992. Mangrove sediments and geomorphology. In: Robertson, A.I., Alongi, D.M. (Eds.), Tropical Mangrove Ecosystems. American Geophysical Union, Washington, DC, pp. 7–42.

Woodroffe, C., 2002. Coasts: Form, Process and Evolution. Cambridge University Press, Cambridge.

Woodroffe, C.D., Rogers, K., McKee, K.L., Lovelock, C.E., Mendelssohn, I.A., Saintilan, N., 2016. Mangrove sedimentation and response to relative sea-level rise. Ann. Rev. Mar. Sci. 8, 243–266. https://doi.org/10.1146/annurev-marine-122414-034025.

Worthington, T.A., zu Ermgassen, P.S.E., Friess, D.A., Krauss, K.W., Lovelock, C.E., Thorley, J., Tingey, R., Woodroffe, C.D., Bunting, P., Cormier, N., Lagomasino, D., Lucas, R., Murray, N.J., Sutherland, W., Spalding, M., et al., 2020. A global biophysical typology of mangroves and its relevance for ecosystem structure and deforestation. Sci. Rep. 10, 14652. https://doi.org/10.1038/s41598-020-71194-5.

Xu, X., Hirata, E., Enoki, T., Tokashiki, Y., 2004. Leaf litter decomposition and nutrient dynamics in a subtropical forest after typhoon disturbance. Plant Ecol. 173, 161–170. https://doi.org/10.1023/B:VEGE.0000029319.05980.70.

Yin, S., An, S., Deng, Q., Zhang, J., Ji, H., Cheng, X., 2015. Spartina alterniflora invasions impact CH_4 and N_2O fluxes from a salt marsh in eastern China. Ecol. Eng. 81 (Suppl. C), 192–199. https://doi.org/10.1016/j.ecoleng.2015.04.044.

Youssef, T., Saenger, P., 1996. Anatomical adaptive strategies to flooding and rhizosphere oxidation in mangrove seedlings. Aust. J. Bot. 44 (3). https://doi.org/10.1071/BT9960297.

Zhang, C., Denka Durgan, S., Lagomasino, D., 2019. Modeling risk of mangroves to hurricanes: a case study of Hurricane Irma. Estuar. Coast. Shelf Sci. 224. https://doi.org/10.1016/j.ecss.2019.04.052.

Zhang, D., Zhang, H., Zheng, J., Cheng, X., Tian, D., Chen, D., 2020. Changes in tropical-cyclone translation speed over the western north pacific. Atmosphere 11, 93. https://doi.org/10.3390/atmos11010093.

The role of biogenic structures for greenhouse gas balance in vegetated intertidal wetlands

Erik Kristensen, Cintia Organo Quintana, and Susan Guldberg Graungård Petersen

Department of Biology, University of Southern Denmark, Odense, Denmark

8.1 Introduction

There is currently much focus in both public media and scientific literature on the global greenhouse gas (GHG) balance, and particularly how the massive anthropogenic emission of these gasses (i.e., CO_2, CH_4, and N_2O) can be reduced. Unfortunately, even a substantial reduction of human GHG emissions is not enough to avoid excessive future global warming and associated climate change impacts (IPCC, 2019). Efficient mitigation approaches are therefore needed to reduce the excess GHG inventory in the atmosphere. An obvious approach is to exploit the carbon (C) sequestration capacity of natural ecosystems. There is a broad consensus that vegetated intertidal wetlands are among the most valuable blue C ecosystems (Donato et al., 2011; Kelleway et al., 2016; Rogers et al., 2019). In fact, such vegetated ecosystems (mangrove forests, saltmarshes, and seagrass beds) that cover just 0.2% of the ocean surface are estimated to contribute nearly 50% of the total C sequestration by marine ecosystems (Mcleod et al., 2011; Duarte et al., 2013). The highly productive vegetated coastal areas rapidly convert CO_2 into structural plant biomass (Donato et al., 2011; Alongi, 2014), which eventually is stored under anoxic conditions in sediments where microbial degradation is very slow (Kristensen and Holmer, 2001; Mcleod et al., 2011). Intertidal mangrove forests and saltmarshes are also efficient depositional ecosystems with potential to sequester allochthonous C by trapping particulate matter from river catchments and marine sources via tidal actions (Chen et al., 2018). It is, though, not yet fully clear how the abundant biogenic structures (e.g., roots and burrows) formed by the plants and animals inhabiting intertidal ecosystems affect the C storage capacity and the GHG balance.

Most vegetated intertidal wetland sediments have a dense network of roots and host a variety of burrow- and tube-dwelling benthic animals (Alongi, 2016; Aschenbroich et al., 2016; Redelstein et al., 2018). These biogenic structures may

Carbon Mineralization in Coastal Wetlands. https://doi.org/10.1016/B978-0-12-819220-7.00001-7

augment C storage by subducting substantial amounts of organic matter deep into anoxic sediments where microbial activity is slow (Lee et al., 2008; Dunn et al., 2019; Thomson et al., 2020). However, the biogenic structures are also maintained oxic by diffusional and active downward transport of oxygen, which may increase microbial organic matter degradation and CO_2 release (Kristensen and Holmer, 2001; Middelburg, 2018). Furthermore, the biogenically induced oxic conditions may inhibit CH_4 generation and promote coupled nitrification-denitrification and associated N_2O generation (Chen and Gu, 2017; Mehring et al., 2017). Nevertheless, there are also strong indications that deep biogenic structures may reach zones with high methanogenesis (Bonaglia et al., 2017; Jeffrey et al., 2019) and act as conduits for rapid release of the potent GHG CH_4 to the atmosphere (Kristensen et al., 2008; Chen et al., 2015). The extent by which these latter effects of biogenic structures counteract the otherwise efficient CO_2 sequestration by wetland ecosystems is not yet fully clear.

The purpose of this chapter is therefore to elucidate in detail the current knowledge on the role of biogenic structures for GHG biogeochemistry and dynamics in vegetated intertidal wetlands. We will start describing these wetlands according to our definition in the present context. This will be followed by an overview of the relevant biogeochemical processes affecting GHG in wetland sediments. Then we will portray the major types of biogenic structures formed by plants and animals inhabiting wetlands. The functioning of these biogenic structures for C dynamics and GHG exchange in intertidal wetlands will be demonstrated by assessing and compiling the current knowledge from densely vegetated and bioturbated tropical mangrove forests and temperate *Spartina* marshes. The stage is then set for evaluating the climate perspective with respect to the role of biogenic structures on ecosystem functioning of coastal wetlands in a world with rising sea level and increasing temperatures.

8.2 Types of vegetated intertidal wetlands

Coastal intertidal wetlands located at the interface between land and sea (Bruland, 2008) are here defined as vegetated areas between the low- and high water mark where soil/sediment surfaces are covered temporarily with seawater by tidal cycles. The substratum in these intertidal areas will in the present marine ecological context be denoted sediment according to the guidelines given by Kristensen and Rabenhorst (2015). The extent of water cover and physical exposure as well as sediment characteristics and climate determine the structure of plant and animal communities present in coastal wetlands (Duke et al., 1998; Loebl et al., 2006; Raabe and Stumpf, 2016). These ecosystems are often located in estuaries and along physically protected coasts with large tidal amplitudes, forming environments with oscillating temperatures and salinities. The sediments are typically fine-grained due to the physical protection and even muddy where the vegetation efficiently traps suspended particles transported from the outside ocean by tides or from land by rivers (e.g., Chen et al., 2018). Plants and animals living under these conditions must possess special adaptations to withstand not only unpredictable variations in water submergence and temperature but also to cope with simultaneous stress from sediment anoxia and sulfidic conditions as well as osmotic stress due to salinity extremes (Lee, 2003; Welker et al., 2013; Naidoo, 2016). Most species in these communities are of marine origin, although freshwater or semi and fully terrestrial species may intrude in the less marine influenced upper intertidal zone (Nagelkerken et al., 2008; Rog and Cook, 2017). The vegetated wetlands dealt with here, mangrove forests and *Spartina* marshes, typically have an unvegetated outer boundary consisting of intertidal sand- or mudflats.

Mangrove forests are dominant wetlands along subtropical and tropical coasts (Spalding et al., 2010) with a current global cover of about 140,000 km^2 (Hamilton and Friess, 2018). They are composed of intertidal trees and scrubs growing primarily in sheltered muddy areas where quiescent conditions support their establishment (Chapman, 1976; Duke et al., 1998). There are worldwide about 70 true mangrove tree species within 32 genera from 18 families with highest species richness in Southeast Asia (Duke, 2017). Mangrove trees therefore represent an ecological rather than a taxonomic assemblage of plant species having common attributes that enable them to occupy harsh environments subjected to large variations in temperature, salinity, and oxygen levels. The trees and their root systems trap particles brought in by tides and rivers, consolidating the deposits on which they grow. The intertidal mangrove forest and its biota must be highly dynamic because they are affected by considerable stress associated with the periodic seawater inundation (Duke et al., 1998). A high abundance of benthic animals, particularly crabs, live on and in mangrove sediments with substantial bioturbation impact through their feeding and burrow construction activities (Cannicci et al., 2008; Kristensen, 2008). However, the biodiversity of benthic infauna such as crabs and other crustaceans associated with mangrove sediments varies considerably among continents; from over hundred species in Southeast Asia to less than half of that in East Africa and the Americas (Gillikin and Schubart, 2004; Lee et al., 2008). Nevertheless, the functional types of bioturbation are similar in all biogeographic regions, as are the typical densities of burrows (Kristensen et al., 2017).

Mangrove forests are in temperate climates replaced with saltmarshes dominated by a range of *Spartina* species. The genus *Spartina* was in 2014 subsumed into the genus *Sporobolus* and reassigned to the taxonomic status of a section (Petersen et al., 2014). However, in accord with Bortolus et al. (2019), we will here adhere to the opinion that *Spartina* is a genus, although relatively paucal with only approximately 15 species. *Spartina anglica* is today an invasive and dominating species in Europe and many other parts of the world. It is a fertile polyploid derived from a hybrid, *Spartina x townsendii*, of the North American *Spartina alterniflora* and the European *Spartina maritima* (Gray et al., 1991). Because of its phenotypic plasticity, *S. anglica* can tolerate a wider range of environmental conditions than its ancestors (Gray et al., 1991). It occurs in coherent saltmarsh meadows between the spring and neap high tide line and is seaward bordered by unvegetated tidal flats (Reise, 2005). Macrobenthic assemblages in saltmarshes show inconsistent responses to *S. anglica* invasions around the world. The infauna in European marshes is limited to few species of polychaetes, while ocypodid and sesarmid crabs are abundant in other parts of the world (Gribsholt and Kristensen, 2003; Raposa et al., 2018; Guimond et al., 2020). However, *Spartina* marshes generally appear inhospitable for benthic infauna (Gribsholt and Kristensen, 2003; Neira et al., 2005; Tang and Kristensen, 2010; Cutajar et al., 2012). Negative relationships between infauna and below-ground plant biomass suggests space limitation due to densely entangled roots and rhizomes below the sediment surface (Gribsholt and Kristensen, 2002; Brusati and Grosholz, 2006; Cutajar et al., 2012). Most burrowing activity in saltmarsh areas therefore occurs in nearby unvegetated patches and along creek banks (Gribsholt and Kristensen, 2002; Wasson et al., 2019; Guimond et al., 2020).

8.3 Greenhouse gas biogeochemistry

Biogeochemistry is defined as all biological, geological, chemical, and physical properties and processes within an ecosystem. These drive the circulation and transport of elements between the biotic and abiotic sphere and the exchange of organic and inorganic compounds within and among ecosystems (Canfield et al., 2005; Aller, 2014; Middelburg, 2018). The controlling role of sediment biogeochemistry for the formation, mitigation, and release of GHG in wetland sediments will briefly be outlined below.

Interstices of wetland sediments are often saturated with seawater both when inundated and air exposed. Temporal coverage with seawater defines intertidal biogeochemistry as it favors certain physical, chemical, biological, and morphological properties that differentiate these sediments from subtidal sediments, freshwater sediments, and upland soils. Water saturation promotes anoxic conditions within sediments, and the oxic layer where aerobic microbial respiration takes place is typically only a few mm thick at the sediment surface in coastal areas (Glud, 2008). This oxygen limitation is partly caused by the sluggish diffusive transport of oxygen in water-saturated sediments, which is more than 10,000 times slower than in air. Furthermore, oxygen penetration is precluded in intertidal wetland sediments due to high oxygen demand driven by rapid microbial degradation of deposited organic matter from primary producers such as benthic microalgae, macroalgae, mangrove trees, and saltmarsh plants, as well as accreted allochthonous material (Michaels and Zieman, 2013). The low oxygen penetration provides space for anoxic conditions and promotes facultative and obligate anaerobic microorganisms near the sediment surface.

Intertidal wetland sediments support a range of microbial communities and associated metabolic pathways that not only play a key role in the degradation of organic matter but also control the dynamics of GHGs. Most sediment microbes gain energy by breaking C bonds in organic matter. The easily degradable parts of organic C are decomposed to inorganic elements, whereas a large fraction of the refractory pools is often buried in the sediment. While aerobic respiration can readily degrade organic matter to inorganic products in one step, anaerobic degradation in wetland sediments is typically performed much slower and through at least three steps by mutualistic consortia of bacteria (Canfield et al., 2005):

(Step 1) Hydrolysis, where particulate organic matter is split and converted to large-molecular dissolved organic carbon (L-DOC) by hydrolyzing exoenzymes.
(Step 2) Fermentation, where L-DOC is fermented to smaller dissolved organic carbon moieties (S-DOC), such as acetate and other organic acids. This step of decomposition is performed both extracellularly and intracellularly.
(Step 3) Respiration, where a range of anaerobic microorganisms with specific respiration processes degrades S-DOC to CO_2. These processes are redox sensitive and depends on the availability of electron acceptors. This leads to a sequence of respiration processes competing for organic matter with an outcome depending on the energy yield of the process (Fig. 8.1).

The most important respiration processes in intertidal sediments are typically aerobic respiration and anaerobic sulfate reduction (SR), with a typical share of 30%–50% each (Alongi et al., 2000; Kristensen et al., 2011). Other pathways such as denitrification, Mn, and Fe respiration are in many cases unimportant for C cycling in wetlands due to limited availability of the electron acceptors NO_3^-, Mn(IV), and Fe(III) (Rivera-Monroy and Twilley, 1996; Kristensen et al., 2000). Yet, evidence suggests that the role of Fe respiration may be comparable to or higher than that of SR in Fe-rich wetland sediments (Kostka et al., 2002; Kristensen et al., 2017). The partitioning of electron acceptors within intertidal wetland sediments is also dependent on other

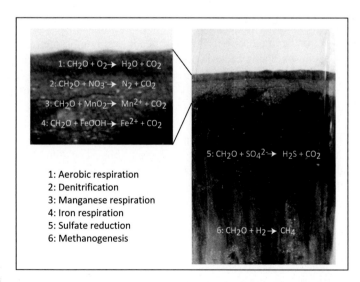

1: $CH_2O + O_2 \rightarrow H_2O + CO_2$
2: $CH_2O + NO_3 \rightarrow N_2 + CO_2$
3: $CH_2O + MnO_2 \rightarrow Mn^{2+} + CO_2$
4: $CH_2O + FeOOH \rightarrow Fe^{2+} + CO_2$

5: $CH_2O + SO_4^{2-} \rightarrow H_2S + CO_2$

1: Aerobic respiration
2: Denitrification
3: Manganese respiration
4: Iron respiration
5: Sulfate reduction
6: Methanogenesis

6: $CH_2O + H_2 \rightarrow CH_4$

FIG. 8.1

Sediment column from an intertidal wetland with idealized biogeochemical zonation of microbial respiration processes. The name of each process with decreasing energy yield in the vertical profile is indicated in the lower left corner.

factors. These include vegetation type and density, extent of water logging and flooding duration, and the intensity of faunal burrowing activities (Kristensen et al., 2017). Methanogenesis is the least energy-yielding step in the sedimentary metabolic pathway, and in marine environments, it is considered unimportant for the C cycling (Canfield et al., 2005). Thus, the generally slow degradation of refractory organic C (i.e., leaf litter) rich in structural components by the dominant anaerobic microorganisms is believed to be the primary reason for the high capacity of wetlands to retain and store C (Kristensen and Holmer, 2001; Alongi, 2012). Deposition and accretion of organic C in these sediments may therefore exceed the microbial degradation capacity.

CO_2 is quantitatively the most significant GHG with respect to global warming due to its high concentration in the atmosphere (IPCC, 2018). It has a global warming potential (GWP) of 1 and is the global warming reference gas. CO_2 is released as a product of almost all steps in the organic C degradation, including respiration and most fermentation processes (Canfield et al., 2005). Methane (CH_4) has a GWP of about 34 on a 100-year scale when climate carbon feedbacks are included (Gillett and Matthews, 2010; Myhre et al., 2013) and is the second-most important anthropogenic GHG (Whalen, 2005; Bloom et al., 2010). CH_4 is produced in sediments by methanogenesis, which is the ultimate step in anaerobic C degradation. The rarer nitrous oxide (N_2O) is the third most important GHG due to its very high 100-year GWP of about 298 including climate carbon feedbacks (Gillett and Matthews, 2010; Myhre et al., 2013). This gas is also produced in intertidal sediments as a by-product of microbial processes, primarily nitrification and denitrification (Davidson et al., 2000).

While CO_2 is generated by most microbial processes in wetlands, the production of N_2O and CH_4 is confined to specific processes in restricted redox zones of sediments. Ammonium-oxidizing bacteria produce N_2O near the sediment surface when O_2 availability and pH is low (Adviento-Borbe et al., 2006). Denitrifiers also generate N_2O in near-surface sediment, but only during periods of anoxia combined with high concentrations of NO_3^- and low pH (Lind et al., 2013). Since nitrification and denitrification can occur simultaneously in sediment micro-niches with varying O_2 availability, it is often difficult to identify the origin of N_2O, but denitrification is generally considered to be the largest contributor (Abbasi and Adams, 2000). The hydrological pulsation in intertidal wetlands rapidly modify the abovementioned environmental parameters (such as O_2 content, pH, and redox potential) in surface sediments and thereby modulates the biogeochemical processes responsible for the formation of N_2O.

Sulfate-reducing bacteria and methanogens compete for similar S-DOC compounds in deeper anoxic sediment. When S-DOC is limiting in the presence of sulfate, SR-producing CO_2 is superior due to higher energy yield and inhibits CH_4 production by methanogenesis (Mitsch and Gosselink, 2007). SR is usually more dominating in marine than freshwater wetlands due to the high level of sulfate in seawater (28 mM at a salinity of 35). Methanogenesis can therefore only occur in intertidal wetland sediments where sulfate is exhausted (Fig. 8.2).

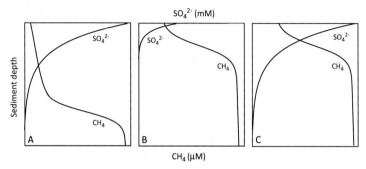

FIG. 8.2

Vertical profiles of sulfate (SO_4^{2-}) and methane (CH_4) in three types of intertidal wetlands. (A) An environment with full strength seawater where methanogenesis is hampered by sulfate reduction in the upper SO_4^{2-} containing part of the sediment. CH_4 produced below is consumed by methanogens while diffusing upwards. (B) CH_4 approaches the sediment surface in brackish intertidal areas with lower SO_4^{2-} concentration in the overlying water. (C) An environment with full strength seawater where the presence of excess reactive organic carbon in the sediment allows coexistence of sulfate reduction and methanogenesis leading to high near-surface CH_4 concentrations.

This typically occurs deep (>0.5 m) into the sediments with little to no overlap between the zone of SR and the zone of methanogenesis (Lovley and Phillips, 1987; Kuivila et al., 1989). However, SR and methanogenesis may occasionally "co-exist" in organic-rich sediments when S-DOC is in surplus (Holmer and Kristensen, 1994; Fonseca et al., 2019). CH_4 is released from intertidal wetlands to the atmosphere through two major pathways: Either by porewater diffusion from the deep anoxic sediment or by diffusion through air-filled biogenic structures during low tide. Transport via porewater diffusion is slow and a large proportion, if not all, of the produced CH_4 is oxidized by methanotrophs during transit in the sediment column (Segers, 1998). Methanotrophs cover a wide range of aerobic and anaerobic bacteria and archaea, including sulfate reducers. The emission of CH_4 from intertidal wetland sediments is therefore the net result of methanogenesis and methanotrophy, with biogenic structures as the major emission conduits in most cases.

8.4 Biogenic structures and their function

The term biogenic structure is in this chapter defined as any modification or extension of the sediment fabric by the presence of plant roots and benthic fauna burrows/tubes. These structures are dynamic in the sense that they are conduits for material exchange (i.e., GHGs) driven by the life processes of the organisms they host and the external environmental conditions. There are obvious morphological and functional differences between the two major types of biogenic structures, roots, and burrows/tubes (Fig. 8.3). Roots are an integrated living part of plants with essential functions as anchors to prevent uprooting as well as for absorption and vertical transport of gases, solutes, and water (Hogarth, 2015). Burrows/tubes, on the other hand, are external nonliving sheaths formed by their infaunal inhabitants for

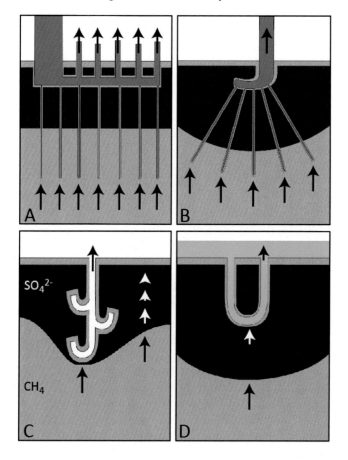

FIG. 8.3

Schematic drawing of biogenic structures in intertidal wetlands. *Brown* color indicates oxidized sediment; *black* color indicates the sulfate reduction zone; and *gray* color indicates the zone of methanogenesis. Arrows show the transport direction of CH_4. A shows a mangrove tree trunk with horizontal cable roots and pneumatophores. The deep roots reach the zone of methanogenesis. B shows a *Spartina* stem with horizontal rhizome and deep roots reaching the zone of methanogenesis. C shows an air-filled crab burrow at low tide. The lower part of the burrow approaches the zone of methanogenesis and allows rapid transport of CH_4 that otherwise would have been lost by oxidation during diffusive transit in the sulfate reduction zone. D shows an open-ended polychaete burrow while ventilated during inundation. The sulfate reduction zone extends deep into the sediment and allows only traces of CH_4 to enter the burrow.

protection, feeding, and respiration purposes (Kristensen et al., 2012). Both plant roots and burrows/tubes have confounding impact on rates and distribution of sediment biogeochemistry, and as such strongly affect C and N (nitrogen) cycling as well as GHG dynamics (Kristensen and Alongi, 2006; Ferreira et al., 2007; Bertics et al., 2012).

8.4.1 Plant roots

Submerged plants have roots specially adapted for the anoxic and often sulfidic conditions in sediment flooded with seawater. Roots of mangrove and saltmarsh plants have well-developed aerenchyma tissue that may account for up to 70% of the root volume (Hogarth, 2015). This air-filled tissue facilitates a rapid supply of oxygen from aboveground sources to the belowground tissues, where it may leak across root surfaces and oxidize the surrounding sediment with strong effects on biogeochemical functioning (Koop-Jakobsen and Wenzhöfer, 2015). Mangrove forests have trees with a variety of aerial and subsurface root morphology (Srikanth et al., 2016) providing functionalities related to, for example, sedimentation potential (Krauss et al., 2003) and gas exchange (Kitaya et al., 2002). Thus, *Kandelia* trees have roots that mostly grow under the sediment surface, while *Rhizophora* trees have extended prop or stilt roots that provide physical support in soft mud. Other trees, like *Avicennia* and *Sonneratia*, have distinctive pneumatophores that arise vertically at high density from subsurface cable roots spreading far away from the tree (Fig. 8.3A). Knee roots of *Ceriops* and *Bruguiera*, on the other hand, are cable roots that bend upward and curve down forming a loop above the sediment. Prop roots, pneumatophores, and knee roots connect the subsurface root system directly with the atmosphere through lenticels at the surface allowing rapid exchange of gases, such as O_2, CO_2, and CH_4. Diffusive transport of these gases into and out of deep sediment layers through the air-filled aerenchyma tissue of roots is much faster than in porewater of the surrounding sediment (Purvaja et al., 2004; Purnobasuki and Suzuki, 2005; Penha-Lopes et al., 2010). Previous studies have indicated that pneumatophores are more efficient in exchanging gases than other forms of mangrove roots (Fig. 8.3A, Kreuzwieser et al., 2003), and the presence of pneumatophores can increase the exchange of gases several fold in some tropical mangrove wetlands (Kitaya et al., 2002; Kristensen et al., 2008; Penha-Lopes et al., 2010).

Spartina and other saltmarsh plants (e.g., *Phragmites*, *Suaeda*, and *Cyperus*) also have stems and roots with aerenchyma that allows rapid exchange of gases between the atmosphere and sediments. The fibrous and relatively thick roots of *Spartina* typically extend 40–50 cm into the sediment from the vertical stem with horizontal rhizomes used for vegetative reproduction (Fig. 8.3B) (Wang et al., 2015; Bernal et al., 2017). This root structure has high capacity for transporting oxygen to underground roots and rhizospheres supporting rapid aerobic respiration and CO_2 production (Tong et al., 2012). Accordingly, Koop-Jakobsen and Wenzhöfer (2015) showed that *S. anglica* supports a $\sim$1.5-mm oxic zone radially around its roots. The oxygen is partly supplied by photosynthetic oxygen production and partly by downward diffusive transport of atmospheric oxygen. Concurrently, the CO_2 generated by root respiration and microbial respiration in the surrounding sediment are rapidly transported upward through the aerenchyma back to the atmosphere (Fig. 8.3B; Koop-Jakobsen et al., 2018). In locations where the roots penetrate deep into sediment layers with methanogenesis, CH_4 emission is also enhanced considerably by diffusion through the root aerenchyma (Laanbroek, 2010).

8.4.2 Infaunal burrows and tubes

Numerous benthic fauna species are burrow or tube dwellers in sheltered intertidal wetland environments. These include many polychaete and crustacean species with decapods being particularly important in mangrove environments (Kristensen, 2008). Burrows or tubes are constructed by benthic animals for feeding, respiration, or as a refuge against adverse environmental conditions and predators (Botto and Iribarne, 2000; Thongtham and Kristensen, 2003). The biogeochemical consequences of infaunal organisms strongly depend on the morphology of the structures they construct and their burrow-dwelling behavior. Burrows and tubes can be blind- or open-ended with galleries divided into one or several branches (Fig. 8.3C and D), and depending on the particle reworking activity, the inhabitants can be defined as biodiffusors, upward conveyors, downward conveyors, and regenerators (Kristensen et al., 2012). Blind-ended burrows with one opening are not well ventilated when inundated with water, particularly in nonpermeable sediments, while open-ended burrows with two or more openings can be flushed easily with overlying water from one end to the other (Kristensen et al., 2012). The ventilation rate therefore controls the biogeochemical influence of the burrow structure and associated solute and gas transport (Kristensen et al., 2011; Quintana et al., 2018). When burrows are exposed to air at low tide, the galleries may drain in permeable sediments, providing direct connectivity and rapid gas transport in air between the atmosphere and deep sediment layers (De la Iglesia et al., 1994; Stieglitz et al., 2013).

Sesarmid and ocypodid crabs dominate the benthic fauna in mangrove environments worldwide (Kristensen, 2008), while mud lobsters (*Thalassina* sp.) are ubiquitous features in Southeast Asia (Moh et al., 2015). They are all regenerators inhabiting blind-ended and nonventilated burrows that commonly drain during low tide. Burrow morphology and depth vary strongly among crab species where sesarmids, like *Neoepisesarma versicolor* in southeast Asia and *Sesarma rectum* in the Americas, build a network consisting of simple burrows with several branches and openings (Fig. 8.3C), while ocypodids, such as fiddler crabs and the swamp ghost crab (*Ucides cordatus*), build single shafted "J" shaped burrows (Kristensen, 2008). The burrow depth ranges from 10 to 40 cm for *Uca* spp. and sesarmid crabs and up to 2 m for *U. cordatus* (Pülmanns et al., 2016) and *Thalassina* sp. (Moh et al., 2015). These decapods not only modify the mangrove topography and biogeochemistry by their burrow excavation but also the exchange of solutes and gases from deep sediments to the overlying water and atmosphere by tidal flushing of burrows (Stieglitz et al., 2013).

Since temperate *Spartina* marshes in Europe generally have few large burrow-dwelling infauna (Cottet et al., 2007; Tang and Kristensen, 2010), the role of bioturbation for sediment biogeochemistry and exchange of solutes may be limited in these environments. The roots and rhizomes of *Spartina* occupy most of the space and prevent the activities of burrow-dwelling animals (Pillay et al., 2011). However, the polychaete *Hediste diversicolor* is common in many European saltmarsh environments, particularly in unvegetated patches within the marshes. It typically lives in relatively shallow burrows, often open-ended, that are intensely ventilated during high tide and contain almost stagnant water when nondraining flats are exposed to air (Fig. 8.3D, Gribsholt and Kristensen, 2003). The worm has adaptations to survive by anaerobic metabolism for extended periods of anoxia when ventilation is prevented (Schöttler, 1979; Wohlgemuth et al., 2000; Welker et al., 2013). When the burrows are ventilated during inundation, oxygen and other electron acceptors, such as SO_4^-, are rapidly transported to deep subsurface sediment and metabolites, such as H_2S, are flushed out of the sediment (Banta et al., 1999). Burrows with 2–3 mm oxidized walls may then extent to at least 10–20 cm depth driving SO_4^{2-} penetration substantially deeper (Kristensen et al., 2011). Some saltmarshes in South and North America, Africa, Asia, and Australia instead support sesarmid and ocypodid crabs as well as other decapods, and they form burrow structures of similar shape and function as those in mangrove forests (Escapa et al., 2008; Wang et al., 2015; Guimond et al., 2020). These may, as in mangrove forests, affect the biogeochemistry by enhancing sediment aeration as well as transport of solutes and gases to and from the overlying water and atmosphere (Fanjul et al., 2015; Xiao et al., 2019). However, it must be emphasized that crab and polychaete burrowing activity in saltmarshes is by far highest in unvegetated patches and along creekbanks (Gribsholt and Kristensen, 2003; Raposa et al., 2018; Guimond et al., 2020).

8.5 Mangrove carbon cycling and biogenic greenhouse gas dynamics

Vegetated wetlands, such as mangrove forests, typically increase sediment elevations by vertical accretion through belowground root production as well as aboveground autochthonous litter and allochthonous sediment deposition. By doing so, they facilitate continued burial of organic C (Lovelock et al., 2014; Kelleway et al., 2016; Lamont et al., 2020). The reported long-term sequestration capacity of C in mangrove ecosystems generally exceeds that of most other marine and terrestrial ecosystems (Ouyang and Lee, 2020). However, many studies do not fully consider the role of biogenic structures and GHG emissions when formulating the net climate effect of mangrove forests. Based on a large range of published reports on mangrove net primary production (NPP), C sequestration, and GHG emissions, this chapter will provide new global average C and GHG budgets that fully include the impact of biogenic structures.

The global average living mangrove tree biomass of 104 Mg C ha^{-1} with a range of 93–115 Mg C ha^{-1} (Table 8.1; Hutchison et al., 2013; Kauffman et al., 2020; Ouyang and Lee, 2020) supports a global average NPP of 13.7 Mg C ha^{-1} yr^{-1} with a range from 11 to 16 Mg C ha^{-1} yr^{-1} (Table 8.1; Bouillon et al., 2008; Alongi, 2012; Alongi and Mukhopadhyay, 2015). Trees are the dominant primary producers in mangrove forests with only limited contribution from macroalgae and microphytobenthos at the shaded sediment surface under the canopy (Alongi, 2012; Peer et al., 2019). The high NPP combined with the ability of mangrove forests to trap allochthonous sediment particles both from water discharged upstream and the coastal ocean by tides promote a long-term C storage in the anoxic sedimentary environment. Past compilations of living biomass and sediment organic matter to a depth of about 1 m have shown a wide range of mangrove ecosystem C stocks (72–2139 Mg C ha^{-1}) (Kauffman et al., 2020; Ouyang and Lee, 2020) that are controlled by geographic, climatic, and environmental variations. However, recent global estimates of C stocks provide a much narrower range of 455 to 956 Mg C ha^{-1} with an average of 663 Mg C ha^{-1} (Table 8.1; Hutchison et al., 2013; Alongi and Mukhopadhyay, 2015; Atwood et al., 2017; Kauffman et al., 2020; Ouyang and Lee, 2020). These and other C budget estimates are, however, strongly dependent on the methodology applied with respect to sampling design, collection methods, data interpretation, and reporting protocols (Owers et al., 2018; Ouyang and Lee, 2020). There are only few reliable estimates of allochthonous C captured as fine particles, but Alongi (2014) reasoned that up to 1 Mg C ha^{-1} yr^{-1} of the global average burial may originate from allochthonous sources. Thus, to balance the budget, NPP and allochthonous input at a total rate of 14.7 Mg C ha^{-1} yr^{-1} (=335 mmol m^{-2} d^{-1}) must be counterbalanced by C sequestration, ecosystem respiration by fauna, and through microbial degradation as well as export of C in the form of particulate organic C (POC), dissolved organic C (DOC), and dissolved inorganic C (DIC).

Bouillon et al. (2008) and Alongi (2009) found that known C sinks (sequestration, sediment CO_2 efflux, and export) together accounted for less than half of the estimated NPP, thus leaving a surprisingly large part of the mangrove production unaccounted for. Various budget estimates have provided global net sequestration in the form of buried organic C by mangrove ecosystems consistently within the range of

Table 8.1 Global carbon stocks (total and living to about 1 m sediment depth) as well as annual net primary production (NPP) and carbon sequestration in vegetated intertidal wetlands.

Wetland type	Total C stock (Mg C ha^{-1})	Living C stock (Mg C ha^{-1})	NPP (Mg C ha^{-1} yr^{-1})	C sequestration (Mg C ha^{-1} yr^{-1})
Mangrove forest	663 (229)	104 (11)	13.7 (1.6)	1.66 (0.37)
Saltmarsh	186 (82)	12 (10)	13.5 (3.5)	2.08 (0.34)

Data are compiled from the sources reporting global estimates as mentioned in the text and given as average ±(SD).

1.2 to 2.0 Mg C ha^{-1} yr^{-1} with an average of 1.66 Mg C ha^{-1} yr^{-1} (Table 8.1; Bouillon et al., 2008; Alongi, 2012; Hutchison et al., 2013; Alongi, 2014; Alongi and Mukhopadhyay, 2015; Ouyang et al., 2017; Pérez et al., 2018), which is equivalent to ~11% of the net C input (Fig. 8.4). Respiration by sediments devoid of biogenic structures, as compiled in this chapter, provide an average benthic CO_2 emission of 68 mmol m^{-2} d^{-1} (Table 8.2). This is equivalent to 3.0 Mg C ha^{-1} yr^{-1} or about 20% of the presently compiled C input (Fig. 8.4) and comparable to that reported by Bouillon et al. (2008). Microphytobenthos assimilation may temporarily offset CO_2 emissions in mangrove habitats, but only to a limited extent when long-term integrated over light-dark periods (Alongi, 2012; Chen et al., 2019). However, CO_2 sink components of the global mangrove C budget are severely underestimated when ignoring emissions via burrows and pneumatophores (Kristensen et al., 2008). Crab burrows can greatly enhance the surface area of the sediment-air or sediment-water interface where excessive aerobic activity and release of microbially- and faunal-derived CO_2 can occur (Thongtham and Kristensen, 2003). Similarly, emission of microbially derived CO_2 from oxidized rhizospheres is greatly enhanced due to the chimney effect of pneumatophores (Kristensen et al., 2008). A fraction (i.e., ~30%) of the contribution from pneumatophores is derived from root

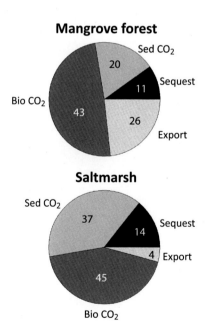

FIG. 8.4

Pie charts showing the carbon budget of vegetated wetland sediments. "Sed CO_2" is diffusive CO_2 release; "Bio CO_2" is CO_2 released via biogenic structures (roots and burrows/tubes); "Export" is the tidal loss of organic and inorganic carbon; "Sequest" is the long-lasting carbon burial within the sediment.

Table 8.2 GHG emission from air exposed and darkened mangrove sediments without biogenic structures (−) and with pneumatophores (+Pneu) and burrows (+Burr).

Location	Vegetation	Biogenic	CO_2 emission (mmol m^{-2} d^{-1})	CH_4 emission (µmol m^{-2} d^{-1})	N_2O emission (µmol m^{-2} d^{-1})	Ref.
Luggage P, Australia	*Avicennia*	−		101	8.6	1
Myora Spr, Australia	*Avicennia*	−		33	0.7	1
Port Douglas, Australia	*Avicennia*	−		36	1.0	1
Jiulong, China	*Kandelia*	−	37	58	19.0	2
Zhangjiang, China	*Kandelia*	−	16	59		3
Bhitarkanika, India	*Avicennia*	−		135	32.0	4
Mtoni, Tanzania	*Sonneratia*	−	40	55		5
Ras Dege, Tanzania	*Avicennia*	−	45	12		5
Nobrega, Brazil	*Laguncularia*	−	156	252		6
Olaria, Brazil	*Laguncularia*	−	111	218		6
avg			**68**	**96**	**12.3**	
sd			**54**	**82**	**13.3**	
Luggage P, Australia	*Avicennia*	+Pneu		586	3.7	1
Myora Spr, Australia	*Avicennia*	+Pneu		205	1.3	1
Port Douglas, Australia	*Avicennia*	+Pneu		204	1.3	1
Jiulong, China	*Sonneratia*	+Pneu	65	117	24	2
Zhangjiang, China	*Avicennia*	+Pneu	13	647		3
Muthupet, India	*Avicennia*	+Pneu		1569	14	7
Pichavaram, India	*Avicennia*	+Pneu		463		8
Mtoni, Tanzania	*Sonneratia*	+Pneu	193	182		5
Ras Dege, Tanzania	*Avicennia*	+Pneu	156	221		5
Nobrega, Brazil	*Laguncularia*	+Pneu	334	483		6
Olaria, Brazil	*Laguncularia*	+Pneu	247	517		6
avg			**168**	**472**	**8.9**	
sd			**118**	**408**	**10.0**	

Sulawesi, Indonesia	Ceriops	+Burr	174	573	84	8
Sulawesi, Indonesia	Rhizophora	+Burr	50	8692	17	8
Mtoni, Tanzania	Sonneratia	+Burr	193			5
Ras Dege, Tanzania	Avicennia	+Burr	103	78		5
Nobrega, Brazil	Laguncularia	+Burr	233	1781		6
Olaria, Brazil	Laguncularia	+Burr	212	847		6
avg			*161*	*2394*	*50.5*	
sd			*70*	*3575*	*47.4*	

Data are from various studies in Australia, Asia, Africa and South America from mangrove forests with different dominant tree species.
1. Kreuzwieser et al. (2003); 2. Chen et al. (2015); 3. Gao et al. (2018); 4. Chauhan et al. (2015); 5. Kristensen et al. (2008); 6. Kristensen et al. (unpublished); 7. Krithika et al. (2008); 8. Purvaja et al. (2004).

respiration (Ouyang et al., 2018) and should not be considered a C sink as it has already been accounted for in the NPP calculations. By considering these contributions and considerations, the presently compiled data show that sediment with pneumatophores and crab burrows emits 118 and 161 mmol CO_2 m^{-2} d^{-1}, respectively (Table 8.2). This is 1.7- to 2.4-fold increase by these biogenic structures and in total they return about 6.3 Mg C ha^{-1} yr^{-1} extra to the atmosphere, explaining about 43% of the global average C input (Fig. 8.4). The remaining 26% must consist of up to 3.8 Mg C ha^{-1} yr^{-1} export of POC, DOC, and DIC to the adjacent ocean (Bouillon et al., 2008; Adame and Lovelock, 2011; Alongi, 2014). The benthic CO_2 emission is therefore the sink that returns about ⅔ of the C input, leaving the rest for net C sequestration and export (Sasmito et al., 2020). The C export fraction is more than twice as high as sequestration by organic C burial in mangrove forests. Most of the C export is in the form of DIC (Maher et al., 2018; Taillardat et al., 2018), which may remain dissolved as total alkalinity in the ocean for millennia (Reithmaier et al., 2020). Total alkalinity therefore represents an additional long-term mangrove sink for atmospheric C of comparable magnitude to sedimentary C burial.

The emission of other GHG than CO_2 from mangrove sediments devoid of biogenic structures is rather modest due to the slow diffusive transport and rapid removal by microbial reactions within the sediment (Kreuzwieser et al., 2003; Kristensen et al., 2008). Accordingly, the global average CH_4 emission of 96 μmol m^{-2} d^{-1} from sediments without biogenic structures is about 700 times lower than CO_2 emission (Table 8.2). However, the global average covers a wide range of published rates (Table 8.2) indicating the important role of local environmental conditions, such as salinity, organic matter availability, and sediment permeability (Al-Haj and Fulweiler, 2020). Thus, the concentration and penetration depth of sulfate in wetland sediments may strongly influence methanogenesis (Fig. 8.5), leading to highly variable levels of CH_4 within and emissions from the porewater. Conversely, CH_4 emission from mangrove sediments inhabited by biogenic structures is stimulated considerably, i.e., by a factor of more than 5 via pneumatophores and a factor of up to 25 via crab burrows (Table 8.2). These biogenic structures act as conduits for rapid CH_4 transport via aerenchyma tissue in roots and air-filled burrows when they extend into or near the zone of methanogenesis in deep sediment layers devoid of sulfate (Fig. 8.3). The importance of these biogenic structures is clearly substantiated from significant correlations between, for example, the abundance of pneumatophores and CH_4 emission (Table 8.3). In some cases, even shallower structures (e.g., *Uca* burrows) may reach near-surface CH_4 hotspots (Fig. 8.2) due to high availability of labile organic C that allows coexistence of methanogens and sulfate reducers (Chuang et al., 2016). The larger impact on CH_4 emission by burrows than roots may be related to a free passage through cm-wide burrows compared with narrow aerenchyma tissues and small lenticels in roots.

N_2O emission from mangrove sediments is low (Table 8.2) due to the general scarcity of nitrogen in many mangrove environments (Feller et al., 2007; Murray et al., 2020). A requirement for generation of N_2O is the presence of NH_4^+ and NO_3^-, which are often very low in sediment porewaters of nitrogen-limited mangrove

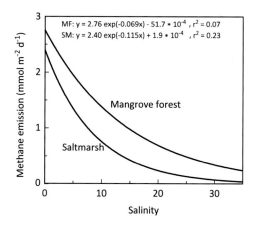

FIG. 8.5

The CH_4 emission from mangrove forest and saltmarsh sediments as a function of overlying water salinity.

Redrawn from Al-Haj, A.N., Fulweiler, R.W., 2020. A synthesis of methane emissions from shallow vegetated coastal ecosystems. Glob. Chang. Biol. 26, 2988–3005.

forests (Kristensen et al., 1998; Murray et al., 2020). Accordingly, the impact of pneumatophores on N_2O emission is weak (Tables 8.2 & 8.3), while burrows may enhance the flux by a factor of up to 5. The larger stimulation by burrows is probably driven by the intimate contact between the temporarily air-filled burrow lumen and surrounding sediment, where N_2O is generated by coupled nitrification and denitrification along a large surface area of burrow walls (Stief, 2013). Nevertheless, the emission of N_2O from mangrove sediments is only 10%–20% of the CH_4 emission, and in the presence of burrows it may even drop down to 1% (Table 8.2).

Table 8.3 Linear relationships ($y = ax + b$) reported between abundance of *Avicennia* pneumatophores (x, m^{-2}) and sediment emission of CH_4 and N_2O (y, $\mu mol\, m^{-2}\, d^{-1}$).

Location	a	b	r^2	Ref.
CH_4				
Moreton Bay, Australia	7.1	486	0.45	Kreuzwieser et al. (2003)
Pichavaram, India	28.1	−900	0.70	Purvaja et al. (2004)
Muthupet, India	5.3	1464	0.96	Krithika et al. (2008)
N_2O				
Moreton Bay, Australia	0.0055	2.67	0.04	Kreuzwieser et al. (2003)
Muthupet, India	0.0562	−4.14	0.81	Krithika et al. (2008)

The present data compilation allows a rough estimate of all net transfers involving C and GHG in mangrove forests (Fig. 8.6), and thus the overall effect on global warming. It must be emphasized that the wide range and large variability of the data used in the calculations and the fact that most rates were measured during exposure at low tide renders the outcome with a high degree of uncertainty. Anyway, C burial in mangrove sediments is equivalent to a storage of around 6.1 Mg CO_2 ha^{-1} yr^{-1}, which has led to the general opinion that mangrove forests have a strong mitigation effect on climate change (e.g., Mcleod et al., 2011; Cusack et al., 2018). However, when the contribution from biogenic structures are considered and the compiled CH_4 and N_2O emission are converted to CO_2 using the global warming potential of these gases reported above, the GHG balance of mangrove forests actually turns negative by 1.8 Mg CO_2 ha^{-1} yr^{-1} (Table 8.5). C sequestration through burial into mangrove forests therefore cannot keep pace with the GHG emission. However, due to the large variability in the data, there is a wide

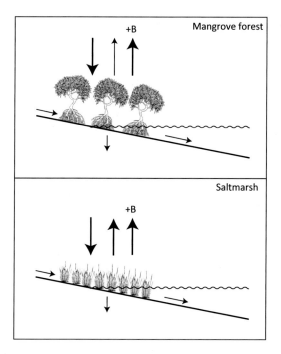

FIG. 8.6

Schematic drawing of mangrove forest and saltmarsh intertidal wetlands with indications of major carbon transfers to and from the atmosphere and adjacent ecosystems. Arrow thickness is proportional to global average rates of transfer in CO_2 equivalents. Inputs are net primary production and particle import, and outputs are emission of GHG through diffusive release by bare sediment plus excess release by biogenic structures (+B), carbon sequestration by burial in the sediment, and export to the adjacent ocean.

range in the mitigation effects among mangrove forests worldwide. Furthermore, the recent discovery that total alkalinity export to the ocean from mangrove ecosystems represents a long-term sink for atmospheric carbon (Reithmaier et al., 2020) may balance the GHG budget. In any case, many of these forests should be considered more or less climate neutral, as also noted recently by Al-Haj and Fulweiler (2020).

8.6 Saltmarsh carbon cycling and biogenic greenhouse gas dynamics

Vegetation cover is important for the capacity of intertidal ecosystems, like salt-marshes, to retain organic C, but the actual C dynamics and storage is also dependent on vegetation type (Kelleway et al., 2016). Since *Spartina* marshes are the most widespread vegetated wetlands in temperate and subtropical zones, this chapter will, based on published reports, focus on C and GHG budgets in *Spartina*-vegetated ecosystems, with emphasis on the role of biogenic structures.

At first glance, it is remarkable that the global average saltmarsh NPP of $13.5\,Mg\ C\ ha^{-1}\ yr^{-1}$ with a range of $11–16\,Mg\ C\ ha^{-1}\ yr^{-1}$ (Table 8.1; Duarte et al., 2013; Alongi, 2014) is similar to that of mangrove forests when the saltmarsh global average standing living biomass of $12\,Mg\ C\ ha^{-1}$ with a range of $4–26\,Mg\ C\ ha^{-1}$ (Table 8.1; Couto et al., 2013; Van de Broek et al., 2017; Zhang et al., 2017; Santini et al., 2019) is less than 10% of than in mangrove forests. However, part of the explanation is that most mangrove biomass consists of photosynthetically inactive wood mass, while all aboveground parts of marsh grasses, like *Spartina* spp., contribute to their primary production. Like in mangrove forests, the role of micro-phytobenthos is generally limited in *Spartina*-marshes due to light limitation under the dense canopy (Gribsholt and Kristensen, 2002). Saltmarshes grow in a variety of sedimentary landscapes and may accumulate C-rich deposits of autochthonous and allochthonous origin (Temmerman et al., 2005; Chen et al., 2018). However, the global average total C stocks in saltmarshes of $186\,Mg\ C\ ha^{-1}$ with a range from 100 to $300\,Mg\ C\ ha^{-1}$ are generally lower than those in mangrove forests (Table 8.1; Duarte et al., 2013; Kelleway et al., 2017; Van Ardenne et al., 2018; Santini et al., 2019). This is caused by an overall higher C content in deposits under mangrove forests than saltmarshes (Duarte et al., 2013) combined with a shorter longevity of saltmarshes than mangrove forests and a generally shallower depth of saltmarsh deposits (Kelleway et al., 2017). Saltmarshes have higher sediment particle trapping ability than mangrove forests due to both the hydraulic sediment trapping and the direct trapping by vegetation surface (Chen et al., 2018). However, there are to our knowledge no published data that can support a global average estimate of allochthonous C captured as fine particles. Nevertheless, by assuming a particle capture of $1–2\,Mg\ C\ ha^{-1}\ yr^{-1}$, the global average organic C input will be in the order of $\sim15\,Mg\ C\ ha^{-1}\ yr^{-1}$ ($=350\,mmol\,m^{-2}\,d^{-1}$). C sinks in the form of sequestration within sediments, CO_2 emission via benthic respiration, and export of POC, DOC, and DIC from saltmarshes to adjacent environments must be of the same or slightly higher magnitude than that in mangrove forests to counterbalance all organic C sources. Any differences between the effects of mangrove forests and saltmarshes on the capacity of C storage must therefore be related to the physical structure of the vegetation and environmental conditions.

C sinks in saltmarshes are highly dependent on geomorphological, climatic, and hydrographical settings (Cornell et al., 2007; Bu et al., 2019). Thus, vegetation dynamics and activity vary both spatially with intertidal elevation (Silvestri et al., 2005) and temporally with the seasons in temperate and subtropical saltmarshes. Nevertheless, the global estimates of C sequestration in saltmarshes are well confined with an average $2.08\,\mathrm{Mg\,C\,ha^{-1}\,yr^{-1}}$ (range from 1.5 to $2.4\,\mathrm{Mg\,C\,ha^{-1}\,yr^{-1}}$) (Table 8.1; Chmura et al., 2003; Alongi, 2012; Duarte et al., 2013; Alongi, 2014; Ouyang et al., 2017) corresponding to 14% of the C input (Fig. 8.4), which is slightly higher than that of mangrove forests. On the other hand, the capricious environmental conditions are evident from the large variability of measured CO_2 emissions in *Spartina* marshes, ranging from 50 to $500\,\mathrm{mmol\,m^{-2}\,d^{-1}}$ (Table 8.4). For example, the higher oxidation level in rarely inundated sandy environments where sediments are well drained may lead to the rapid loss of organic C as CO_2 (Saintilan et al., 2013; Kelleway et al., 2016), whereas lower oxidation in fine-grained impermeable sediments with more frequent inundation hampers organic C oxidation and CO_2 emission.

CO_2 emission in saltmarshes is also dependent on the vegetation density and, not the least, on how measurements were performed. Dark rates of CO_2 release are 2.6 times higher when the measuring units (e.g., benthic chambers) are large enough to include the vegetation canopy. Furthermore, the general lack of information about the abundance of crab and polychaete burrows in saltmarshes hampers a reliable evaluation of burrow contribution to CO_2 emissions (Guimond et al., 2020). The results available suggest that organic C is efficiently oxidized in crab bioturbated sediment and may quickly be exported out of the sediment as CO_2 (Fanjul et al., 2015). It is anticipated, though, that the role of burrows is relatively small in the densely vegetated part of saltmarshes due to the limited space for infauna (Gribsholt and Kristensen, 2002, 2003; Raposa et al., 2018). In any case, the global average sediment release is $5.6\,\mathrm{Mg\,C\,ha^{-1}\,yr^{-1}}$ via porewater diffusion of CO_2 to the atmosphere (~37% of the C input; Fig. 8.4), while CO_2 channeled by the vegetation delivers another $9.6\,\mathrm{Mg\,C\,ha^{-1}\,yr^{-1}}$ (Table 8.4). About 30% or more of this vegetation mediated emission is respiration by the plants themselves (Dai and Wiegert, 1996; Neubauer et al., 2000), and as such should not be included as a C sink because it is included in the NPP estimate. Nevertheless, a large proportion of the CO_2 emitted via plants, $6.7\,\mathrm{Mg\,C\,ha^{-1}\,yr^{-1}}$ (~45% of the C input; Fig. 8.4) is derived from microbial C oxidation within the sediment. Some CO_2 must be emitted via the few burrows present, but the extent of this contribution is presently unknown (Guimond et al., 2020). It is remarkable, though, that the global average C sequestration (Table 8.3) and CO_2 emission (Table 8.4) estimated for the vegetated saltmarshes are similar to the abovementioned C inputs. The loss of POC, DOC, and DIC from saltmarshes must therefore be limited and considerably lower than

Table 8.4 GHG emission from air exposed and darkened *Spartina* sp. saltmarsh sediments with (Yes) and without (No) the canopy present in incubation chambers.

Location	Canopy	CO_2 emission (mmol m^{-2} d^{-1})	CH_4 emission (µmol m^{-2} d^{-1})	N_2O emission (µmol m^{-2} d^{-1})	Ref.
Dongtang marsh, China	Yes	444	12,280	7.8	1
Shanyutan, China	Yes		16,378		2
Sheyang, China	Yes		3355		3
Wangang, China	Yes		2070		4
Yancheng, China	Yes		2145		5
Yancheng, China	Yes	523	59	11.9	6
Yancheng, China	Yes	368	280	9.0	7
Yancheng, China	Yes		291	−3.2	8
Yangtze Estuary, China	Yes		3316		9
Dovey Estuary, Wales	Yes	268	103	93.6	10
Bay of Fundy, Canada	Yes	240	41	3.0	11
Bay of Fundy, Canada	Yes	308	18	−1.4	12
Dauphin Island, USA	Yes	243	975		14
Waquoit Bay, USA	Yes	279	99		15
St. Jones Reserve, USA	Yes	467	5884		16
avg		*348*	*3152*	*17.2*	
sd		*106*	*4905*	*34.1*	
Jiulong Estuary, China	No	56	89	17.2	17
Westerschelde, Holland	No	203			18
Blackwater, England	No		36	2.0	19
Blackwater, England	No	144	30	−10.9	20
Wadden Sea, Denmark	No	108			21
Cocodrie, USA	No		5505		22
Provincetown, USA	No		−26	0.5	23
Rowley, USA	No		15	0.9	23
avg		*128*	*942*	*1.9*	
sd		*62*	*2236*	*10.0*	

Data are from various studies in Asia, Europe and North America.

1. Sheng et al. (2014); 2. Tong et al. (2012); 3. Xiang et al. (2015); 4. Zhang and Ding (2011); 5. Wang and Wang (2017); 6. Xinwanghao et al. (2017); 7. Xu et al. (2014); 8. Yuan et al. (2015); 9. Liu et al. (2019); 10. Dausse et al. (2012); 11. Chmura et al. (2011); 12. Chmura et al. (2016); 13. Weston et al. (2014); 14. Wilson et al. (2015); 15. Abdul-Aziz et al. (2018); 16. Trifunovic et al. (2020); 17. Chen et al. (2015); 18. Gribsholt and Kristensen (2003); 19. Adams et al. (2012); 20. Burden et al. (2013); 21.; Gribsholt and Kristensen (2002); 22. Rietl et al. (2017); 23. Emery et al. (2019)

reported for mangrove forests. Accordingly, Hyndes et al. (2014) found that the global average C export from saltmarshes only accounts for $0.6\,\mathrm{Mg\,C\,ha^{-1}\,yr^{-1}}$, which is equivalent to only 4% of the C input (Fig. 8.4). However, the export varies considerably among wetlands depending on environmental conditions driven by river discharge, tidal amplitude, geomorphology, creek bank bioturbation, and hydrodynamics as well as age of the wetland.

The emission of GHG other than CO_2 from saltmarshes varies strongly and is probably dependent on climate, vegetation type, sediment type, and tidal elevation. Thus, the CH_4 emissions range from 15 to more than $16{,}000\,\mathrm{\mu mol\,m^{-2}\,d^{-1}}$ in *Spartina* marshes (Table 8.4). The underlying mechanisms for these large differences are not well known, but the methodology applied seems to explain some of the variability, since the vegetation canopy is not always included in the measurements. The role of vegetation is clearly evident from the significantly increased CH_4 emissions after *Spartina* invasion of Chinese coastal wetlands (Tong et al., 2012; Chen et al., 2015; Yuan et al., 2015). Accordingly, the compiled data in Table 8.4 show that measurements from saltmarshes where the canopy is excluded provide a global average CH_4 emission of $942\,\mathrm{\mu mol\,m^{-2}\,d^{-1}}$, while 3.3 times higher emissions are evident in the presence of the canopy. Another possible cause for the highly variable CH_4 emissions is the salinity of tidal waters. Low saline water in river-dominated estuaries may promote excessive CH_4 generation and emission when SR is hampered by SO_4^{2-} limitation within the sediment (Fig. 8.2; Poffenbarger et al., 2011). Furthermore, wetlands are spatially heterogeneous within the intertidal zone, with low marsh areas experiencing higher frequencies of tidal inundation than high marshes (Tang and Kristensen, 2010). Such differences in water cover may affect sediment oxidation and control methanogenesis (Niedermeier and Robinson, 2007). The relative partitioning of oxic and anoxic sediment can therefore directly control CH_4 dynamics, leaving the oxidized upper marsh with lower CH_4 emission than the reduced lower marsh (Ding et al., 2010).

N_2O emission is also highly variable in saltmarshes with global average rates in vegetated sediments almost 10 times higher than in unvegetated sediment (Table 8.4). The high variability is caused by other mechanisms than those for CH_4 emission. A major controlling factor is the nitrogen availability in the form of NH_4^+ and NO_3^-. High eutrophication level may enhance N_2O production via stimulated nitrification and denitrification in surface sediment (Moseman-Valtierra et al., 2011; Chmura et al., 2016; Roughan et al., 2018). Salinity in tidal water may also affect N_2O emissions when salt and sulfide hamper both nitrification and denitrification (Pathak and Rao, 1998; Ardon et al., 2018). However, the positive role of vegetation on N_2O emissions is probably a consequence of enhanced nitrification-denitrification in the extended oxic rhizospheres of the densely vegetated marshes (Sun et al., 2013; Xu et al., 2014). This will extend N_2O production deeper in the sediment and lead to excess emission through the canopy after transport from the roots through aerenchyma tissue.

C sequestration in saltmarsh sediments corresponds to a global average storage of $7.6\,\mathrm{Mg\,CO_2\,ha^{-1}\,yr^{-1}}$, pointing toward a strong mitigation effect on climate change (Chmura et al., 2003; Mcleod et al., 2011; Cusack et al., 2018; Burden et al., 2019).

Table 8.5 Global warming potential of vegetated intertidal wetlands.

Mg CO_2 ha^{-1} yr^{-1}	Mangrove forest	Saltmarsh
C-sequestration	6.1	7.6
CH_4 emission	−5.5	−5.9
N_2O emission	−2.4	−0.9
Balance	*−1.8*	*0.8*

The balance between carbon sequestration and $CH_4 + N_2O$ emission is presented as CO_2 equivalents based on the compiled data given in the text. Positive values indicate removal of CO_2 equivalents from the atmosphere.

By converting the reported CH_4 and N_2O emissions to CO_2 equivalents using the global warming potential of these gases, saltmarshes still have a global warming mitigation effect of about 0.8 Mg CO_2 ha^{-1} yr^{-1} (Table 8.5). The exact magnitude of this mitigation is uncertain due to the large variability among saltmarshes in various parts of the world and the lack of data on a possible fauna stimulation of GHG emissions. Conversely, the export of C in the form of persistent alkalinity in the ocean may not contribute significantly to saltmarsh-derived C sequestration due to the generally low export (Fig. 8.6). Thus, the data currently available suggest, in line with the observations from mangrove forests, that most saltmarshes may be considered close to climate neutrality.

8.7 **Perspectives**

It is compelling that the examined vegetated wetlands exhibit relatively consistent patterns in the global estimates of C sequestration and GHG emissions despite the inherent uncertainties caused by, for example, climate, vegetation type, sediment type, and tidal elevation. The global average NPP by mangrove forests and saltmarshes are remarkably similar (Fig. 8.6), despite the larger standing biomass of mangrove forests combined with generally higher temperatures and insolation in climates supporting these forests. The data compilation in this chapter emphasizes the important role of biogenic (i.e., roots structures and infaunal burrows/tubes) in C cycling and GHG dynamics of intertidal wetlands. There is no doubt that vegetated wetlands sequester large amounts of C and that diffusive emission of GHG from the sediment surface cannot balance the vegetation capacity for retaining CO_2 from the atmosphere. However, when the available data on the effect of biogenic structures as conduits for rapid CH_4 and N_2O emission are considered, the global warming mitigation capacity of these wetlands, particularly mangrove forests, is offset (Fig. 8.6). Although this conclusion may appear discouraging in a climate perspective, climate neutrality of vegetated wetlands is still beneficial compared with the massive net GHG emissions experienced from most anthropogenically impacted terrestrial, freshwater, and marine environments. This is particularly evident at locations where the natural vegetation is lost due to deforestation and eutrophication.

The large uncertainties and variability of the data used in the present compilation of global averages strongly advocate for more studies applying state-of-the-art approaches to substantiate the exact role of all C and GHG processes in vegetated intertidal wetlands. It is imperative from the present chapter that these studies first of all must strengthen our knowledge on the quantitative impact of biogenic structures. It is also urgently needed with focus on, for example, the role of microphytobenthos biofilms for the sediment-air exchange of CO_2 and the long-term C storage capacity of exported alkalinity to the adjacent ocean. Future studies must also provide detailed knowledge on the impact of environmental conditions, including climate, vegetation type, sediment type, salinity, and tidal elevation for the ecosystem functioning of vegetated wetlands. We cannot provide a true and reliable estimate on the role of these wetlands as "blue carbon" depositories before the appropriate scientific evidence is available.

In any case, it is important to focus on future regulations for protection of vegetated wetlands. These ecosystems are threatened by human coastal development and eutrophication leading to loss of vegetation cover and thus ecosystem functionality and services. Furthermore, climate change with higher temperatures and sea-level rise will accelerate coastal squeeze due to human reinforcement of coastal protection structures, whereby the areal extension of intertidal wetlands rapidly diminishes. Loss of these ecosystems will raise concerns about the biodiversity of intertidal flora and fauna and their role for the ecological balance and food web structure in the adjacent marine and terrestrial ecosystems. Despite the apparent climate neutrality of vegetated intertidal wetlands, it is still vital to stimulate and expand conservation efforts to protect these valuable and unique ecosystems.

References

Abbasi, M.K., Adams, W.A., 2000. Gaseous N emission during simultaneous nitrification-denitrification associated with mineral N fertilization to a grassland soil under field conditions. Soil Biol. Biochem. 32, 1251–1259.

Abdul-Aziz, O.I., Ishtiaq, K.S., Tang, J., Moseman-Valtierra, S., Kroeger, K.D., Gonneea, M. E., Mora, J., Morkeski, K., 2018. Environmental controls, emergent scaling, and predictions of greenhouse gas (GHG) fluxes in coastal salt marshes. J. Geophys. Res. Biogeo. 123, 2234–2256.

Adame, M.F., Lovelock, C.E., 2011. Carbon and nutrient exchange of mangrove forests with the coastal ocean. Hydrobiologia 663, 23–50.

Adams, C.A., Andrews, J.E., Jickells, T., 2012. Nitrous oxide and methane fluxes vs. carbon, nitrogen and phosphorous burial in new intertidal and saltmarsh sediments. Sci. Total Environ. 434, 240–251.

Adviento-Borbe, M.A.A., Doran, J.W., Drijber, R.A., Dobermann, A., 2006. Soil electrical conductivity and water content affect nitrous oxide and carbon dioxide emissions in intensively managed soils. J. Environ. Qual. 35, 1999–2010.

Al-Haj, A.N., Fulweiler, R.W., 2020. A synthesis of methane emissions from shallow vegetated coastal ecosystems. Glob. Chang. Biol. 26, 2988–3005.

Aller, R.C., 2014. Sedimentary diagenesis, depositional environments, and benthic fluxes. In: Holland, H.D., Turekian, K.K. (Eds.), Treatise on Geochemistry, Second ed. vol. 8. Elsevier, Oxford, pp. 293–334.

Alongi, D.M., 2009. The Energetics of Mangrove Forests. Springer Science, Dordrecht.

Alongi, D.M., 2012. Carbon sequestration in mangrove forests. Carbon Manage. 3, 313–322.

Alongi, D.M., 2014. Carbon cycling and storage in mangrove forests. Ann. Rev. Mar. Sci. 6, 195–219.

Alongi, D.M., 2016. Mangroves. In: Kennish, M.J. (Ed.), Encyclopedia of Estuaries. Springer Dordrecht, Heidelberg, New York, London, pp. 393–404.

Alongi, D.M., Mukhopadhyay, S.K., 2015. Contribution of mangroves to coastal carbon cycling in low latitude seas. Agric. For. Meteorol. 213, 266–272.

Alongi, D.M., Tirendi, F., Clough, B.F., 2000. Below-ground decomposition of organic matter in forests of the mangroves *Rhizophora stylosa* and *Avicennia marina* along the arid coast of Western Australia. Aquat. Bot. 68, 97–122.

Ardon, M., Helton, A.M., Bernhardt, E.S., 2018. Salinity effects on greenhouse gas emissions from wetland soils are contingent upon hydrologic setting: a microcosm experiment. Biogeochemistry 140, 217–232.

Aschenbroich, A., Michaud, E., Stieglitz, T., Fromard, F., Gardel, A., Tavares, M., Thouzeau, G., 2016. Brachyuran crab community structure and associated sediment reworking activities in pioneer and young mangroves of French Guiana, South America. Estuar. Coast. Shelf Sci. 182, 60–71.

Atwood, T.B., Connolly, R.M., Almahasheer, H., Carnell, P.E., Duarte, C.M., Lewis, C.J.E., Irigoien, X., Kelleway, J.J., Lavery, P.S., Macreadie, P.I., Serrano, O., Sanders, C.J., Santos, I., Steven, A.D.L., Lovelock, C.E., 2017. Global patterns in mangrove soil carbon stocks and losses. Nat. Clim. Chang. 7, 523–528.

Banta, G.T., Holmer, M., Jensen, M.H., Kristensen, E., 1999. Effects of two polychaete worms, *Nereis diversicolor* and *Arenicola marina*, on aerobic and anaerobic decomposition in sandy marine sediment. Aquat. Microb. Ecol. 19, 189–204.

Bernal, B., Megonigal, J.P., Mozdzer, T.J., 2017. An invasive wetland grass primes deep soil carbon pools. Glob. Chang. Biol. 23, 2104–2116.

Bertics, V.J., Sohm, J.A., Magnabosco, C., Ziebis, W., 2012. Denitrification and nitrogen fixation dynamics in the area surrounding an individual ghost shrimp (*Neotrypaea californiensis*) burrow system. Appl. Environ. Microbiol. 78, 3864–3872.

Bloom, A.A., Palmer, P.I., Fraser, A., Reay, D.S., Frankenberg, C., 2010. Large-scale controls of methanogenesis inferred from methane and gravity spaceborne data. Science 327, 322–325.

Bonaglia, S., Brüchert, V., Callac, N., Vicenzi, A., Fru, E.C., Nascimento, F.J.A., 2017. Methane fluxes from coastal sediments are enhanced by macrofauna. Sci. Rep. 7, 13145. https://doi.org/10.1038/s41598-017-13263-w.

Bortolus, A., Adam, P., Adams, J.B., Ainouche, M.L., Ayres, D., Bertness, M.D., Bouma, T.J., Bruno, J.F., Cacador, I., Carlton, J.T., Castillo, J.M., Costa, C.S.B., Davy, A.J., Deegan, L., Duarte, B., Figueroa, E., Gerwein, J., Gray, A.J., Grosholz, E.D., Hacker, S.D., Hughes, A. R., Mateos-Naranjo, E., Mendelssohn, I.A., Morris, J.T., Munoz-Rodriguez, A.F., Nieva, F.J.J., Levin, L.A., Li, B., Liu, W., Pennings, S.C., Pickart, A., Redondo-gomez, S., Richardson, D.M., Salmon, A., Schwindt, E., Silliman, B.R., Sotka, E.E., Stace, C., Sytsma, M., Temmerman, S., Turner, R.E., Valiela, I., Weinstein, M.P., Weis, J.S., 2019. Supporting *Spartina*: interdisciplinary perspective shows *Spartina* as a distinct solid genus. Ecology 100, e02863.

Botto, F., Iribarne, O., 2000. Contrasting effects of two burrowing crabs (*Chasmagnathus granulata* and *Uca uruguayensis*) on sediment composition and transport in estuarine environments. Estuar. Coast. Shelf Sci. 51, 141–151.

Bouillon, S., Borges, A.V., Castañeda-Moya, E., Diele, K., Dittmar, T., Duke, N.C., Kristensen, E., Lee, S.Y., Marchand, C., Middelburg, J.J., Rivera-Monroy, V.H., Smith III, T.J., Twilley, R.R., 2008. Mangrove production and carbon sinks: a revision of global budget estimates. Global Biogeochem. Cycles 22, GB2013. https://doi.org/10.1029/2007GB003052.

Bruland, G.L., 2008. Coastal wetlands: function and role in reducing impact of land-based management. In: Fares, A., El-Kadi, A.I. (Eds.), Coastal Watershed Management. WIT Press, Southampton, UK, pp. 85–124.

Brusati, E.D., Grosholz, E.D., 2006. Native and introduced ecosystem engineers produce contrasting effects on estuarine infaunal communities. Biol. Invasions 8, 683–695.

Bu, N., Wu, S., Yang, X., Sun, Y., Chen, Z., Ma, X., Song, Y., Ma, F., Yan, Z., 2019. *Spartina alterniflora* invasion affects methane emissions in the Yangtze River estuary. J. Soil. Sediment. 19, 579–587.

Burden, A., Garbutt, A., Evans, C.D., 2019. Effect of restoration on saltmarsh carbon accumulation in Eastern England. Biol. Lett. 15, 20180773. https://doi.org/10.1098/rsbl.2018.0773.

Burden, A., Garbutt, A., Evans, C.D., Jones, D.L., Cooper, D.M., 2013. Carbon sequestration and biogeochemical cycling in a saltmarsh subject to coastal managed realignment. Estuar. Coast. Shelf Sci. 120, 12–20.

Canfield, D.E., Kristensen, E., Thamdrup, B., 2005. Aquatic Geomicrobiology. Academic Press, San Diego.

Cannicci, S., Burrows, D., Fratini, S., Smith III, T.J., Offenberg, J., Dahdouh-Guebas, F., 2008. Faunal impact on vegetation structure and ecosystem function in mangrove forests: a review. Aquat. Bot. 89, 186–200.

Chapman, V.J., 1976. Mangrove Vegetation. Cramer, Vaduz, Liechtenstein.

Chauhan, R., Datta, A., Ramanathan, A.L., Adhya, T.K., 2015. Factors influencing spatiotemporal variation of methane and nitrous oxide emission from a tropical mangrove of eastern coast of India. Atmos. Environ. 107, 95–106.

Chen, Y., Chen, G., Ye, Y., 2015. Coastal vegetation invasion increases greenhouse gas emission from wetland soils but also increases soil carbon accumulation. Sci. Total Environ. 526, 19–28.

Chen, S., Chmura, G.L., Wang, Y., Yu, D., Ou, D., et al., 2019. Benthic microalgae offset the sediment carbon dioxide emission in subtropical mangrove in cold seasons. Limnol. Oceanogr. 64, 1297–1308.

Chen, J., Gu, J.-D., 2017. Faunal burrows alter the diversity, abundance, and structure of AOA, AOB, anammox and n-damo communities in coastal mangrove sediments. Microb. Ecol. 74, 140–156.

Chen, Y., Li, Y., Thompson, C., Wang, X., Cai, T., Chang, Y., 2018. Differential sediment trapping abilities of mangrove and saltmarsh vegetation in a subtropical estuary. Geomorphology 318, 270–282.

Chmura, G.L., Anisfeld, S.C., Cahoon, D.R., Lynch, J.C., 2003. Global carbon sequestration in tidal, saline wetland soils. Global Biogeochem. Cycles 17, 1111. https://doi.org/10.1029/2002GB001917.

Chmura, G.L., Kellman, L., Guntenspergen, G.R., 2011. The greenhouse gas flux and potential global warming feedbacks of a northern macrotidal and microtidal salt marsh. Environ. Res. Lett. 6. https://doi.org/10.1088/1748-9326/6/4/044016, 044016.

Chmura, G.L., Kellman, L., Van Ardenne, L., Guntenspergen, G.R., 2016. Greenhouse gas fluxes from salt marshes exposed to chronic nutrient enrichment. PLoS One 11. https://doi.org/10.1371/journal.pone.0149937, e0149937.

Chuang, P.-C., Young, M.B., Dale, A.W., Miller, L.G., Herrera-Silveira, J.A., Paytan, A., 2016. Methane and sulfate dynamics in sediments from mangrove-dominated tropical coastal lagoons, Yucatán, Mexico. Biogeosciences 13, 2981–3001.

Cornell, J.A., Craft, C.B., Megonigal, J.P., 2007. Ecosystem gas exchange across a created salt marsh chronosequence. Wetlands 27, 240–250.

Cottet, M., de Montaudouin, X., Blanchet, H., Lebleu, P., 2007. *Spartina anglica* eradication experiment and in situ monitoring assess structuring strength of habitat complexity on marine macrofauna at high tidal level. Estuar. Coast. Shelf Sci. 71, 629–640.

Couto, T., Duarte, B., Cacador, I., Baeta, A., Marques, J.C., 2013. Salt marsh plants carbon storage in a temperate Atlantic estuary illustrated by a stable isotopic analysis based approach. Ecol. Indic. 32, 305–311.

Cusack, M., Saderne, V., Arias-Ortiz, A., Masqué, P., Krishnakumar, P.K., Rabaoui, L., Qurban, M.A., Qasem, A.M., Prihartato, P., Loughland, R.A., Elyas, A.A., Duarte, C.M., 2018. Organic carbon sequestration and storage in vegetated coastal habitats along the western coast of the Arabian Gulf. Environ. Res. Lett. 13. https://doi.org/10.1088/1748-9326/aac899, 074007.

Cutajar, J., Shimeta, J., Nugegoda, D., 2012. Impacts of the invasive grass *Spartina anglica* on benthic macrofaunal assemblages in a temperate Australian saltmarsh. Mar. Ecol. Prog. Ser. 464, 107–120.

Dai, T., Wiegert, R.G., 1996. Estimation of the primary productivity of *Spartina alterniflora* using a canopy model. Ecography 19, 410–423.

Dausse, A., Garbutt, A., Norman, L., Papadimitriou, S., Jones, L.M., Robins, P.E., Thomas, D.N., 2012. Biogeochemical functioning of grazed estuarine tidal marshes along a salinity gradient. Estuar. Coast. Shelf Sci. 100, 83–92.

Davidson, E.A., Keller, M., Erickson, H.E., Verchot, L.V., Veldkamp, E., 2000. Testing a conceptual model of soil emissions of nitrous and nitric oxides: using two functions based on soil nitrogen availability and soil water content, the hole-in-the-pipe model characterizes a large fraction of the observed variation of nitric oxide and nitrous oxide emissions from soils. Bioscience 50, 667–680.

De la Iglesia, H., Rodriguez, M.E., Dezi, R.E., 1994. Burrow plugging in the fiddler crab *Uca uruguayensis* and its synchronization with two environmental cycles. Physiol. Behav. 55, 913–919.

Ding, W., Zhang, Y., Cai, Z., 2010. Impact of permanent inundation on methane emissions from a *Spartina alterniflora* coastal salt marsh. Atmos. Environ. 44, 3894–3900.

Donato, D.C., Kauffman, J.B., Murdiyarso, D., Kurnianto, S., Stidham, M., Kanninen, M., 2011. Mangroves among the most carbon-rich forests in the tropics. Nat. Geosci. https://doi.org/10.1038/NGEO1123.

Duarte, C.M., Losada, I.J., Hendriks, I.E., Mazarrasa, I., Marbà, N., 2013. The role of coastal plant communities for climate change mitigation and adaptation. Nat. Clim. Chang. 3. https://doi.org/10.1038/NCLIMATE1970.

Duke, N.C., 2017. Mangrove floristics and biogeography revisited: further deductions from biodiversity hot spots, ancestral discontinuities and common evolutionary processes. In: Rivera-Monroy, V.H., Lee, S.Y., Kristensen, E., Twilley, R.R. (Eds.), Mangrove Ecosystems: A Global Biogeographic Perspective. Structure, Function and Ecosystem Services. Springer Nature, pp. 17–53.

Duke, N.C., Ball, M.C., Ellison, J.C., 1998. Factors influencing biodiversity and distributional gradients in mangroves. Glob. Ecol. Biogeogr. Lett. 7, 27–47.

Dunn, R.J.K., Welsh, D.T., Teasdale, P.R., Gilbert, F., Poggiale, J.-C., Waltham, N.J., 2019. Effects of the bioturbating marine yabby *Trypaea australiensis* on sediment properties in sandy sediments receiving mangrove leaf litter. J. Mar. Sci. Eng. 7, 426. https://doi.org/10.3390/jmse7120426.

Emery, H.E., Angell, J.H., Fulweiler, R.W., 2019. Salt marsh greenhouse gas fluxes and microbial communities are not sensitive to the first year of precipitation change. J. Geophys. Res. Biogeo. 124, 1071–1087.

Escapa, M., Perillo, G.M.E., Iribarne, O., 2008. Sediment dynamics modulated by burrowing crab activities in contrasting SW Atlantic intertidal habitats. Estuar. Coast. Shelf Sci. 80, 365–373.

Fanjul, E., Escapa, M., Montemayor, D., Addino, M., Alvarez, M.F., Grela, M.A., Iribarne, O., 2015. Effect of crab bioturbation on organic matter processing in South West Atlantic intertidal sediments. J. Sea Res. 95, 206–216.

Feller, I.C., Lovelock, C.E., McKee, K.L., 2007. Nutrient addition differentially affects ecological processes of *Avicennia germinans* in nitrogen versus phosphorus limited mangrove ecosystems. Ecosystems 10, 347–359.

Ferreira, T.O., Otero, X.L., Vidal-Torrado, P., Macías, F., 2007. Effects of bioturbation by root and crab activity on iron and sulfur biogeochemistry in mangrove substrate. Geoderma 142, 36–46.

Fonseca, A.L.S., Marinho, C.C., Esteves, F.A., 2019. Acetate and sulphate as regulators of potential methane production in a tropical coastal lagoon. J. Soil. Sediment. 19, 2604–2612.

Gao, G.F., Li, P.F., Shen, Z.J., Qin, Y.Y., Zhang, X.M., Ghoto, K., Zhu, X.Y., Zheng, H.L., 2018. Exotic *Spartina alterniflora* invasion increases CH_4 while reduces CO_2 emissions from mangrove wetland soils in southeastern China. Sci. Rep. 8, 9243. https://doi.org/10.1038/s41598-018-27625-5.

Gillett, N., Matthews, H., 2010. Accounting for carbon cycle feedbacks in a comparison of the global warming effects of greenhouse gases. Environ. Res. Lett. 5, 034011.

Gillikin, D.P., Schubart, C.D., 2004. Ecology and systematics of mangrove crabs of the genus *Perisesarma* (Crustacea: Brachyura: Sesarmidae) from East Africa. Zool. J. Linn. Soc. 141, 435–445.

Glud, R.N., 2008. Oxygen dynamics of marine sediments. Mar. Biol. Res. 4, 243–289.

Gray, A.J., Marshall, D.F., Raybould, A.F., 1991. A century of evolution in *Spartina anglica*. Adv. Ecol. Res. 21, 1–62.

Gribsholt, B., Kristensen, E., 2002. Effects of bioturbation and plant roots on salt marsh biogeochemistry: a mesocosm study. Mar. Ecol. Prog. Ser. 241, 71–87.

Gribsholt, B., Kristensen, E., 2003. Benthic metabolism and sulfur cycling along an inundation gradient in a tidal *Spartina anglica* salt marsh. Limnol. Oceanogr. 48, 2151–2162.

Guimond, J.A., Seyfferth, A.L., Moffett, K.B., Michael, H., 2020. A physical-biogeochemical mechanism for negative feedback between marsh crabs and carbon storage. Environ. Res. Lett. 15. https://doi.org/10.1088/1748-9326/ab60e2, 034024.

Hamilton, S.E., Friess, D.A., 2018. Global carbon stocks and potential emissions due to mangrove deforestation from 2000 to 2012. Nat. Clim. Chang. 8, 240–244.

Hogarth, P.J., 2015. The Biology of Mangroves and Seagrasses, third ed. Oxford University Press, Oxford, UK.

Holmer, M., Kristensen, E., 1994. Coexistence of sulfate reduction and methane production in an organic-rich sediment. Mar. Ecol. Prog. Ser. 107, 177–184.

Hutchison, J., Manica, A., Swetnam, R., Balmford, A., Spalding, M., 2013. Predicting global patterns in mangrove forest biomass. Conserv. Lett. 7, 233–240.

Hyndes, G.A., Nagelkerken, I., McLeod, R.J., Connolly, R.M., Lavery, P.S., Vanderklift, M. A., 2014. Mechanisms and ecological role of carbon transfer within coastal seascapes. Biol. Rev. 89, 232–254.

IPCC, 2018. Global Warming of 1.5°C. An IPCC Special Report on the Impacts of Global Warming of 1.5°C above Pre-Industrial Levels and Related Global Greenhouse Gas Emission Pathways, in the Context of Strengthening the Global Response to the Threat of Climate Change, Sustainable Development, and Efforts to Eradicate Poverty. Geneva.

IPCC, 2019. Summary for policymakers. In: IPCC Special Report on the Ocean and Cryosphere in a Changing Climate. Geneva.

Jeffrey, L.C., Maher, D.T., Johnston, S.G., Kelaher, B.P., Steven, A., Tait, D.R., 2019. Wetland methane emissions dominated by plant-mediated fluxes: contrasting emissions pathways and seasons within a shallow freshwater subtropical wetland. Limnol. Oceanogr. 9999, 1–18. https://doi.org/10.1002/lno.11158.

Kauffman, J.B., Adame, M.F., Arifanti, V.B., Schile-Beers, L.M., Bernardino, A.F., Bhomia, R.K., Donato, D.C., Feller, I.C., Ferreira, T.O., Garcia, M.D.J., MacKenzie, R.A., Megonigal, J.P., Murdiyarso, D., Simpson, L., Trejo, H.H., 2020. Total ecosystem carbon stocks of mangroves across broad global environmental and physical gradients. Ecol. Monogr. 90, e01405.

Kelleway, J.J., Saintilan, N., Macreadie, P.I., Baldock, J.A., Heijnis, H., Zawadzki, A., Gadd, P., Jacobsen, G., Ralph, P.J., 2017. Geochemical analyses reveal the importance of environmental history for blue carbon sequestration. J. Geophys. Res. Biogeo. 122, 1789–1805.

Kelleway, J.J., Saintilan, N., Macreadie, P.I., Ralph, P.J., 2016. Sedimentary factors are key predictors of carbon storage in SE Australian saltmarshes. Ecosystems 19, 865–880.

Kitaya, Y., Yabuki, K., Kiyota, M., Tani, A., Hirano, T., Aiga, I., 2002. Gas exchange and oxygen concentration in pneumatophores and prop roots of four mangrove species. Trees 16, 155–158.

Koop-Jakobsen, K., Mueller, P., Meier, R.J., Liebsch, G., Jensen, K., 2018. Plant-sediment interactions in salt marshes—an optode imaging study of O_2, pH, and CO_2 gradients in the rhizosphere. Front. Plant Sci. 9, 541. https://doi.org/10.3389/fpls.2018.00541.

Koop-Jakobsen, K., Wenzhöfer, F., 2015. The dynamics of plant-mediated sediment oxygenation in *Spartina anglica* rhizospheres—a planar optode study. Estuar. Coasts 38, 951–963.

Kostka, J.E., Gribsholt, B., Petrie, E., Dalton, D., Skelton, H., Kristensen, E., 2002. The rates and pathways of carbon oxidation in bioturbated saltmarsh sediments. Limnol. Oceanogr. 47, 230–240.

Krauss, K.W., Allen, J.A., Cahoon, D.R., 2003. Differential rates of vertical accretion and elevation change among aerial root types in Micronesian mangrove forests. Estuar. Coast. Shelf Sci. 56, 251–259.

Kreuzwieser, J., Buchholz, J., Rennenberg, H., 2003. Emission of methane and nitrous oxide by Australian mangrove ecosystems. Plant Biol. 5, 423–431.

Kristensen, E., 2008. Mangrove crabs as ecosystem engineers, with emphasis on sediment processes. J. Sea Res. 59, 30–43.

Kristensen, E., Alongi, D.M., 2006. Control by fiddler crabs (*Uca vocans*) and plant roots (*Avicennia marina*) on carbon, iron and sulfur biogeochemistry in mangrove sediment. Limnol. Oceanogr. 51, 1557–1571.

Kristensen, E., Andersen, F.Ø., Holmboe, N., Holmer, M., Thongtham, N., 2000. Carbon and nitrogen mineralization in sediment of the Bangrong mangrove area, Phuket, Thailand. Aquat. Microb. Ecol. 22, 199–213.

Kristensen, E., Connolly, R.M., Otero, X.L., Marchand, C., Ferreira, T.O., Rivera-Monroy, V. H., 2017. Biogeochemical cycles: Global approaches and perspectives. In: Rivera-Monroy, V.H., Lee, S.Y., Kristensen, E., Twilley, R.R. (Eds.), Mangrove Ecosystems: A Global Biogeographic Perspective. Structure, Function and Ecosystem Services. Springer Nature, pp. 163–209.

Kristensen, E., Flindt, M.R., Borges, A.V., Bouillon, S., 2008. Emission of CO_2 and CH_4 to the atmosphere by sediments and open waters in two Tanzanian mangrove forests. Mar. Ecol. Prog. Ser. 370, 53–67.

Kristensen, E., Holmer, M., 2001. Decomposition of plant materials in marine sediment exposed to different electron acceptors (O_2, NO_3^- and SO_4^{2-}), with emphasis on substrate origin, degradation kinetics and the role of bioturbation. Geochim. Cosmochim. Acta 65, 419–434.

Kristensen, E., Jensen, M.H., Banta, G.T., Hansen, K., Holmer, M., King, G.M., 1998. Transformation and transport of inorganic nitrogen in sediments of a Southeast Asian mangrove forest. Aquat. Microb. Ecol. 15, 165–175.

Kristensen, E., Mangion, P., Tang, M., Flindt, M.R., Ulomi, S., 2011. Benthic metabolism and partitioning of electron acceptors for microbial carbon oxidation in sediments of two Tanzanian mangrove forests. Biogeochemistry 103, 143–158.

Kristensen, E., Penha-Lopes, G., Delefosse, M., Valdemarsen, T., Quintana, C.O., Banta, G.T., 2012. What is bioturbation? Need for a precise definition for fauna in aquatic sciences. Mar. Ecol. Prog. Ser. 446, 285–302.

Kristensen, E., Rabenhorst, M.C., 2015. Do marine rooted plants grow in sediment or soil? A critical appraisal on definitions, methodology and communication. Earth Sci. Rev. 145, 1–8.

Krithika, K., Purvaja, R., Ramesh, R., 2008. Fluxes of methane and nitrous oxide from an Indian mangrove. Curr. Sci. 94, 218–224.

Kuivila, K.M., Murray, J.W., Devol, A.H., Novelli, P.C., 1989. Methane production, sulfate reduction and competition for substrates in the sediments of Lake Washington. Geochim. Cosmochim. Acta 53, 409–416.

Laanbroek, H.J., 2010. Methane emission from natural wetlands: interplay between emergent macrophytes and soil microbial processes. A mini-review. Ann. Bot. 105, 141–153.

Lamont, K., Saintilan, N., Kelleway, J.J., Mazumder, D., Zawadzki, A., 2020. Thirty-yearrepeat measures of mangrove above- and below-ground biomass reveals unexpectedly high carbon sequestration. Ecosystems 23, 370–382.

Lee, R.W., 2003. Physiological adaptations of the invasive cordgrass *Spartina anglica* to reducing sediments: rhizome metabolic gas fluxes and enhanced O_2 and H_2S transport. Mar. Biol. 143, 9–15.

Lee, R.Y., Porubsky, W.P., Feller, I.C., McKee, K.L., Joye, S.B., 2008. Porewater biogeochemistry and soil metabolism in dwarf red mangrove habitats (Twin Cays, Belize). Biogeochemistry 87, 181–198.

Lind, L.P.D., Audet, J., Tonderski, K., Hoffmann, C.C., 2013. Nitrate removal capacity and nitrous oxide production in soil profiles of nitrogen loaded riparian wetlands inferred by laboratory microcosms. Soil Biol. Biochem. 60, 156–164.

Liu, L., Wang, D., Chen, S., Yu, Z., Xu, Y., Li, Y., Ge, Z., Chen, Z., 2019. Methane emissions from estuarine coastal wetlands: implications for global change effect. Soil Sci. Soc. Am. J. 83, 1368–1377.

Loebl, M., van Beusekom, J.E.E., Reise, K., 2006. Is spread of the neophyte *Spartina anglica* recently enhanced by increasing temperatures? Aquat. Ecol. 40, 315–324.

Lovelock, C.E., Adame, M.F., Bennion, V., Hayes, M., O'Mara, J., Reef, R., Santini, N.S., 2014. Contemporary rates of carbon sequestration through vertical accretion of sediments in mangrove forests and saltmarshes of South East Queensland, Australia. Estuar. Coasts 37, 763–771.

Lovley, D.R., Phillips, E.J.P., 1987. Competitive mechanisms for inhibition of sulfate reduction and methane production in the zone of ferric iron reduction in sediments. Appl. Environ. Microbiol. 53, 2636–2641.

Maher, D.T., Call, M., Santos, I.R., Sanders, C.J., 2018. Beyond burial: lateral exchange is a significant atmospheric carbon sink in mangrove forests. Biol. Lett. 14, 20180200. https://doi.org/10.1098/rsbl.2018.0200.

Mcleod, E., Chmura, G.L., Bouillon, S., Salm, R., Björk, M., Duarte, C.M., Lovelock, C.E., Schlesinger, W.H., Silliman, B.R., 2011. A blueprint for blue carbon: toward an improved understanding of the role of vegetated coastal habitats in sequestering CO_2. Front. Ecol. Environ. 9, 552–560.

Mehring, A.S., Cook, P.L.M., Evrard, V., Grant, S.B., Levin, L.A., 2017. Pollution-tolerant invertebrates enhance greenhouse gas flux in urban wetlands. Ecol. Appl. 27, 1852–1861.

Michaels, R.E., Zieman, J.C., 2013. Fiddler crab (*Uca* spp.) burrows have little effect on surrounding sediment oxygen concentrations. J. Exp. Mar. Biol. Ecol. 448, 104–113.

Middelburg, J.J., 2018. Reviews and syntheses: to the bottom of carbon processing at the seafloor. Biogeosciences 15, 413–427.

Mitsch, W.J., Gosselink, J.G., 2007. Wetlands, fourth ed. Wiley, Hoboken.

Moh, H.H., Chong, V.C., Sasekumar, A., 2015. Distribution and burrow morphology of three sympatric species of *Thalassina* mud lobsters in relation to environmental parameters on a Malayan mangrove shore. J. Sea Res. 95, 75–83.

Moseman-Valtierra, S.M., Gonzalez, R., Kroeger, K., Tang, J., Chun, W., Crusius, J., Bratton, J., Green, A., Shelton, J., 2011. Short-term nitrogen additions can shift a coastal wetland from a sink to a source of N_2O. Atmos. Environ. 45, 4390–4397.

Murray, R., Erler, D.V., Rosentreter, J., Wells, N.S., Eyre, B.D., 2020. Seasonal and spatial controls on N_2O concentrations and emissions in low-nitrogen estuaries: evidence from three tropical systems. Mar. Chem. 221, 103779.

Myhre, G., Shindell, D., Bréon, F.-M., Collins, W., Fuglestvedt, J., Huang, J., Koch, D., Lamarque, J.-F., Lee, D., Mendoza, B., Nakajima, T., Robock, A., Stephens, G., Takemura, T., Zhang, H., 2013. Anthropogenic and natural radiative forcing. In: Stocker, T.F., Qin, D., Plattner, G.-K., Tignor, M., Allen, S.K., Boschung, J., Midgley, P.M. (Eds.), Climate Change 2013: The Physical Science Basis. Contribution of Working Group I to the Fifth Assessment Report of the Intergovernmental Panel on Climate Change. Cambridge University Press, Cambridge, pp. 659–740.

Nagelkerken, I., Blaber, S.J.M., Bouillon, S., Green, P., Haywood, M., Kirton, L.G., Meynecke, J.-O., Pawlik, J., Penrose, H.M., Sasekumar, A., Somerfield, P.J., 2008. The habitat function of mangroves for terrestrial and marine fauna: a review. Aquat. Bot. 89, 155–185.

Naidoo, G., 2016. The mangroves of South Africa: an ecophysiological review. S. Afr. J. Bot. 107, 101–113.

Neira, C., Levin, L.A., Grosholz, E.D., 2005. Benthic macrofaunal communities of three sites in San Francisco Bay invaded by hybrid *Spartina*, with comparison to uninvaded habitats. Mar. Ecol. Prog. Ser. 292, 111–126.

Neubauer, S.C., Miller, W.D., Anderson, I.C., 2000. Carbon cycling in a tidal freshwater marsh ecosvstem: a carbon gas flux studv. Mar. Ecol. Prog. Ser. 199, 13–30.

Niedermeier, A., Robinson, J.S., 2007. Hydrological controls on soil redox dynamics in a peat-based, restored wetland. Geoderma 137, 318–326.

Ouyang, X., Lee, S.Y., 2020. Improved estimates on global carbon stock and carbon pools in tidal wetlands. Nat. Commun. 11, 317. https://doi.org/10.1038/s41467-019-14120-2.

Ouyang, X., Lee, S.Y., Connolly, R.M., 2017. The role of root decomposition in global mangrove and saltmarsh carbon budgets. Earth Sci. Rev. 166, 53–63.

Ouyang, X., Lee, S.Y., Connolly, R.M., 2018. Using isotope labeling to partition sources of CO2 efflux in newly established mangrove seedlings. Limnol. Oceanogr. 63, 731–740.

Owers, C.J., Rogers, K., Woodroffe, C.D., 2018. Spatial variation of above-ground carbon storage in temperate coastal wetlands. Estuar. Coast. Shelf Sci. 210, 55–67.

Pathak, H., Rao, D.L.N., 1998. Carbon and nitrogen mineralization from added organic matter in saline and alkali soils. Soil Biol. Biochem. 30, 695–702.

Peer, N., Miranda, N.A.F., Perissinotto, R., 2019. Impact of fiddler crab activity on microphytobenthic communities in a South African mangrove forest. Estuar. Coast. Shelf Sci. 227. https://doi.org/10.1016/j.ecss.2019.106332.

Penha-Lopes, G., Kristensen, E., Flindt, M., Mangion, P., Bouillon, S., Paula, J., 2010. The role of biogenic structures on biogeochemical functioning of mangrove constructed wetlands sediments—a mesocosm approach. Mar. Pollut. Bull. 60, 560–572.

Pérez, A., Libardoni, B.G., Sanders, C.J., 2018. Factors influencing organic carbon accumulation in mangrove ecosystems. Biol. Lett. 14, 20180237. https://doi.org/10.1098/rsbl.2018.0237.

Petersen, P.M., Romaschenko, K., Arriet, Y.H., Saarela, J.M., 2014. A molecular phylogeny and new subgeneric classification of *Sporobolus* (Poaceae: Chloridoideae: Sporobolinae). Taxon 63, 1212–1234.

Pillay, D., Branch, G.M., Dawson, J., Henry, D., 2011. Contrasting effects of ecosystem engineering by the cordgrass *Spartina maritima* and the sandprawn *Callianassa kraussi* in a marine-dominated lagoon. Estuar. Coast. Shelf Sci. 91, 169–176.

Poffenbarger, H.J., Needelman, B.A., Megonigal, J.P., 2011. Salinity influence on methane emissions from tidal marshes. Wetlands 31, 831–842.

Pülmanns, N., Mehlig, U., Nordhaus, I., Saint-Paul, U., Diele, K., 2016. Mangrove crab *Ucides cordatus* removal does not affect sediment parameters and stipule production in a one year experiment in Northern Brazil. PLoS One 11. https://doi.org/10.1371/journal.pone.0167375, e0167375.

Purnobasuki, H., Suzuki, M., 2005. Aerenchyma tissue development and gas-pathway structure in root of *Avicennia marina* (Forsk.) Vierh. J. Plant Res. 118, 285–294.

Purvaja, R., Ramesh, R., Frenzel, P., 2004. Plant-mediated methane emission from an Indian mangrove. Glob. Chang. Biol. 10, 1825–1834.

Quintana, C.O., Raymond, C., Nascimento, F., Bonaglia, S., Forster, S., Gunnarson, J.S., Kristensen, E., 2018. Functional performance of three invasive *Marenzelleria* species under contrasting ecological conditions within the Baltic Sea. Estuar. Coasts 41, 1766–1781.

Raabe, E.A., Stumpf, R.P., 2016. Expansion of tidal marsh in response to sea-level rise: Gulf Coast of Florida, USA. Estuar. Coasts 39, 145–157.

Raposa, K.B., McKinney, R.A., Wigand, C., Hollister, J.W., Lovall, C., Szura, K., Gurak Jr., J. A., McNamee, J., Raithel, C., Watson, E.B., 2018. Top-down and bottom-up controls on southern New England salt marsh crab populations. PeerJ 6. https://doi.org/10.7717/peerj.4876, e4876.

Redelstein, R., Dinter, T., Hertel, D., Leuschner, C., 2018. Effects of inundation, nutrient availability and plant species diversity on fine root mass and morphology across a salt-marsh flooding gradient. Front. Plant Sci. 9. https://doi.org/10.3389/fpls.2018.00098.

Reise, K., 2005. Coast of change: habitat loss and transformations in the Wadden Sea. Helgol. Mar. Res. 59, 9–21.

Reithmaier, G.M.S., Ho, D.T., Johnston, S.G., Maher, D.T., 2020. Mangroves as a source of greenhouse gases to the atmosphere and alkalinity and dissolved carbon to the coastal ocean: a case study from the Everglades National Park, Florida. J. Geophys. Res. Biogeosci. 125. https://doi.org/10.1029/2020JG005812, e2020JG005812.

Rietl, A.J., Nyman, J.A., Lindau, C.W., Jackson, C.R., 2017. Gulf ribbed mussels (*Geukensia granosissima*) increase methane emissions from a coastal *Spartina alterniflora* marsh. Estuar. Coasts 40, 832–841.

Rivera-Monroy, V.H., Twilley, R.R., 1996. The relative role of denitrification and immobilization in the fate of inorganic nitrogen in mangrove sediments (Terminos Lagoon, Mexico). Limnol. Oceanogr. 41, 284–296.

Rog, S.M., Cook, C.N., 2017. Strengthening governance for intertidal ecosystems requires a consistent definition of boundaries between land and sea. J. Environ. Manage. 197, 694–705.

Rogers, K., Macreadie, P.I., Kelleway, J.J., Saintilan, N., 2019. Blue carbon in coastal landscapes: a spatial framework for assessment of stocks and additionality. Sustain. Sci. 14, 453–467.

Roughan, B.L., Kellman, L., Smith, E., Chmura, G.L., 2018. Nitrous oxide emissions could reduce the blue carbon value of marshes on eutrophic estuaries. Environ. Res. Lett. 13. https://doi.org/10.1088/1748-9326/aab63c, 044034.

Saintilan, N., Rogers, K., Mazumder, D., Woodroffe, C., 2013. Allochthonous and autochthonous contributions to carbon accumulation and carbon store in southeastern Australian coastal wetlands. Estuar. Coast. Shelf Sci. 128, 84–92.

Santini, N.S., Lovelock, C.E., Hua, Q., Zawadzki, A., Mazumder, D., Mercer, T., Munoz-Rojas, M., Hardwick, S., Madala, B.S., Cornwell, W., Thomas, T., Marzinelli, E.M., Adam, P., Paul, S., Vergés, A., 2019. Natural and regenerated saltmarshes exhibit similar soil and belowground organic carbon stocks, root production and soil respiration. Ecosystems 22, 1803–1822.

Sasmito, S.D., Kuzyakov, Y., Lubis, A.A., Murdiyarso, D., Hutley, L.B., Bachri, S., Friess, D. A., Martius, C., Borchard, N., 2020. Organic carbon burial and sources in soils of coastal mudflat and mangrove ecosystems. Catena 187, 104414.

Schöttler, U., 1979. On the anaerobic metabolism of three species of *Nereis* (Annelida). Mar. Ecol. Prog. Ser. 1, 249–254.

Segers, R., 1998. Methane production and methane consumption: a review of processes underlying wetland methane fluxes. Biogeochemistry 41, 23–51.

Sheng, Q., Zhao, B., Huang, M., Wang, L., Quan, Z., Fang, C., Li, B., Wua, J., 2014. Greenhouse gas emissions following an invasive plant eradication program. Ecol. Eng. 73, 229–237.

Silvestri, S., Defina, A., Marani, M., 2005. Tidal regime, salinity and salt marsh plant zonation. Estuar. Coast. Shelf Sci. 62, 119–130.

Spalding, M., Kainuma, M., Collins, L., 2010. World Atlas of Mangroves. Earthscan Ltd., London.

Srikanth, S., Kaihekulani, S., Lum, Y., Chen, Z., 2016. Mangrove root: adaptations and ecological importance. Trees 30, 451–465.

Stief, P., 2013. Stimulation of microbial nitrogen cycling in aquatic ecosystems by benthic macrofauna: mechanisms and environmental implications. Biogeosciences 10, 7829–7846.

Stieglitz, T.C., Clark, J.F., Hancock, G.J., 2013. The mangrove pump: the tidal flushing of animal burrows in a tropical mangrove forest determined from radionuclide budgets. Geochim. Cosmochim. Acta 102, 12–22.

Sun, Z., Wanga, L., Tian, H., Jiang, H., Moua, X., Sun, W., 2013. Fluxes of nitrous oxide and methane in different coastal *Suaeda salsa* marshes of the Yellow River estuary, China. Chemosphere 90, 856–865.

Taillardat, P., Willemsen, P., Marchand, C., Friess, D.A., Widory, D., Baudron, P., Truong, V. V., Nguyen, T.N., Ziegler, A.D., 2018. Assessing the contribution of porewater discharge in carbon export and CO_2 evasion in a mangrove tidal creek (Can Gio, Vietnam). J. Hydrol. 563, 303–318.

Tang, M., Kristensen, E., 2010. The impact of cordgrass, *Spartina anglica*, on macrobenthos distribution in the Danish Wadden Sea. Helgol. Mar. Res. 64, 321–329.

Temmerman, S., Bouma, T.J., Govers, G., Wang, Z.B., De Vries, M.B., Herman, P.M.J., 2005. Impact of vegetation on flow routing and sedimentation patterns: three-dimensional modeling for a tidal marsh. J. Geophys. Res. 110, F04019. https://doi.org/10.1029/2005JF000301.

Thomson, A.C.G., Kristensen, E., Valdemarsen, T., Quintana, C.O., 2020. Short-term fate of seagrass and macroalgal detritus in *Arenicola marina* bioturbated sediments. Mar. Ecol. Prog. Ser. 639, 21–35.

Thongtham, N., Kristensen, E., 2003. Physical and chemical characteristics of mangrove crab (*Neoepisesarma versicolor*) burrows in the Bangrong mangrove forest, Phuket, Thailand; with emphasis on behavioural response to changing environmental conditions. Vie et Milieu 53, 141–151.

Tong, C., Wang, W.-Q., Huang, J.-F., Gauci, V., Zhang, L.-H., Zeng, C.-S., 2012. Invasive alien plants increase CH_4 emissions from a subtropical tidal estuarine wetland. Biogeochemistry 111, 677–693.

Trifunovic, B., Vázquez-Lule, A., Capooci, M., Seyfferth, A.L., Moffat, C., Vargas, R., 2020. Carbon dioxide and methane emissions from temperate salt marsh tidal creek. J. Geophys. Res. Biogeosci. 125. https://doi.org/10.1029/2019JG005558, e2019JG005558.

Van Ardenne, L.B., Jolicouer, S., Bérubé, D., Burdick, D., Chmura, G.L., 2018. The importance of geomorphic context for estimating the carbon stock of salt marshes. Geoderma 330, 264–275.

Van de Broek, M., Vandendriessche, C., Poppelmonde, D., Merckx, R., Temmerman, S., Govers, G., 2017. Long-term organic carbon sequestration in tidal marsh sediments is dominated by old-aged allochthonous inputs in a macrotidal estuary. Glob. Chang. Biol. 24, 2498–2512.

Wang, J.-Q., Bertness, M.D., Li, B., Chen, J.-K., Lü, W.-G., 2015. Plant effects on burrowing crab morphology in a Chinese salt marsh: native vs. exotic plants. Ecol. Eng. 74, 376–384.

Wang, J., Wang, J., 2017. *Spartina alterniflora* alters ecosystem DMS and CH_4 emissions and their relationship along interacting tidal and vegetation gradients within a coastal salt marsh in Eastern China. Atmos. Environ. 167, 346–359.

Wasson, K., Raposa, K., Almeida, M., Beheshti, K., Crooks, J.A., Decks, A., Dix, N., Garvey, C., Goldstein, J., Johnson, D.S., Lerberg, S., Marcum, P., Peter, C., Puckett, B., Schmitt, J., Smith, E., St Laurent, K., Swanson, K., Tyrrell, M., Guy, R., 2019. Pattern and scale: evaluating generalities in crab distributions and marsh dynamics from small plots to a national scale. Ecology 100. https://doi.org/10.1002/ecy.2813.

Welker, A.F., Moreira, D.C., Campos, É.G., Hermes-Lima, M., 2013. Role of redox metabolism for adaptation of aquatic animals to drastic changes in oxygen availability. Comp. Biochem. Physiol. 165A, 384–404.

Weston, N.B., Neubauer, S.C., Velinsky, D.J., Vile, M.A., 2014. Net ecosystem carbon exchange and the greenhouse gas balance of tidal marshes along an estuarine salinity gradient. Biogeochemistry 120, 163–189.

Whalen, S.C., 2005. Biogeochemistry of methane exchange between natural wetlands and the atmosphere. Environ. Eng. Sci. 22, 73–94.

Wilson, B.J., Mortazavi, B., Kiene, R.P., 2015. Spatial and temporal variability in carbon dioxide and methane exchange at three coastal marshes along a salinity gradient in a northern Gulf of Mexico estuary. Biogeochemistry 123, 329–347.

Wohlgemuth, S.E., Taylor, A.C., Grieshaber, M.K., 2000. Ventilatory and metabolic responses to hypoxia and sulphide in the lugworm *Arenicola marina* (L.). J. Exp. Biol. 203, 3177–3188.

Xiang, J., Liu, D., Ding, W., Yuan, J., Lin, Y., 2015. Invasion chronosequence of *Spartina alterniflora* on methane emission and organic carbon sequestration in a coastal salt marsh. Atmos. Environ. 112, 72–80.

Xiao, K., Wilson, A.M., Li, H., Ryan, C., 2019. Crab burrows as preferential flow conduits for groundwater flow and transport in salt marshes: a modeling study. Adv. Water Resour. 132. https://doi.org/10.1016/j.advwatres.2019.103408.

Xinwanghao, X., Guanghe, F., Xinqing, Z., Chendong, G., Yifei, Z., 2017. Diurnal variations of carbon dioxide, methane, and nitrous oxide fluxes from invasive *Spartina alterniflora* dominated coastal wetland in northern Jiangsu Province. Acta Oceanol. Sin. 36, 105–113.

Xu, X., Zou, X., Cao, L., Zhamangulova, N., Zhao, Y., Tang, D., Liu, D., 2014. Seasonal and spatial dynamics of greenhouse gas emissions under various vegetation covers in a coastal saline wetland in Southeast China. Ecol. Eng. 73, 469–477.

Yuan, J., Ding, W., Liu, D., Kang, H., Freeman, C., Xiang, J., Lin, Y., 2015. Exotic *Spartina alterniflora* invasion alters ecosystem-atmosphere exchange of CH_4 and N_2O and carbon sequestration in a coastal salt marsh in China. Glob. Chang. Biol. 21, 1567–1580.

Zhang, T., Chen, H., Cao, H., Ge, Z., Zhang, L., 2017. Combined influence of sedimentation and vegetation on the soil carbon stocks of a coastal wetland in the Changjiang estuary. Chinese J. Oceanol. Limnol. 35, 833–843.

Zhang, Y., Ding, W., 2011. Diel methane emissions in stands of *Spartina alterniflora* and Suaeda salsa from a coastal salt marsh. Aquat. Bot. 95, 262–267.

Greenhouse gas emissions from intertidal wetland soils under anthropogenic activities

9

Guangcheng Chen[a,b], Nora F.Y. Tam[c,d], Yong Ye[e], and Bin Chen[a,b]

[a]*Third Institute of Oceanography, Ministry of Natural Resources, Xiamen, China,*
[b]*Observation and Research Station of Coastal Wetland Ecosystem in Beibu Gulf,
Ministry of Natural Resources, Beihai, China,*
[c]*School of Science and Technology, The Open University of Hong Kong, Hong Kong Special
Administrative Region, China,*
[d]*Department of Chemistry, City University of Hong Kong, Hong Kong Special Administrative
Region, China,*
[e]*Key Laboratory of the Ministry of Education for Coastal and Wetland Ecosystems,
College of the Environment and Ecology, Xiamen University, Xiamen, China*

9.1 Introduction

Intertidal wetlands, including mangroves and salt marshes, are located at the interface between the terrestrial and marine systems and are subject to serious anthropogenic activities, including agriculture, aquaculture, fisheries, tourism, urbanization, and industrial development (Newton et al., 2020). Despite their importance as habitats for marine biota, coastal protection, and carbon sinks (Barbier et al., 2011; Mcleod et al., 2011; Lee et al., 2014), these wetlands have been subjected to massive loss and degradation due to land conversions for agriculture, mariculture, urbanization activities, and climate change (Alongi, 2002; Valiela et al., 2009). Since intertidal wetlands are intermittently flooded by incoming tides, their soil often has alternating anoxic and oxic conditions (Alongi, 2009). These conditions are favorable for microbial nitrification, denitrification, and aerobic and anaerobic respiration processes, leading to the production of different greenhouse gases (GHGs), such as carbon dioxide (CO_2), methane (CH_4), and nitrous oxide (N_2O). The production and emission of GHGs from intertidal wetland soils are controlled by a variety of environmental and physiological factors (e.g., soil pH, redox potential, temperature, soil organic matter content; Allen et al., 2007; Chen et al., 2010; Chmura et al., 2011). Changes in these factors due to anthropogenic activities will likely affect microbial metabolism in wetland soils and the emission of GHGs. In this chapter, we summarize the impacts of various anthropogenic activities, in particular nutrient enrichment, tidal restriction, deforestation, and restoration on soil biogeochemical characteristics and GHG fluxes in vegetated intertidal wetlands.

9.2 Anthropogenic activities in intertidal wetlands

Mangroves and salt marshes have been adversely impacted by human activities to varying degrees. Substantial losses of both salt marshes and mangroves have been recorded globally, with considerable portions of the remaining habitats being degraded (Valiela et al., 2009). These intertidal wetlands have undergone diking and drainage over the past several centuries of human civilization for grazing, agriculture, land creation, aquaculture, mosquito control, rice production, and wildfowl management (Gedan et al., 2009; Roman and Burdick, 2012). These practices have continued in recent decades along many coastlines around the world where drained wetlands are used for modern industrial and urbanization purpose (Byun et al., 2004; Stedman and Dahl, 2008; Goldberg et al., 2020). Diking and drainage cause partial restriction or even complete blockage of tides, reducing the tidal range and inevitably altering the ecosystem structure, processes, and functions (summarized by Portnoy, 1999; Emery and Fulweiler, 2017). In some places, roads and railroads are crossing tidal marshes without adequately sized bridges and culverts, leading to the restriction of tides and the reduction of tidal ranges. Diked salt marshes with little drainage capacity remain waterlogged with freshwater for most of the time and are effectively impounded, while diked marshes that are intentionally drained by ditching, creek channelization, and the provision of large culverts develop more aerobic soils due to the lowered water table (Portnoy, 1999).

High population pressure in the coastal areas has led to the conversion of many mangrove areas to other uses, causing an alarming 20% loss of the global mangrove area (3.6 million hectares) from 1980 to 2005 (FAO, 2007). Although global mangrove loss has slowed down considerably since 2000 (at a rate of 0.13% per year between 2000 and 2016), deforestation is still continuing in many countries, especially in the largest mangrove-holding nations in Southeast Asia, that results in mangrove destruction (Goldberg et al., 2020). Indonesia has experienced the largest loss of mangrove area during 2000–16, followed by Brazil and Malaysia (Table 9.1). (Goldberg et al., 2020) estimated that nearly 80% of the global loss of mangrove area was driven by anthropogenic activities that occurred in the six Southeast Asian nations, including Indonesia, Myanmar, Malaysia, Philippines, Thailand, and Vietnam. Marine aquaculture is identified as a dominant driver of mangrove deforestation, apart from the conversion of deforested mangroves to agricultural farms, urbanized lands (Richards and Friess, 2016; Goldberg et al., 2020), or abandoned lands (Fig. 9.1). In some regions, mangroves also experience small-scale but widespread degradation from indiscriminate wood harvesting for construction and fuelwood purposes (Goldberg et al., 2020). For instance, in Lamu, Kenya, the harvest of mangroves for fuelwood for domestic uses has been legalized through licensing procedures, with a licensed removal rate of 43 poles ha^{-1} $year^{-1}$ in the mangrove forest

Table 9.1 The change in mangrove area from 2000 to 2016 globally and in the top 20 mangrove-holding nations identified by Giri et al. (2011).

No.	Region	Area loss (km^2)			
		2000–05	2005–10	2010–16	2000–16
1	Global	1809.51	990.48	562.61	3362.59
2	Indonesia	687.95	225.05	147.40	1060.41
3	Australia	39.30	31.15	9.59	80.04
4	Brazil	98.68	75.08	49.62	223.39
5	Mexico	33.13	18.43	11.95	63.50
6	Nigeria	66.70	17.56	6.50	90.76
7	Malaysia	54.09	36.39	36.71	127.19
8	Myanmar (Burma)	261.16	170.57	56.42	488.15
9	Papua New Guinea	26.16	26.78	11.78	64.72
10	Bangladesh	49.01	30.54	9.23	88.78
11	Cuba	60.46	74.90	37.77	173.13
12	India	38.28	33.24	12.98	84.50
13	Guinea Bissau	2.90	5.39	1.99	10.29
14	Mozambique	16.74	19.14	13.92	49.80
15	Madagascar	32.22	29.06	33.19	94.47
16	Philippines	20.90	12.01	5.42	38.32
17	Thailand	22.10	8.00	4.70	34.80
18	Colombia	12.70	5.19	4.72	22.61
19	United States	4.32	5.05	0.19	9.57
20	Venezuela	14.04	8.87	9.65	32.56
21	Guinea	16.20	23.16	13.46	52.83

Area losses of the mangrove were derived from Goldberg, L., Lagomasino, D., Thomas, N., Fatoyinbo, T., 2020. Global declines in human-driven mangrove loss. Glob. Chang. Biol. 26, 5844–5855.

(Abuodha and Kairo, 2001). It is reported that this removal rate is recommended based mainly on the national demand of mangrove wood products rather than the natural resilience of mangroves. Therefore, the degradation of mangrove canopies is of great concern for mangrove conservation in Kenya.

In addition to land conversion, massive economic growth, and urban development in the coastal regions have also led to the release of excessive nutrients and toxic pollutants into the coastal areas (Malone and Newton, 2020). There has been a progressive increase in the discharge of untreated, nutrient-rich, domestic, and industrial wastewaters into the coastal ecosystems, leading to the accumulation of nutrients in the soils of intertidal wetlands. Numerous intertidal wetlands around the world, especially the mangroves in Asian countries, are under increasing threat from the wastewater discharged from the surrounding mariculture ponds (Vaiphasa et al., 2007; Chen et al., 2020). During maricultural activities, pond waters are routinely and frequently discharged into adjacent coastal areas. At the end of each culture cycle, the pond sediments containing harmful pathogenic bacteria, uneaten

(A)

(B)

FIG. 9.1

Small-scale mangrove deforestation is still widespread in developing countries. The sites shown above remain abandoned after the removal of canopy. Photos were taken in North Sulawesi, Indonesia (A), and Jiulong River Estuary, China (B).

food, and fish feces (Gräslund et al., 2003; Naylor and Burke, 2005) are completely dredged and dumped in the form of sludge into the coastal areas (Fig. 9.2). Chen et al. (2020) found that the discharge of wastewater resulted in a higher level of ammonia (up to ~200 μg g^{-1}) in the surface mangrove soil, which was more than 10 times higher than that in soil that is not subjected to wastewater discharge. In addition, fertilization practices that are often carried out in the nitrogen-limited tidal salt marshes

FIG. 9.2

Mariculture is one of the most important activities carried out in the tropical and subtropical coastal areas. During culture, harmful pathogenic bacteria, and nutrients derived from feed additives accumulate in the bottom sediment of farm ponds. Pond sediments are often completely dredged and dumped into adjacent intertidal areas in the form of sludge.

(Pennings et al., 2002) to facilitate the growth of wetland plants can also increase the nutrient availability in wetland soils for microbial activities. Various types of waste-waters, such as shrimp pond effluent, livestock wastewater, and municipal sewage, are enriched in carbon, nitrogen, and phosphorus, which can be retained in mangrove soils and enhance plant growth (Ye et al., 2001; Gautier et al., 2001; Yang et al., 2008; Barcellos et al., 2019; Queiroz et al., 2020). These wastewater discharges can also lead to elevated levels of toxic compounds, including heavy metals (Sámano et al., 2014) and antibiotics (Liu et al., 2016) in coastal wetlands, subsequently affecting GHG emissions from soils through their toxic effects on soil microbial processes (Chen et al., 2014; Huang et al., 2017; Yin et al., 2016).

The loss and degradation of mangrove wetlands, along with the significant rise in public consciousness of environment protection, have prompted a worldwide effort to mitigate mangrove loss, mostly through the rehabilitation of mangroves along the coastlines where mangroves have been damaged, or on newly accreted lands, non-vegetated mudflats, and abandoned shrimp ponds in recent decades (Worthington and Spalding, 2018; Lewis III et al., 2019). Many Asian countries have conducted large-scale rehabilitations of mangrove forests. For example, approximately $1200\,km^2$ of mangroves had been planted in Bangladesh Sundarbans between the 1960s and the early 1990s (Saenger and Siddiqi, 1993), while nearly $530\,km^2$ of mangroves had been planted in Vietnam between 1975 and the early 1990s (FAO, 2007).

There has been considerable interest in the restoration of tidal flows to diked salt marshes in order to re-establish the important intertidal wetlands over the past few decades (Portnoy, 1999). Restoration of intertidal wetlands occurs where hydrologic modifications to reverse drainage or remove impoundments and other obstructions to hydrologic flow take place (Roman and Burdick, 2012). The typical methods for tidal flow restoration include manually opening or abandoning the tide gates, enlarging culverts, and removing fills from former marshes that have been buried. Such restoration efforts can lead to a desirable shift in the floral and faunal communities of a restricted marsh toward preimpact reference conditions and cause changes in the biogeochemical processes in wetland soils.

9.3 Greenhouse gas emissions due to tidal restriction and restoration

The emissions of CH_4 and CO_2 from wetland soils depend largely on water salinity, the degree of flooding and water saturation, and organic matter contents (Bartlett et al., 1987; Olsson et al., 2015). The flux of CH_4 from coastal wetland soils decreases with salinity following a log-linear function (Poffenbarger et al., 2011), with an increase in the oxidation of CH_4 to CO_2 by chemoautotrophic methanotrophs in response to the lowering of water table (Bridgham et al., 2013). Reduced flooding and water saturation of wetland soils due to drainage can expose the formerly water-logged and anoxic soils to oxygen (Table 9.2), thereby increasing the aerobic zone and decreasing the anaerobic zone of organic matter decomposition (Portnoy and Giblin, 1997). Previous studies have reported an increase in CO_2 emissions but a decrease in CH_4 emissions in response to wetland drainage (Nyman and DeLaune, 1991; Salm et al., 2012). A reduction in water supply in the arid areas will result in hypersalinization and further limit CH_4 emissions because sulfate reduction can outcompete methanogenesis, and sulfate can also serve as the terminal electron acceptor in the oxidation of CH_4 to CO_2 (Megonigal and Schlesinger, 2002).

Rewetting drained soils by the re-establishment of tidal connectivity promotes oxygen depletion in the wetland soils and favors anaerobic microbial activity,

Table 9.2 Impacts of tidal restriction and restoration on soil biogeochemical processes involved in the production of greenhouse gases.

Restriction	Potential impacts	Restoration	Potential impacts	References
Drainage	• Wetland water table is lowered, and soil is aerated • Aerobic carbon decomposition is enhanced • Nitrification is promoted and produces nitrate under long-term drought conditions • Hypersalinization occurs in arid area and limits methanogenesis	Rewetting	• Tidal inundation is achieved with a reduction in soil redox potential • Anaerobic decomposition is enhanced in the drained soil • Exposure to saline water limits methanogenesis • Denitrification occurs with rapid nitrate depletion • Ammonia availability increases with sulfate reduction and cation exchange • Sulfite produced from sulfate reduction hinders nitrification and denitrification	Ardón et al. (2013), Joye and Hollibaugh (1995), Megonigal and Schlesinger (2002), Nyman and DeLaune (1991), Osborne et al. (2015), Portnoy (1999), Portnoy and Giblin (1997), Senga et al. (2006), Steinmuller and Chambers (2017), and Weston et al. (2010)
Impoundment	• Wetland soil becomes more reduced • Decreased sulfate supply promotes methanogenesis as a major pathway of anaerobic respiration • Gas diffusion from the substrate to the atmosphere is slow	Tide re-establishment	• Exposure of the soil surface during low tides may favor aerobic carbon decomposition • Seawater intrusion reduces methanogenesis • Ammonia availability increases with sulfate reduction and cation exchange • Sulfite produced from sulfate reduction hinders nitrification and denitrification	Ardón et al. (2013), Helton et al. (2014), Joye and Hollibaugh (1995), Osborne et al. (2015), Steinmuller and Chambers (2017), Senga et al. (2006), and Weston et al. (2010)

leading to lower redox potentials in soils (Portnoy, 1999). Microbial decomposition is predicted to shift from aerobic to anaerobic regimes, increasing the potential of CH_4 emissions from rewetted wetland soils. However, the rates of methanogenesis and fluxes of CH_4 from wetland soils under anoxic conditions would likely decrease when exposed to tidal water (Marton et al., 2012; Chambers et al., 2011; Helton et al., 2014), owing to an alleviation of competition by the energetically more efficient sulfate- and nitrate-reducing bacteria than the methanogenic bacteria (Biswas et al., 2007). Helton et al. (2019) reported insignificant effect of tidal restoration on the flux of CH_4 from wetland soils.

A decrease in sulfate supply as a result of seawater exclusion in impounded wetlands can cause a shift in the predominant pathway of anaerobic respiration from sulfate reduction to methanogenesis, with concomitant changes in pore-water chemistry (Portnoy, 1999). The reduced soil conditions due to flooding and poor pore water exchange are favorable for methanogenesis and CH_4 emissions (Olsson et al., 2015 Emery and Fulweiler, 2017; Yang et al., 2020), while the restoration of impounded wetlands to intermittently tidal-flooded wetlands reverses the effect. Ding et al. (2010) reported that the permanent inundation of a *Spartina alterniflora* salt marsh by blackish water (salinity 5 psu) significantly reduced the soil redox potential and stimulated soil CH_4 production as compared to the intermittently inundated soil, resulting in higher CH_4 emissions from the soil. Helton et al. (2019) also found that wetland soils intermittently flooded by seawater had a higher CO_2 flux but lower CH_4 flux than soils permanently flooded by fresh water. This might be attributed to the tidal water as a likely inexhaustible source of sulfate for the soils in the restored wetland, leading to favorable conditions for sulfate reduction that will in turn lower CH_4 emissions. It has also been suggested that increases in the oxygen availability in surface soils can enhance the aerobic decomposition of organic carbon as well as CH_4 consumption by methanotrophs while at the same time suppress methanogenesis (Helton et al., 2014). Methanogenesis is naturally low in the surface soils in mangrove forest because the oxic/suboxic conditions do not favor methanogenesis (Nóbrega et al., 2016; Das et al., 2021). Studies have suggested that even in deeper mangrove soils with favorable conditions to methanogenesis, CH_4 emission is low due to CH_4 consumption by methanotrophs during the upward diffusion of CH_4 in the soil column (Kristensen et al., 2008). Moreover, the intrusion of sulfate has been found to enhance the mineralization of soil carbon and the associated CO_2 emissions from wetland soils through increased sulfate reduction (Portnoy, 1999; Chambers et al., 2011; Morrissey et al., 2014). However, Helton et al. (2019) found no significant effect of sulfate addition on CO_2 emissions in either permanently or intermittently flooded soils. Microbial functions related to oxidation-reduction potential, CO_2 production, porewater NH_4^+-N, and dissolved organic carbon in the brackish soil could be more strongly impacted by increased inundation than increased salinity (Chambers et al., 2016).

Aerobic decomposition is more efficient than anaerobic decomposition (Reddy and Patrick Jr., 1975; Kristensen et al., 1995; Reddy and De Laune, 2008).

Drainage favored aerobic decomposition that results in a higher loss of soil carbon and lower accumulation of soil organic carbon (OC) than in waterlogged wetlands (Portnoy, 1999). On the other hand, reduced tidal intrusion increases the productivity of salt marsh by relieving salt stress on the existing halophytes and allowing more productive brackish and freshwater species to colonize (Roman et al., 1984). The blockage or restriction of tidal exchange also limits the export of plant materials by tides, resulting in the retention of plant-derived organic matter within the wetland soils.

Apart from CO_2 and CH_4 emissions, N_2O fluxes will also respond to tidal restriction and restoration. Soil aeration due to drainage may promote nitrification, which will in turn likely inhibit the anaerobic process of denitrification. Reflooding of drained wetland soils potentially results in N_2O emissions even when the floodwater is nitrogen-limited (Blackwell et al., 2010; Wollenberg et al., 2018). The emission of N_2O is probably due to the reduction of intrinsic soil nitrate produced after an extended period of drought before reflooding. Tidal restoration could lead to a suboxic condition in the surface layer of the wetland soils, which is more favorable for organic nitrogen mineralization than the oxic condition before reflooding, resulting in greater release of inorganic nitrogen and N_2O emission (Queiroz et al., 2019). If an area has an increased nitrogen supply (e.g., fertilized land) prior to tidal reflooding, it can become a substantial source of N_2O during reflooding. Adams et al. (2012) reported higher N_2O emissions in managed realignment marshes, especially those with high anthropogenic nitrogen enrichment before restoration, than in natural marshes. However, when nitrogen substrate is rapidly depleted without continuous replenishment, N_2O emissions from rewetted soils may be negligible. Previous studies have found that exogenous nutrient inputs, in the forms of either nitrogen addition or wastewater discharge from mariculture farms, produce instant N_2O emission from the impacted wetland soils, while the flux decreases when the nitrogen content drops back to the background levels prior to the nutrient input (Moseman-Valtierra et al., 2011; Chen et al., 2020).

Ammonium is likely released from wetland soils upon exposure to tidal water as a result of sulfate reduction, with the enhanced cation exchange also contributing to the mobilization of ammonia (Weston et al., 2010; Ardón et al., 2013; Steinmuller and Chambers, 2017). An increase in porewater ammonium concentration in restored soils by up to tenfold in response to reflooding has been demonstrated in previous studies by flooding soil cores from historically drained salt marshes (Portnoy and Giblin, 1997; Blackwell et al., 2004). The increased nitrogen supply alters the substrate availability for nitrification and the subsequent denitrification, both of which will produce N_2O. However, the toxic effect of sulfite as a product of sulfate reduction on soil microbial nitrification and denitrification (Joye and Hollibaugh, 1995; Senga et al., 2006; Osborne et al., 2015) makes it difficult to predict N_2O emissions following the incursion of nutrient-limited seawater into drained soil. Some studies have found that N_2O emissions are not influenced by tidal restriction and restoration in salt marshes (Emery and Fulweiler, 2017; Yang et al., 2020).

9.4 Soil greenhouse gas emissions from deforested and rehabilitated mangroves

The flux of CO_2 from intact forest soils can be attributed to both root respiration and microbial decomposition of organic matter (Poungparn et al., 2009; Lang'at et al., 2014), while that from soils in cleared mangroves is mainly due to microbial decomposition (Lovelock et al., 2011), which varies as a function of the biotic and abiotic factors such as soil organic carbon concentration, redox potential, flooding, and pH (Chen et al., 2012). Based on field measurements along a chronosequence of sites representing different periods since mangrove clearance, Lovelock et al. (2011) found a transient increase in CO_2 emissions shortly after clearance that was related to the oxidation of relatively labile fractions of soil organic matter, followed by a logarithmic decline in CO_2 flux from soils over time as a result of the slow decomposition of refractory soil organic matter after the depletion of the labile fraction. Some researchers have also proposed that the substantial loss of soil organic carbon may be the main factor accounting for the decreasing rate of soil CO_2 emissions with time in cleared mangrove forest sites (Bulmer et al., 2015; Grellier et al., 2017). Castillo et al. (2017) observed lower CH_4 fluxes in a cleared mangrove site than an intact mangrove site, but no significant difference in N_2O flux between these two sites.

Due to the shade provided by its complex aboveground structure, the mangrove canopy ameliorates the harsh physical conditions caused by high temperatures and high evaporation rates in the intertidal areas (McGuinness, 1994). A loss of the mangrove canopy following vegetation clearance or thinning results in the exposure of the soil surface to insolation, likely leading to an increase in surface temperature and a decrease in soil moisture. These conditions are favorable for the microbial decomposition of soil organic carbon. An increase in soil CO_2 emissions soon after small-scale cutting of mangrove trees in Gazi Bay in Kenya was observed due to the high soil surface temperatures and the decomposition of soil organic matter in the cut plots, although the flux decreased to levels similar to that measured in the intact mangrove sites within a few months (Lang'at et al., 2014). It is also found that the easily fermentable substrates released from the dying roots could provide substrates for anaerobic respiration and stimulate short-term CH_4 emissions from the cut plots. In contrast to an intact mangrove forest where excess oxygen is transferred from the aerial parts to the rhizosphere through aerenchyma cells to ameliorate the anoxic conditions (Pi et al., 2009; Srikanth et al., 2016), a cleared mangrove is more favorable for anaerobic respiration as a result of a lack of oxygen release from mangrove roots.

The term "blue carbon sink" refers to the large capacity of coastal vegetated ecosystems in storing carbon (Alongi, 2014; Chmura et al., 2003). Previous studies have reported that the carbon stored in the soils of coastal ecosystems is equivalent to 20%–30% of the total amount of carbon stored in global soils (Lal, 2008; Mitsch et al., 2013). Continuous accumulation of organic carbon has been observed in rehabilitated mangrove soils, which can be derived from either allochthonous suspended particulate organic matter imported by tides or autochthonous mangrove materials

(Chen et al., 2021). In a comparison of rehabilitated mangrove sites with reference sites that have similar geomorphology as those disturbed sites before rehabilitation, i.e., bare flat or abandoned aquaculture ponds, a higher CO_2 flux is generally found at the rehabilitated mangrove sites, with the flux increasing with the amount of soil organic carbon accumulated following rehabilitation (Table 9.3). CO_2 fluxes gradually increase with forest age as a consequence of organic matter accumulation in the soil, while N_2O and CH_4 fluxes show no clear trends with age along the chronosequence.

Macrobenthos are important component in mangrove forests and their bioturbation activities can influence the biogeochemical cycles, mainly through their burrowing and feeding activities (Werry and Lee, 2005; Kristensen, 2008; Chen et al., 2016b). Their burrowing activities can modify the particle size distribution, the drainage of tidal water, and enhance the exposure of sediment to air (Botto and Iribarne, 2000; Kristensen, 2008). The herbivorous benthos can retain large quantities of mangrove-derived organic matter in the ecosystem during their shredding, grazing, and burying activities (Lee, 1998; Chen et al., 2008). These bioturbation activities therefore have the potential to affect microbial mineralization and GHG emissions in soils, which are related to the community composition of the benthos (Kristensen, 2008; Mehring et al., 2017; Bernardino et al., 2020). Because mangrove clearance or rehabilitation can change the abundance and composition of macrobenthos (Fondo and Martens, 1998; Chen et al., 2007; Bernardino et al., 2020), these anthropogenic activities may impact the bioturbation activities and ultimately the microbial mineralization of soil carbon and soil GHG emissions, with potential involvement in carbon sequestration that deserves further study (Bernardino et al., 2020).

The removal/creation of mangrove canopies may also affect the microphytobenthos via changes in illumination, which may then alter the emissions of GHGs from the soil. Although there is no study directly examining this impact so far, previous studies on the ecology of microphytobenthos in mangrove forests have provided indirect evidence of their impact on GHG emissions. Microphytobenthos, as the primary producer in mangrove forests, grow within the top several millimeters of illuminated soils (Macintyre et al., 1996). Their abundance is generally low or undetectable in mangroves because of the limited light penetration to the surface soil due to shading by the tree canopy, the limitation of nutrients, and inhibition by soil organic compounds such as tannin (Alongi, 1994; Gattuso et al., 1998). The microphytobenthos on the soil surface form conspicuous microbial mats or biofilms, especially during the cold seasons in subtropical or temperate mangrove forests where light penetration is sufficient (Lovelock, 2008; Chen et al., 2019). The presence of microphytobenthic mats or biofilms can change soil GHG fluxes through photosynthesis under light conditions or respiration in the dark (Chen et al., 2019). The mat/biofilm may also act as a barrier to prevent the upward diffusion of gases produced in the deeper soil, which may reduce the net GHG flux to the atmosphere (Bulmer et al., 2015; Leopold et al., 2015). Chen et al. (2019) found that the

Table 9.3 Soil greenhouse gas fluxes in the rehabilitated mangroves and the reference sites that are representative of either natural or prerehabilitated disturbed conditions.

Location	Site condition	CO_2 flux $(mg\,m^{-2}\,h^{-1})$	CH_4 flux $(\mu g\,m^{-2}\,h^{-1})$	N_2O flux $(\mu g\,m^{-2}\,h^{-1})$	OC $(mg\,g^{-1})$	TN $(mg\,g^{-1})$	Reference
Jiulong River Estuary, China	Bare flat	−9.8	30.9	17.2	16.7	1.2	Yu (2014)
	26-year mangrove	51.0	73.9	10.2	18.3	1.4	
	50-year mangrove	79.4	412.2	11.2	30.8	2.0	
	Natural mangrove	181.5	1608.7	22.3	27.2	1.8	
North Sulawesi, Indonesia	Abandoned pond	292.2	2853.9	99.6	ND	ND	Cameron et al. (2019)
	10-year mangrove	256.9–335.6	502.3–2237.4	88.1–149.4	ND	ND	
	Natural mangrove	92.5–319.6	1050.2–15,936.1	30.6–153.2	ND	ND	
Red River Delta, Northern Viet Nam	Bare flat	30.0–161.1 (77.8)	ND	ND	8.0	ND	Hien et al. (2018)
	18-year mangrove	68.1–418.8 (175.1)	ND	ND	12.4	ND	
Central Lombok, Indonesia	Abandoned pond	38.05	0.64	4.66	3.5	0.3	Chen et al., unpublished data
	12-year mangrove	50.84	0.54	11.13	12.5	1.0	
	20-year mangrove	63.09	1.83	1.85	43.1	1.5	
	Natural mangrove	203.02	0.46	3.52	50.0	1.5	

ND, not determined.

microphytobenthic biofilms formed at the soil surface could change the mangrove soil from a source to a sink of CO_2 in the cold seasons with low microbial respiration, but exert no significant effect on the fluxes of both CH_4 and N_2O probably due to their low magnitudes. In the open-canopy mangroves or cleared mangrove sites, microbial mats or biofilms may form and change the soil fluxes of GHGs, especially CO_2.

9.5 Greenhouse gas fluxes in nutrient-enriched wetlands

It has been suggested that the emissions of CH_4 and N_2O from mangroves and other intertidal wetland soils are negligible when compared to those from other wetlands (Sotomayor et al., 1994) as mangrove soils are generally oligotrophic or nutrient-limited (Allen et al., 2007; Pennings et al., 2002), and the high salinity in saline wetlands further reduces CH_4 emissions (Biswas et al., 2007). Under a high degree of saturation in soils, the rate of decomposition of organic matter decreases, leading to low CO_2 emissions from these wetland soils (Armentano and Menges, 1986). However, not all intertidal wetlands are limited by low nutrient availability. Many wetlands receive nutrient inputs from various human activities, which can shift the wetland soil from a GHG sink to a source of or substantially enhance the strength of existing GHG sources (Chen et al., 2011, 2020; Moseman-Valtierra et al., 2011; Chmura et al., 2016). For instance, the annual mean fluxes of CH_4 and N_2O reached $3899\,\mu g\,CH_4\,m^{-2}\,h^{-1}$ and $19.0\,\mu g\,N_2O\,m^{-2}\,h^{-1}$, respectively, in a Brisbane mangrove impacted by discharges from a sewage treatment plant in Queensland, Australia (Allen et al., 2007). Even higher GHG fluxes of $453.6–1047.6\,\mu g\,N_2O\,m^{-2}\,h^{-1}$ and $161.6–82,697.6\,\mu g\,CH_4\,m^{-2}\,h^{-1}$ have been reported from the mangrove soils in a eutrophic Futian mangrove in South China (Chen et al., 2010). Table 9.4 compares the fluxes of CH_4 and N_2O measured in various mangrove forests that are subjected to nutrient enrichment with those in pristine or less-polluted mangrove forests. The stimulatory effect of nutrient enrichment on the fluxes of CH_4 and N_2O in intertidal wetland soils is evident, with the fluxes ranging from $−15.4$ to $50.0\,\mu g\,m^{-2}\,h^{-1}$ for N_2O and from $−96.8$ to $5360.1\,\mu g\,m^{-2}\,h^{-1}$ for CH_4 in the nonpolluted or less-polluted mangrove forests, while those in nutrient-enriched forests were much higher ranging from $−4$ to $1047.6\,\mu g\,N_2O\,m^{-2}\,h^{-1}$ and from 13.3 to $82,697.6\,\mu g\,CH_4\,m^{-2}\,h^{-1}$. Therefore, the global warming impacts of these two GHGs cannot be ignored.

The composition of nutrient loads affects the quantity and relative proportion of different GHGs released from wetland soils (Chen et al., 2011). The addition of ammonia is more effective in stimulating N_2O flux from wetland soils than nitrate addition under the same load (Munoz-Hincapie et al., 2002), because the N_2O emission under nitrate addition is attributed to denitrification alone, while both nitrification and denitrification of the nitrate/nitrite produced through nitrification can contribute to the total N_2O flux under ammonium addition (Munoz-Hincapie et al., 2002). Chen et al. (2011, 2020) found that wetland soils receiving ammonia-rich wastewater exhibited enhanced N_2O emissions which increased

Table 9.4 Soil-atmosphere fluxes of N_2O and CH_4 from nutrient-enriched and nonpolluted or less-polluted mangrove soils.

Location	N_2O flux ($\mu g\,m^{-2}\,h^{-1}$)	CH_4 flux ($\mu g\,m^{-2}\,h^{-1}$)	References
Nutrient-enriched mangroves[a]			
Brisbane River and Moreton Bay, Queensland, Australia	-4–202.0	3–17370	Allen et al. (2007, 2011)
Bay of La Parguera, Southwest Puerto Rico	5.3–43.0	154.2–3400.0	Sotomayor et al. (1994) and Corredor et al. (1999)
Bird Island, Southwest Puerto Rico	343.2	ND	Corredor et al. (1999)
Maipo, Hong Kong, South China	32.1–533.7	547.2–4390.4	Chen et al. (2010, 2012)
Futian, Shenzhen, South China	453.6–1047.6	161.6–82,697.6	Chen et al. (2010)
Mtoni, Tanzania	ND	13.3–58.4	Kristensen et al. (2008)
Muthupet, South India	17.9–33.8	778.8–1561.3	Krithika et al. (2008)
Pichavaram, South India	39.2–79.2	308.3–13,520.0	Purvaja and Ramesh (2001), Purvaja et al. (2004), and Chauhan et al. (2008)
Indian Sundarbangs, East India	ND	136–507	Dutta et al. (2013)
Bhitarkanika, eastern coast of India	9.2–207.9	91.7–3230.0	Chauhan et al. (2008)
Jaguaribe River Estuary, Ceará, NE Brazil	90.9	200	Queiroz et al. (2019)
Nonpolluted or less-polluted mangroves[a]			
Myora Spring, Australia	1.5–3.0	20–130	Kreuzwieser et al. (2003)
Enrique Reef, Southwest Puerto Rico	11	ND	Corredor et al. (1999)
Bahia La Parguera, Southwest Puerto Rico	ND	12.5–79.2	Sotomayor et al. (1994)
Magueyes Island, Southwest Puerto Rico	16.72	ND	Corredor et al. (1999)
Ras Dege, Tanzania	ND	0–16.7	Kristensen et al. (2008)
Ranong Biosphere Reserve, Thailand	ND	7.9–37.9	Lekphet et al. (2005)
Yung Shue O, Hong Kong	11–32.56	211.2–1025.6	Chen et al. (2010)

Table 9.4 Soil-atmosphere fluxes of N_2O and CH_4 from nutrient-enriched and nonpolluted or less-polluted mangrove soils—cont'd

Location	N_2O flux ($\mu g\,m^{-2}\,h^{-1}$)	CH_4 flux ($\mu g\,m^{-2}\,h^{-1}$)	References
North Sulawesi, Indonesia	−15.4–26.84	−96.8–210.2	Chen et al. (2014)
Jiulong River Estuary, China	−1.6–50.0	−1.4–5360.1	Chen et al. (2016a)
Andaman Island, Bay of Bengal, India	12.3–29.0	ND	Jennifer et al. (2013)
Jaguaribe River Estuary, Ceará, NE Brazil	40.4	800	Queiroz et al. (2019)
Honda Bay, Puerto Princesa, Palawan, Philippines	~0.44	72.8–122.24	Castillo et al. (2017)

ND, no determination of the in situ gas flux.
[a]*Nutrient-enriched mangroves are those sites impacted by wastewater discharge or eutrophic coastal water, while nonpolluted or less-polluted mangroves are pristine mangroves and/or mangroves that have not received any direct wastewater discharge or eutrophic coastal water.*

exponentially with the soil ammonia content, while the increase in soil nitrate was insignificant or small that indicated a loss of nitrate through microbial denitrification. In the presence of high levels of dissolved organic carbon, enhanced microbial respiration reduces the availability of oxygen in soils, leading to anoxic conditions that promote the emissions of CH_4 and N_2O from anaerobic respiration and denitrification, respectively (Chen et al., 2011). This, in turn, limits nitrification even if high levels of ammonia are present (Strauss and Lamberti, 2000; Chen et al., 2011). Strauss and Lamberti (2000) found that the addition of organic carbon increased microbial respiration rates but decreased nitrification rates in a hydric ecosystem. Dissimilatory nitrate reduction to ammonium (also known as nitrate ammonification) in extremely reduced soil zones has also been reported (Buresh and Patrick, 1981). Anaerobic ammonium oxidation in anoxic soil under the availability of both reduced and oxidized inorganic nitrogenous bases (Thamdrup and Dalsgaard, 2002) can outcompete the nitrogen substrate for gas production and regulate gas emissions from wetland soils.

Numerous studies have demonstrated that nitrogen enrichment, in some cases along with other nutrients, results in higher N_2O emissions from wetland soils (Moseman-Valtierra et al., 2011; Chmura et al., 2016). The nitrogen compounds can also accelerate organic matter decomposition and CO_2 emission from the wetland soils (Queiroz et al., 2019). However, the effect of nitrogen enrichment on CH_4 flux shows inconsistent patterns. The soil ammonia content is positively related to soil CH_4 fluxes in many intertidal wetlands (Allen et al., 2007; Chen et al., 2010; Irvine et al., 2012) because of the competitive inhibition of methane monooxygenase enzymes by ammonia and nitrite produced during ammonium oxidation (Bosse et al., 1993, Schnell and King, 1995). The addition of both organic (urea, Irvine et al., 2012) and inorganic (NH_4NO_3, Kim et al., 2020) nitrogen has been found to significantly enhance CH_4 emissions from marsh soils (Irvine et al., 2012; Kim et al., 2020), with the flux increasing linearly with the level of nitrogen applied (Irvine et al., 2012). On the other hand, mangrove soils that received effluents from mariculture ponds had CH_4 fluxes comparable to those at an unimpacted site (without effluent discharge), although the latter had lower fluxes of CO_2 and N_2O (Queiroz et al., 2019). This varied effect of nitrogen enrichment on CH_4 may be partially attributed to the concentration and species of nitrogen. Van der Nat et al. (1997) reported the existence of ammonia inhibition, which is only effective when the concentration of ammonium exceeds that of CH_4 by at least 30-fold. Nitrate can also largely inhibit CH_4 production (Wang et al., 2010) because of the intensive oxidation of CH_4 and/or out-competition of methanogenic bacteria by nitrate-reducing bacteria (Biswas et al., 2007). These results show that the production/consumption of CH_4 in intertidal wetland soils is very complicated, requiring more research on this aspect.

Most studies have reported the stimulation of GHG emissions from coastal wetland soils receiving exogenous nutrients as tentative, usually for approximately 2 days (Lindau and Delaune, 1991; Moseman-Valtierra et al., 2011; Chen et al., 2014). The short duration of the enhanced N_2O fluxes is related to the rapid microbial assimilation of nitrogen and the reduction of N_2O to N_2 in intertidal wetland soils, as well as the reversion of the microbial population back to the background level prior to nitrogen addition (Lindau and Delaune, 1991; Moseman-Valtierra et al., 2011). In a nutrient-enriched site, the rapid decline in gas fluxes and soil nitrogen may also be attributed to the loss of nitrogen with tidal flushing, thus reducing the substrate availability for microbial N_2O production. In a study of the effect of the discharge of wastewater from dredging shrimp pond sediments on N_2O emissions from mangrove soils, Chen et al. (2020) found that a longer time was required for the N_2O flux to recover to the previous discharge level when the discharge was performed on a neap tide day than a spring tide day, possibly because of a longer time required for the gas flux and nutrient content to decrease to the level before N addition under a higher nitrogen load. Lindau and Delaune (1991) found that the application of $10\,g$ $NO_3^--N\,m^{-2}$ to marsh soil produced instant N_2O soil emissions that reached $31\,\mu mol\,m^{-2}\,h^{-1}$, which were higher than those produced from the addition of 1.4 $NO_3^--N\,m^{-2}$ to soil, as reported by Moseman-Valtierra et al. (2011). Also, it took a longer time of approximately 11 and 2 days for the flux and nutrient content to decrease to the levels before N addition under the two NO_3^--N loadings, respectively. The GHG emissions and their potential contribution to global warming in mangroves and other intertidal wetlands under the influence of nutrient enrichment deserve more in-depth research.

The import of toxic pollutants, like heavy metals and antibiotics, along with nutrients into the wetland soils has been demonstrated to affect the microbial processes involved in the GHG production in coastal soils (Magalhães et al., 2007; Hou et al., 2015). For instance, thiamphenicol has been shown to exert greater inhibition effect on nitrite reduction in denitrification and nitrite oxidation in nitrification process (Yin et al., 2016). Some studies have found that heavy metals or antibiotics under low levels could enhance nitrification under suboxic conditions (Chen et al., 2014), promote denitrification and methanogenesis under anaerobic conditions (DeVries et al., 2015; Ma et al., 2021), but inhibit these processes under high contamination levels. Soil microorganisms could adapt to these stresses by developing a variety of resistance mechanisms, which occur from less than 1 to 2 years after contamination, with ammonium addition accelerating microbial adaptation to such toxicity (as summarized by Chen et al., 2014). Our previous study found that high zinc concentration in the NH_4^+-N rich wastewater caused significantly lower N_2O emission from wetland soil after discharge relative to the soil receiving wastewater without zinc, while the inhibiting effects on N_2O emission could be relieved rapidly in less than 12 h (Chen et al., 2014). The responses of microbial metabolisms of carbon and nitrogen and GHG production in soils are stronger dependent on various factors, including the metabolism steps, the species and level of toxic pollutants, soil texture, and treatment time (Chen et al., 2014, 2017; DeVries et al., 2015; Yin et al., 2016). However, currently the impacts of these pollutants on soil microbial processes and GHG production have rarely been reported in intertidal wetlands, and our knowledge on their effects on the soil GHG emissions is even scarcer.

9.6 Conclusion

Intertidal wetlands are effective in carbon sequestration and play an important role in mitigating climate change. In this chapter, we review various studies on GHG emissions from intertidal wetland soils, which are found to vary in response to anthropogenic impacts. Anthropogenic activities may affect soil physicochemical properties and environmental settings, as well as introduce toxic pollutants, which in turn lead to a change in the magnitude and patterns of GHG emissions. Among different activities, nutrient enrichment or nutrient pollution is a driving force that largely stimulates GHG emissions (especially N_2O) from wetland soils, or shift the soils from a net sink to a net source of GHGs. In some cases, an increase in soil GHG emissions can offset the amount of carbon sequestered in coastal wetlands and may even counterbalance any benefit of carbon storage in the mitigation of climate change (Suarez-Abelenda et al., 2014; Chen et al., 2016a,b; Roughan et al., 2018), making the ecosystem a source of GHGs. Pollution control and management in the coastal areas are essential to reduce GHG emissions to the atmosphere. Other anthropogenic activities also substantially affect the production and emission of CH_4 and CO_2 gases from wetland soils, mostly by changing the oxygen availability and salinity of the soil.

As anthropogenic activities can impact various biogeochemical processes responsible for the production of GHGs, they may have concurrent effects on other biogeochemical processes that can either increase or decrease GHG production in soils. The effects of these activities on GHG emissions therefore vary from site to site and over time. However, the biogeochemical processes and the role of microbial community in the production of GHGs in intertidal wetland soils under different anthropogenic activities are still poorly understood. Further studies investigating GHG emissions and the underlying mechanisms of GHG production in impacted wetland soils are desperately needed.

Acknowledgments

The work described in this paper was funded by the National Natural Science Foundation of China (41776102, 41976161) and a grant from the Research Grants Council of the Hong Kong Special Administrative Region, China (UGC/IDS(R) 16/19). We appreciate Dr. Gail Chmura for knowledge sharing of the anthropogenic impacts on greenhouse gas emissions from coastal wetland soils and the comments of two anonymous reviewers, which helped improve the manuscript. The authors are grateful to Mr. Jiahui Chen for his assistance in data collection.

References

Abuodha, P., Kairo, J.G., 2001. Human-induced stresses on mangrove swamps along Kenya coast. Hydrobiologia 458, 255–265.

Adams, C.A., Andrews, J.E., Jickells, T., 2012. Nitrous oxide and methane fluxes vs. carbon, nitrogen and phosphorous burial in new intertidal and saltmarsh sediments. Sci Total Environ 434, 240–251.

Allen, D.E., Dalal, R.C., Rennenberg, H., Louise Meyer, R., Reeves, S., Schmidt, S., 2007. Spatial and temporal variation of nitrous oxide and methane flux between subtropical mangrove soils and the atmosphere. Soil Biol. Biochem. 39, 622–631.

Allen, D.E., Dalal, R.C., Rennenberg, H., Schmidt, S., 2011. Seasonal variation in nitrous oxide and methane emissions from subtropical estuary and coastal mangrove sediments, Australia. Plant Biol. 13, 126–133.

Alongi, D.M., 1994. Zonation and seasonality of benthic primary production and community respiration in tropical mangrove forests. Oecologia 98, 320–327.

Alongi, D.M., 2002. Present state and future of the world's mangrove forests. Environ. Conserv. 29, 331–349.

Alongi, D.M., 2009. The Energetics of Mangrove Forests. Springer, Dordrecht.

Alongi, D.M., 2014. Carbon cycling and storage in mangrove forests. Annu. Rev. Mar. Sci. 6, 195–219.

Ardón, M., Morse, J.L., Colman, B., Bernhardt, E.S., 2013. Drought induced saltwater incursion leads to increased wetland nitrogen export. Glob. Chang. Biol. 19, 2976–2985.

Armentano, T.V., Menges, E.S., 1986. Patterns of change in the carbon balance of organic soil-wetlands of the temperate zone. J. Ecol. 74, 755–774.

Barbier, E.B., Hacker, S.D., Kennedy, C., Koch, E.W., Stier, A.C., Silliman, B.R., 2011. The value of estuarine and coastal ecosystem services. Ecol. Monogr. 8, 169–193.

Barcellos, D., Queiroz, H.M., Nóbrega, G.N., de Oliveira Filho, R.L., Santaella, S.T., Otero, X.L., Ferreira, T.O., 2019. Phosphorus enriched effluents increase eutrophication risks for mangrove systems in northeastern Brazil. Mar. Pollut. Bull. 142, 58–63.

Bartlett, K.B., Bartlett, D.S., Harriss, R.C., Sebacher, D.I., 1987. Methane emissions along a salt-marsh salinity gradient. Biogeochemistry 4, 183–202.

Bernardino, A.F., Sanders, C.J., Bissoli, L.B., de Gomes, L.E.O., Kauffman, J.B., Ferreira, T.O., 2020. Land use impacts on benthic bioturbation potential and carbon burial in Brazilian mangrove ecosystems. Limnol. Oceanogr. 65, 2366–2376.

Biswas, H., Mukhopadhyay, S.K., Sen, S., Jana, T.K., 2007. Spatial and temporal patterns of methane dynamics in the tropical mangrove dominated estuary, NE coast of Bay of Bengal, India. J. Mar. Syst. 68, 55–64.

Blackwell, M.S.A., Hogan, D.V., Maltby, E., 2004. The short-term impact of managed realignment on soil environmental variables and hydrology. Estuar. Coast. Shelf Sci. 59, 687–701.

Blackwell, M., Yamulki, S., Bol, R., 2010. Nitrous oxide production and denitrification rates in estuarine intertidal saltmarsh and managed realignment zones. Estuar. Coast. Shelf Sci. 87, 591–600.

Bosse, U., Frenzer, P., Conrad, R., 1993. Inhibition of methane oxidation by ammonium in the surface layer of a littoral sediment. FEMS Microbiol. Ecol. 13, 123–134.

Botto, F., Iribarne, O., 2000. Contrasting effects of two burrowing crabs (*Chasmagnathus granulata* and *Uca uruguayensis*) on sediment composition and transport in estuarine environments. Estuar. Coast. Shelf Sci. 5, 141–151.

Bridgham, S.D., Cadillo-Quiroz, H., Keller, J.K., Zhuang, Q.L., 2013. Methane emissions from wetlands: biogeochemical, microbial, and modeling perspectives from local to global scales. Glob. Chang. Biol. 19, 1325–1346.

Bulmer, R.H., Lundquist, J., Schwendenmann, L., 2015. Sediment properties and CO_2 efflux from intact and cleared temperate mangrove forests. Biogeosciences 12, 6169–6180.

Buresh, R.J., Patrick, W.H., 1981. Nitrate reduction to ammonium and organic nitrogen in an estuarine sediment. Soil Biol. Biochem. 13, 279–283.

Byun, D.S., Wanga, X.H., Holloway, P.E., 2004. Tidal characteristic adjustment due to dyke and seawall construction in the Mokpo Coastal Zone, Korea. Estuar. Coast. Shelf Sci. 59, 185–196.

Cameron, C., Hutley, L.B., Friess, D.A., Munksgaard, N.C., 2019. Hydroperiod, soil moisture and bioturbation are critical drivers of greenhouse gas fluxes and vary as a function of landuse change in mangroves of Sulawesi, Indonesia. Sci. Total Environ. 654, 365–377.

Castillo, J.A.A., Apan, A.A., TMaraseni, T.N., Salmo III, S.G., 2017. Soil greenhouse gas fluxes in tropical mangrove forests and in land uses on deforested mangrove lands. Catena 159, 60–69.

Chambers, L.G., Reddy, K.R., Osborne, T.Z., 2011. Short-term response of carbon cycling to salinity pulses in a freshwater wetland. Soil Sci. Soc. Am. J. 75, 2000–2007.

Chambers, L.G., Guevara, R., Boyer, J.N., Troxler, T.G., Davis, S.E., 2016. Effects of salinity and inundation on microbial community structure and function in a mangrove peat soil. Wetlands 36, 361–371.

Chauhan, R., Ramanathan, A.L., Adhya, T.K., 2008. Assessment of methane and nitrous oxide flux from mangroves along Eastern coast of India. Geofluids 8, 321–332.

Chen, G.C., Ye, Y., Lu, C.Y., 2007. Changes of macro-benthic faunal community with stand age of rehabilitated *Kandelia candel* mangrove in Jiulongjiang Estuary, China. Ecol. Eng. 31, 215–224.

Chen, G.C., Ye, Y., Lu, C.Y., 2008. Seasonal variability of leaf litter removal by crabs in a *Kandelia candel* mangrove forest in Jiulongjiang Estuary, China. Estuar. Coast. Shelf Sci. 79, 701–706.

Chen, G.C., Tam, N.F.Y., Ye, Y., 2010. Summer fluxes of atmospheric greenhouse gases N_2O, CH_4 and CO_2 from mangrove soil in South China. Sci. Total Environ. 408, 2761–2767.

Chen, G.C., Tam, N.F.Y., Wong, Y.S., Ye, Y., 2011. Effect of wastewater discharge on greenhouse gas fluxes from mangrove soils. Atmos. Environ. 45, 1110–1115.

Chen, G.C., Tam, N.F.Y., Ye, Y., 2012. Spatial and seasonal variations of atmospheric N_2O and CO_2 fluxes from a subtropical mangrove swamp and their relationships with soil characteristics. Soil Biol. Biochem. 48, 175–181.

Chen, G.C., Tam, N.F.Y., Ye, Y., 2014. Does zinc in livestock wastewater reduce nitrous oxide (N_2O) emissions from mangrove soils? Water Res. 65, 402–413.

Chen, G., Chen, B., Yu, D., Tam, N.F.Y., Ye, Y., Chen, S.Y., 2016a. Soil greenhouse gas emissions reduce the contribution of mangrove plants to the atmospheric cooling effect. Environ. Res. Lett. 11 (12), 1–11.

Chen, G.C., Lu, C., Li, R., Chen, B., Hu, Q., Ye, Y., 2016b. Effects of foraging leaf litter of *Aegiceras corniculatum* (Ericales, Myrsinaceae) by *Parasesarma plicatum* (Brachyura, Sesarmidae) crabs on properties of mangrove sediment: a laboratory experiment. Hydrobiologia 763, 125–133.

Chen, S., Chmura, G.L., Wang, Y., Yu, D., Ou, D., Chen, B., Ye, Y., Chen, G., 2019. Benthic microalgae offset the sediment carbon dioxide emission in subtropical mangrove in cold seasons. Limnol. Oceanogr. 64, 1297–1308.

Chen, G., Chen, J., Ou, D., Tam, N.F.Y., Chen, S., Zhang, Q., Chen, B., Ye, Y., 2020. Increased nitrous oxide emissions from intertidal soil receiving wastewater from dredging shrimp pond sediments. Environ. Res. Lett. 15, 094015.

Chen, S., Chen, B., Chen, G., Ji, J., Yu, W., Liao, J., Chen, G., 2021. Higher soil organic carbon sequestration potential at a rehabilitated mangrove comprised of *Aegiceras corniculatum* compared to *Kandelia obovata*. Sci. Total Environ. 752, 142279.

Chmura, G.L., Anisfeld, S.C., Cahoon, D.R., Lynch, J.C., 2003. Global carbon sequestration in tidal, saline wetland soils. Glob. Biogeochem. Cycles 17, 1111.

Chmura, G.L., Kellman, L., Guntenspergen, G.R., 2011. The greenhouse gas flux and potential global warming feedbacks of a northern macrotidal and microtidal salt Marsh. Environ. Res. Lett. 6, 044016.

Chmura, G.L., Kellman, L., van Ardenne, L., Guntenspergen, G.R., 2016. Greenhouse gas fluxes from salt marshes exposed to chronic nutrient enrichment. PLoS One 11, e014993.

Corredor, J.E., Morell, J.M., Bauz, J., 1999. Atmospheric nitrous oxide fluxes from mangrove soils. Mar. Pollut. Bull. 38, 473–478.

Das, S., Ganguly, D., De, T.K., 2021. Microbial methane production-oxidation profile in the soil of mangrove and paddy fields of West Bengal, India. Geomicrobiol. J. 38, 220–230.

DeVries, S.L., Loving, M., Li, X., Zhang, P., 2015. The effect of ultralow-dose antibiotics exposure on soil nitrate and N_2O flux. Sci. Rep. 5, 16818.

Ding, W., Zhang, Y., Cai, Z., 2010. Impact of permanent inundation on methane emissions from a Spartina alterniflora coastal salt marsh. Atmos. Environ. 44, 3894–3900.

Dutta, M.K., Chowdhury, C., Jana, T.K., Mukhopadhyay, S.K., 2013. Dynamics and exchange fluxes of methane in the estuarine mangrove environment of the Sundarbans, NE coast of India. Atmos. Environ. 77, 631–639.

Emery, H.E., Fulweiler, R.W., 2017. Incomplete tidal restoration may lead to persistent high CH_4 emission. Ecosphere 8, e01968.

FAO, 2007. The world's mangroves 1980–2005. FAO Forestry Paper. No. 153.

Fondo, E.N., Martens, E.E., 1998. Effects of mangrove deforestation on macrofaunal densities, Gazi Bay, Kenya. Mangrove Salt Marshes 2, 75–83.

Gattuso, J., Frankignoulle, P.M., Wollast, R., 1998. Carbon and carbonate metabolism in coastal aquatic ecosystems. Annu. Rev. Ecol. Syst. 29, 405–434.

Gautier, D., Amador, J., Newmark, F., 2001. The use of mangrove wetland as a biofilter to treat shrimp pond effluents: preliminary results of an experiment on the Caribbean coast of Colombia. Aquac. Res. 32, 787–799.

Gedan, K.B., Silliman, B.R., Bertness, M.D., 2009. Centuries of human-driven change in salt marsh ecosystems. Annu. Rev. Mar. Sci. 1, 117–141.

Giri, C., Ochieng, E., Tieszen, L.L., Zhu, Z., Singh, A., Loveland, T., Masek, J., Duke, N., 2011. Status and distribution of mangrove forests of the world using earth observation satellite data. Global Ecol. Biogeogr. 20, 154–159.

Goldberg, L., Lagomasino, D., Thomas, N., Fatoyinbo, T., 2020. Global declines in human-driven mangrove loss. Glob. Chang. Biol. 26, 5844–5855.

Gräslund, S., Holmstrom, K., Wahstrom, A., 2003. A field survey of chemicals and biological products used in shrimp farming. Mar. Pollut. Bull. 46, 81–90.

Grellier, S., Janeau, J., Nhon, D.H., Cu, N.T.K., Quynh, L.T.P., Thao, P.T.T., Nhu-Trang, T., Marchand, C., 2017. Changes in soil characteristics and C dynamics after mangrove clearing (Vietnam). Sci. Total Environ. 593, 654–663.

Helton, A.M., Bernhardt, E.S., Fedders, A., 2014. Biogeochemical regime shifts in coastal landscapes: the contrasting effects of saltwater incursion and agricultural pollution on greenhouse gas emissions from a freshwater wetland. Biogeochemistry 120, 133–147.

Helton, A.M., Ardón, M., Bernhardt, E.S., 2019. Hydrologic context alters greenhouse gas feedbacks of coastal wetland salinization. Ecosystems 22, 1108–1125.

Hien, H.T., Marchand, C., Aimé, J., Nguyen, T.K.C., 2018. Seasonal variability of CO_2 emissions from sediments in planted mangroves (Northern Viet Nam). Estuar. Coast. Shelf Sci. 213, 28–39.

Hou, L., Yin, G., Liu, M., Zhou, J., Zheng, Y., Gao, J., Zong, H., Yang, Y., Gao, L., Tong, C., 2015. Effects of sulfamethazine on denitrification and the associated N_2O release in estuarine and coastal sediments. Environ. Sci. Technol. 49, 326–333.

Huang, Y., Ou, D., Chen, S., Chen, B., Liu, W., Bai, R., Chen, G., 2017. Inhibition effect of zinc in wastewater on the N2O emission from coastal loam soils. Mar. Pollut. Bull. 116, 434–439.

Irvine, I.C., Vivanco, L., Bentley, P.N., Martiny, J.B.H., 2012. The effect of nitrogen enrichment on C1-cycling microorganisms and methane flux in salt marsh sediments. Front. Microbiol. 3, 00090.

Jennifer, I.D., Neetha, V., Hariharan, G., Kakolee, B., Purvaja, V., Ramesh, R., 2013. N_2O flux from south Andaman mangroves and surrounding creek waters. Int. J. Oceans Oceanogr. 7, 73–82.

Joye, S.B., Hollibaugh, J.T., 1995. Influence of sulfide inhibition of nitrification on nitrogen regeneration in sediments. Science 270, 623–625.

Kim, J., Chaudhary, D.R., Kang, H., 2020. Nitrogen addition differently alters GHGs production and soil microbial community of tidal salt marsh soil depending on the types of halophyte. Appl. Soil Ecol. 150, 103440.

Kreuzwieser, J., Buchholz, J., Rennenberg, H., 2003. Emission of methane and nitrous oxide by Australian mangrove ecosystems. Plant Biol. 5, 423–431.

Kristensen, E., 2008. Mangrove crabs as ecosystem engineers, with emphasis on sediment processes. J. Sea Res. 59, 30–43.

Kristensen, E., Bouillon, S., Dittmar, T., Marchand, C., 1995. Organic carbon dynamics in mangrove ecosystems: a review. Aquat. Bot. 89, 201–219.

Kristensen, E., Flindt, M.R., Ulomi, S., Borges, A.V., Abril, G., Bouillon, S., 2008. Emission of CO_2 and CH_4 to the atmosphere by sediments and open waters in two Tanzanian mangrove forests. Mar. Ecol. Prog. Ser. 370, 53–67.

Krithika, K., Purvaja, R., Ramesh, R., 2008. Fluxes of methane and nitrous oxide from an Indian mangrove. Curr. Sci. 94, 218–224.

Lal, R., 2008. Carbon sequestration. Philos. Trans. R. Soc. B Biol. Sci. 363, 815–830.

Lang'at, J.K.S., Kairo, J.G., Mencuccini, M., Bouillon, S., Skov, M.W., et al., 2014. Rapid losses of surface elevation following tree girdling and cutting in tropical mangroves. PLoS One 9 (9), e107868.

Lee, S.Y., 1998. Ecological role of grapsid crabs in mangrove ecosystems: a review. Mar. Freshw. Res. 49, 335–343.

Lee, S.Y., Primavera, J.H., Dahdouh-Guebas, F., McKee, K., Bosire, J.O., Cannicci, S., et al., 2014. Ecological role and services of tropical mangrove ecosystems: a reassessment. Glob. Ecol. Biogeogr. 23, 726–743.

Lekphet, S., Nitisoravut, S., Adsavakulchai, S., 2005. Estimating methane emission from mangrove area in Ranong Province, Thailand. Songklanakarin J Sci. Technol. 27, 153–163.

Leopold, A., Marchand, C., Deborde, J., Allenbach, M., 2015. Temporal variability of CO_2 fluxes at the sediment-air interface in mangroves (New Caledonia). Sci. Total Environ. 502, 617–626.

Lewis III, R.R., Brown, B.M., Flynn, L.L., 2019. Chapter 24—Methods and criteria for successful mangrove forest rehabilitation. In: Perillo, G.M.E., Wolanski, E., Cahoon, D.R., Hopkinson, C.S. (Eds.), Coastal Wetlands. Elsevier, pp. 863–887.

Lindau, C.W., Delaune, R.D., 1991. Dinitrogen and nitrous oxide emission and entrapment in *Spartina alterniflora* saltmarsh soils following addition of N-15 labeled ammonium and nitrate. Estuar. Coast. Shelf Sci. 32, 161–172.

Liu, X., Liu, Y., Xu, J., Ren, K., Meng, X., 2016. Tracking aquaculture-derived fluoroquinolones in a mangrove wetland, South China. Environ. Pollut. 219, 916–923.

Lovelock, C.E., 2008. Soil respiration and belowground carbon allocation in mangrove forests. Ecosystems 11, 342–354.

Lovelock, C.E., Ruess, R.W., Feller, I.C., 2011. CO_2 efflux from cleared mangrove peat. PLoS One 6, e21279.

Ma, J., Ullah, S., Niu, A., Liao, Z., Qin, Q., Xu, S., Lin, C., 2021. Heavy metal pollution increases CH_4 and decreases CO_2 emissions due to soil microbial changes in a mangrove wetland: microcosm experiment and field examination. Chemosphere 269, 128735.

Macintyre, H.L., Geider, R.J., Miller, D.C., 1996. Microphytobenthos: the ecological role of the "secret garden" of unvegetated, shallow-water marine habitats. I. Distribution, abundance and primary production. Estuaries 19, 186–201.

Magalhães, C., Costa, J., Teixeira, C., Bordalo, A.A., 2007. Impact of trace metals on denitrification in estuarine sediments of the Douro River estuary, Portugal. Mar. Chem. 107, 332–341.

Malone, T.C., Newton, A., 2020. The globalization of cultural eutrophication in the coastal ocean: causes and consequences. Front. Mar. Sci. 7, 670.

Marton, J.M., Herbert, E.R., Craft, C.B., 2012. Effects of salinity on denitrification and greenhouse gas production from laboratory-incubated tidal forest soils. Wetlands 32, 347–357.

McGuinness, K.A., 1994. The climbing behaviour of Cerithidea anticipata (Mollusca: Gastropoda): the roles of physical and biological factors. Aust. J. Ecol. 19, 283–289.

Mcleod, E., Chmura, G.L., Bouillon, S., Salm, R., Björk, M., Duarte, C.M., Lovelock, C.E., Schlesinger, W.H., Silliman, B.R., 2011. A blueprint for blue carbon: toward an improved understanding of the role of vegetated coastal habitats in sequestering CO_2. Front. Ecol. Environ. 9, 552–560.

Megonigal, J.P., Schlesinger, W.H., 2002. Methane-limited methanotrophy in tidal freshwater swamps. Global Biogeochem. Cycles 16, 1088.

Mehring, A.S., Cook, P.L.M., Evrard, V., Grant, S.B., Levin, L.A., 2017. Pollution-tolerant invertebrates enhance greenhouse gas flux in urban estuaries. Ecol. Appl. 27, 1852–1861.

Mitsch, W.J., Bernal, B., Nahlik, A.M., Mander, Ü., Zhang, L., Anderson, C.J., Jørgensen, S. E., Brix, H., 2013. Wetlands, carbon, and climate change. Landsc. Ecol. 28, 583–597.

Morrissey, E.M., Gillespie, J.L., Morina, J.C., Franklin, R.B., 2014. Salinity affects microbial activity and soil organic matter content in tidal wetlands. Glob. Chang. Biol. 20 (4), 1351–1362.

Moseman-Valtierra, S., Gonzalez, R., Kroeger, K.D., Tang, J.W., Chao, W.C., Crusius, J., Bratton, J., Green, A., Shelton, J., 2011. Short-term nitrogen additions can shift a coastal wetland from a sink to a source of N_2O. Atmos. Environ. 45, 4390–4397.

Munoz-Hincapie, M., Morell, J.M., Corredor, J.E., 2002. Increase of nitrous oxide flux to the atmosphere upon nitrogen addition to red mangroves sediments. Mar. Pollut. Bull. 44, 992–996.

Naylor, R., Burke, M., 2005. Aquaculture and ocean resources: raising tigers of the sea. Annu. Rev. Environ. Resour. 30, 185–218.

Newton, A., Icely, J., Cristina, S., Perillo, G.M.E., Turner, R.E., Ashan, D., Cragg, S., Luo, Y., Tu, C., Li, Y., Zhang, H., Ramesh, R., Forbes, D.L., Solidoro, C., Béjaoui, B., Gao, S., Pastres, R., Kelsey, H., Taillie, D., Nhan, N., Brito, A.C., de Lima, R., Kuenzer, C., 2020. Anthropogenic, direct pressures on coastal wetlands. Front. Ecol. Evol. 8, 144.

Nóbrega, G.N., Ferreira, T.O., Siqueira Neto, M., Queiroz, H.M., Artur, A.G., Mendonça, E.D. S., Silva, E.D.O., Otero, X.L., 2016. Edaphic factors controlling summer (rainy season) greenhouse gas emissions (CO2 and CH4) from semiarid mangrove soils (NE Brazil). Sci. Total Environ. 542, 685–693.

Nyman, J.A., DeLaune, R.D., 1991. CO_2 emission and Eh responses to different hydrological conditions in fresh, brackish, and saline marsh soils. Limnol. Oceanogr. 36, 1406–1414.

Olsson, L., Ye, S., Yu, X., Wei, M., Krauss, K.W., Brix, H., 2015. Factors influencing CO_2 and CH_4 emissions from coastal wetlands in the Liaohe Delta, Northeast China. Biogeosciences 12, 4965–4977.

Osborne, R.I., Bernot, M.J., Findlay, S.E., 2015. Changes in nitrogen cycling processes along a salinity gradient in tidal wetlands of the Hudson River, New York. USA. Wetlands 35 (2), 323–334.

Pennings, S.C., Stanton, L.E., Stephen Brewer, J., 2002. Nutrient effects on the composition of salt marsh plant communities along the southern Atlantic and Gulf coasts of the United States. Estuaries 25, 1164–1173.

Pi, N., Tam, N.F.Y., Wu, Y., Wong, M.H., 2009. Root anatomy and spatial pattern of radial oxygen loss of eight true mangrove species. Aquat. Bot. 90, 222–230.

Poffenbarger, H.J., Needelman, B.A., Megonigal, J.P., 2011. Salinity influence on methane emissions from tidal marshes. Wetlands 31, 1–12.

Portnoy, J.W., 1999. Salt marsh diking and restoration: biogeochemical implications of altered wetland hydrology. Environ. Manag. 24, 111–120.

Portnoy, J.W., Giblin, A., 1997. Biogeochemical effects of seawater restoration to dyked salt marshes. Ecol. Appl. 7, 1054–1063.

Poungparn, S., Komiyama, A., Tanaka, A., Sangtiean, T., Maknual, C., Kato, S., Tanapermpool, P., Patanaponpaiboon, P., 2009. Carbon dioxide emission through soil respiration in a secondary mangrove forest of eastern Thailand. J. Trop. Ecol. 25, 393–400.

Purvaja, R., Ramesh, R., 2001. Natural and anthropogenic methane emission from coastal wetlands of South India. Environ. Manag. 27, 547–557.

Purvaja, R., Ramesh, R., Frenzel, P., 2004. Plant-mediated methane emission from Indian mangroves. Glob. Chang. Biol. 10, 1825–1834.

Queiroz, H.M., Arturb, A.G., Taniguchi, C.A.K., Silveira, M.R.S., Nascimento, J.C., Nóbrega, C.N., Otero, X.L., Ferreira, T.O., 2019. Hidden contribution of shrimp farming effluents to greenhouse gas emissions from mangrove soils. Estuar. Coast. Shelf Sci. 221, 8–14.

Queiroz, H.M., Ferreira, T.O., Taniguchi, C.A.K., Barcellos, D., do Nascimento, J.C., Nóbrega, G.N., Otero, X.L., Artur, A.G., 2020. Nitrogen mineralization and eutrophication risks in mangroves receiving shrimp farming effluents. Environ. Sci. Pollut. Res. 27, 34941–34950.

Reddy, K.R., De Laune, R.D., 2008. Biogeochemistry of Wetlands: Science and Applications. CRC Press.

Reddy, K.R., Patrick Jr., W.H., 1975. Effect of alternate aerobic and anaerobic conditions on redox potential, organic matter decomposition and nitrogen loss in a flooded soil. Soil Biol. Biochem. 7, 87–94.

Richards, D.R., Friess, D.A., 2016. Rates and drivers of mangrove deforestation in Southeast Asia, 2000–2012. Proc. Natl. Acad. Sci. U.S.A. 113, 344–349.

Roman, C.T., Burdick, D.M., 2012. Tidal Marsh Restoration A Synthesis of Science and Management. Island Press.

Roman, C.T., Niering, W.A., Warren, R.S., 1984. Salt marsh vegetation change in response to tidal restriction. Environ. Manag. 8, 141–149.

Roughan, B.L., Kellman, L., Smith, E., Chmura, G.L., 2018. Nitrous oxide emissions could reduce the blue carbon value of marshes on eutrophic estuaries. Environ. Res. Lett. 13, 1–7.

Saenger, P., Siddiqi, N.A., 1993. Land from the seas: the mangrove afforestation program of Bangladesh. Ocean Coast. Manag. 20, 23–39.

Salm, J.O., Maddison, M., Tammik, S., et al., 2012. Emissions of CO_2, CH_4 and N_2O from undisturbed, drained and mined peatlands in Estonia. Hydrobiologia 692, 41–55.

Sámano, M.L., García, A., Revilla, J.A., Álvarez, C., 2014. Modeling heavy metal concentration distributions in estuarine waters: an application to Suances estuary (Northern Spain). Environ. Earth Sci. 72, 2931–2945.

Schnell, S., King, G.M., 1995. Mechanistic analysis of ammonium inhibition of atmospheric methane consumption in forest soil. Appl. Environ. Microbiol. 60, 3514–3521.

Senga, Y., Mochida, K., Fukumori, R., Okamoto, N., Seike, Y., 2006. N_2O accumulation in estuarine and coastal sediments: the influence of H_2S on dissimilatory nitrate reduction. Estuar. Coast. Shelf Sci. 67, 231–238.

Sotomayor, D., Corredor, J.E., Morell, J.M., 1994. Methane flux from mangrove soils along the southwestern coast of Puerto Rico. Estuaries 17, 140–147.

Srikanth, S., Lum, S.K.Y., Chen, Z., 2016. Mangrove root: adaptations and ecological importance. Trees 30 (2), 451–465.

Stedman, S., Dahl, T.E., 2008. Status and Trends of Wetlands in the Coastal Watersheds of the Eastern United States 1998 to 2004. National Oceanic and Atmospheric Administration, National Marine Fisheries Service and U.S. Department of the Interior, Fish and Wildlife Service.

Steinmuller, H.E., Chambers, L.G., 2017. Can saltwater intrusion accelerate nutrient export from freshwater wetland soils? An experimental approach. Soil Sci. Soc. Am. J. 82 (1), 283–292.

Strauss, E.A., Lamberti, G.A., 2000. Regulation of nitrification in aquatic sediments by organic carbon. Limnol. Oceanogr. 45, 1854–1859.

Suarez-Abelenda, M., Ferreira, T.O., Camps-Arbestain, M., Rivera-Monroy, V.H., Macias, F., Nobrega, G.N., Otero, X.L., 2014. The effect of nutrient-rich effluents from shrimp farming on mangrove soil carbon storage and geochemistry under semiarid climate conditions in northern Brazil. Geoderma 213, 551–559.

Thamdrup, B., Dalsgaard, T., 2002. Production of N_2 through anaerobic ammonium oxidation coupled to nitrate reduction in marine sediments. Appl. Environ. Microbiol. 68, 1312–1318.

Vaiphasa, C., de Boer, W.F., Skidmore, A.K., Panitchart, S., Vaiphasa, T., Bamrongrugsa, N., Santitamnont, P., 2007. Impact of solid shrimp pond waste materials on mangrove growth and mortality: a case study from Pak Phanang, Thailand. Hydrobiologia 591, 47–57.

Valiela, I., Kinney, E., Culbertson, J., Peacock, E., Smith, S., 2009. Global losses of mangroves and salt marshes. In: Duarte, C.M. (Ed.), Global Loss of Coastal Habitats Rates, Causes and Consequences. Fundación BBVA.

Van der Nat, F.W.A., de Brouwer, J.F.C., Middelburg, J.J., Laanbroek, H.J., 1997. Spatial distribution and inhabitation by ammonium of methane oxidation in intertidal freshwater marshes. Appl. Environ. Microbiol. 63, 4734–4740.

Wang, J., Dodla, S., DeLaune, R., Cook, R.L., 2010. Organic carbon transformation along a salinity gradient in Louisiana wetland soils. In: 19th World Congress of Soil Science, Soil Solutions for a Changing World, 1–6 August, Brisbane, Australia.

Werry, J., Lee, S.Y., 2005. Grapsid crabs mediate link between mangrove litter production and estuarine planktonic food chains. Mar. Ecol. Prog. Ser. 293, 165–176.

Weston, N.B., Giblin, A.E., Banta, G.T., Hopkinson, C.S., Tucker, J., 2010. The effects of varying salinity on ammonium exchange in estuarine sediments of the Parker River, Massachusetts. Estuar. Coasts 33, 985–1003.

Wollenberg, J., Biswas, A., Chmura, G.L., 2018. Greenhouse gas flux with reflooding of a drained salt marsh soil. PeerJ 6, e5659.

Worthington, T., Spalding, M., 2018. Mangrove Restoration Potential: A Global Map highlighting a Critical Opportunity., https:/doi.org/10.17863/CAM.39153.

Yang, Q., Tam, N.F.Y., Wong, Y.S., Luan, T.G., Su, W.S., Lan, C.Y., Shin, P.K.S., Cheung, S. G., 2008. Potential use of mangroves as constructed wetland for municipal sewage treatment in Futian, Shenzhen, China. Mar. Pollut. Bull. 57, 735–743.

Yang, H., Tang, J., Zhang, C., Dai, Y., Zhou, C., Xu, P., Perry, D.C., Chen, X., 2020. Enhanced carbon uptake and reduced methane emissions in a newly restored wetland. J. Geophys. Res. Biogeosci. 125, e2019JG005222.

Ye, Y., Tam, N.F.Y., Wong, Y.S., 2001. Livestock wastewater treatment by a mangrove pot-cultivation system and the effect of salinity on the nutrient removal efficiency. Mar. Pollut. Bull. 42, 513–521.

Yin, G., Hou, L., Liu, M., Zheng, Y., Li, X., Lin, X., Gao, J., Jiang, X., 2016. Effects of thiamphenicol on nitrate reduction and N_2O release in estuarine and coastal sediments. Environ. Pollut. 214, 265–272.

Yu, D., 2014. Studies on Changes in Atmospheric Greenhouse Gas Fluxes from Soils of *Kandelia obovata* Mangrove Forests with the Development of Restored Vegetation in Jiulong River Estuary. Xiamen University (MSc thesis).

Carbon storage and mineralization in coastal wetlands

10

Xiaoguang Ouyang[a,b,c], Derrick Y.F. Lai[d], Cyril Marchand[e], and Shing Yip Lee[f]

[a]*Southern Marine Science and Engineering Guangdong Laboratory (Guangzhou), Guangzhou, China,*
[b]*Guangdong Provincial Key Laboratory of Water Quality Improvement and Ecological Restoration for Watersheds, School of Ecology, Environment and Resources, Guangdong University of Technology, Guangzhou, China,*
[c]*Simon F.S. Li Marine Science Laboratory, School of Life Sciences, The Chinese University of Hong Kong, Hong Kong Special Administrative Region, China,*
[d]*Department of Geography and Resource Management, The Chinese University of Hong Kong, Hong Kong Special Administrative Region, China,*
[e]*University of New Caledonia, ISEA, Noumea, New Caledonia,*
[f]*Institute of Environment, Energy and Sustainability, Simon F.S. Li Marine Science Laboratory, School of Life Sciences, The Chinese University of Hong Kong, Hong Kong Special Administrative Region, China*

10.1 Introduction

Early in 2016, the acquisition editor Ms. Louisa Hutchins at Elsevier Press invited the first author to edit a book in the area of coastal wetlands. By 2018, he managed to organize an editorial team to consider producing a book on carbon mineralization in coastal wetlands, as a component of blue carbon which was undervalued in previous studies. In 2020, we witnessed the outbreak of the coronavirus pandemic and the widespread lockdown brought about additional challenges in completing the book. However, we have overcome the difficulties and brought the book to fruition. During the past few years, there has been a surge of publications on carbon mineralization in coastal wetlands, including some studies on root decomposition and response of greenhouse gas emission to environmental changes at the global scale (e.g., Ouyang et al., 2017a, b; O'Connor et al., 2020). This surge reiterates the importance of this topic in coastal wetland research, especially carbon dynamics. In this book, we focus on carbon mineralization ranging from litter decomposition to greenhouse gas emission in coastal wetlands. While carbon mineralization in terrestrial ecosystems may involve similar processes, the unique characteristics of coastal wetlands, including tidal inundation and the interplay of land and sea processes, make the topic more complex and important. The unprecedented trend of global coastal urbanization has imposed ubiquitous disturbances on coastal wetlands in recent decades. Coastal

Carbon Mineralization in Coastal Wetlands. https://doi.org/10.1016/B978-0-12-819220-7.00005-4

295

wetlands also experience direct influence from climate change, such as sea-level rise, warming, fluctuations in rainfall pattern, and extreme weather events.

In this concluding chapter, we synthesize knowledge on carbon mineralization based on different topics discussed in previous chapters: (1) the importance of studying carbon mineralization from the perspective of carbon storage in coastal wetlands; (2) carbon mineralization pathways and methods for estimating carbon mineralization; (3) the impact of bioturbation and macrofauna consumption on carbon mineralization; and (4) the impact of global changes on carbon mineralization.

10.2 Importance of carbon mineralization to carbon storage

Concerns about carbon mineralization in coastal wetlands arise from their effectiveness in carbon sequestration and storage compared to other ecosystems and the vulnerability of their large carbon stock to both climate change and anthropogenic disturbance. In contrast to their small area, vegetated coastal wetlands have disproportionately high sediment carbon burial capacity, in the range of $138–244.7\,g\ C\ m^{-2}\ yr^{-1}$ (Breithaupt et al., 2012; Chmura et al., 2003; Kennedy et al., 2010; Ouyang and Lee, 2014), which is 30 to 60 times higher than that $(4–5.1\,g\ C\ m^{-2}\ yr^{-1})$ of tropical, temperate, and boreal terrestrial ecosystems (Schlesinger and Bernhardt, 2013; Zehetner, 2010; Asner et al., 2009, Table 10.1). Current estimates on organic carbon stock in vegetated coastal wetlands are $237.4–361\,Mg\,ha^{-1}$ in mangrove top-meter sediments (Jardine and Siikamäki, 2014; Sanderman et al., 2018; Ouyang and Lee, 2020), $410.77\,Mg\,ha^{-1}$ in saltmarsh sediments (based on data from Ouyang and Lee, 2014) and $139.7\,Mg\,ha^{-1}$ in seagrass sediments (Fourqurean et al., 2012). Sediment organic carbon stocks in blue carbon ecosystems therefore at least double those $(37.5–75\,Mg\,ha^{-1})$ of terrestrial ecosystems (Batjes, 2011). However, most syntheses ignored sediment inorganic carbon stock as a component of total sediment carbon stock. Sediment inorganic carbon sink is determined by calcium carbonate deposition and dissolution in blue carbon ecosystems (Macreadie et al., 2017). In addition, mineralization of mangrove-derived organic matter produces dissolved inorganic carbon that can precipitate within the sediment depending on the redox conditions (Marchand et al., 2008). Although there are key uncertainties (e.g., sediment inorganic carbon stocks ranging from 0 to $1506.9\,Mg\,ha^{-1}$ in global mangroves), these C components play important but different roles in sediment carbon stock and burial capacity. Sediment inorganic carbon stocks in mangroves were estimated to be $34.7\,Mg\,ha^{-1}$ (Ouyang and Guo, 2020), which is 14.6% of sediment organic carbon stock. Sediment inorganic carbon burial rate reaches $6\,g\ C\ m^{-2}\ yr^{-1}$ and $87\,g\ C\ m^{-2}\ yr^{-1}$ in mangroves and seagrasses (Saderne et al., 2019), respectively, accounting for 3.7% and 63% of sediment organic carbon burial rate in the two ecosystems. Taking into account of both sediment/soil and biomass carbon stock, mangrove ecosystem carbon stocks $(450.6\,Mg\,ha^{-1}$, Ouyang and Lee, 2020) are almost double those of temperate $(204.7–278.7\,Mg\,ha^{-1})$, boreal $(202.3–279.3\,Mg\,ha^{-1})$, and tropical $(236.9–258.9\,Mg\,ha^{-1})$ terrestrial ecosystems (combined data from Batjes, 2011 and Ogle et al., 2019).

Table 10.1 Global carbon stock and burial capacity of coastal wetlands in comparison with terrestrial ecosystems.

Ecosystems	SOC (Mg ha⁻¹)	SIC (Mg ha⁻¹)	SOAC (g C m⁻² yr⁻¹)	SIAC (g C m⁻² yr⁻¹)	NPP (g C m⁻² yr⁻¹)	ECS (Mg ha⁻¹)	References
Coastal wetlands							
Mangroves	237.4 (6.1–1526.9), $n=235$; 361 (86–729), $n=149$	34.7 (0–1506.9), $n=25$	163 (20–1020), $n=19$	6, $n=22$	400	450.6 (157.9–3260.1), $n=400$	Jardine and Siikamäki (2014); Sanderman et al. (2018); Breithaupt et al. (2012); Saderne et al. (2019); Ouyang and Lee (2020); Duarte et al. (2013)
Tidal marshes	410.77 ± 21.95, $n=143$	NA	244.7 ± 26.1, $n=50$	NA	440	NA	Ouyang and Lee (2014); Duarte et al. (2013)
Seagrasses	139.7 (9.1–628.1)	NA	83–133	87, $n=15$	278	NA	Kennedy et al. (2010); Fourqurean et al. (2012); Saderne et al. (2019); Duarte et al. (2013)
Terrestrial ecosystems							
Temperate	95.4 ± 55.2	NA	5.1	NA	625–779	202.3–279.3[a]	Huston and Wolverton (2009); Batjes (2011); Schlesinger and Bernhardt (2013); Wieder et al. (2013); Ogle et al. (2019)

Continued

Table 10.1 Global carbon stock and burial capacity of coastal wetlands in comparison with terrestrial ecosystems—cont'd

Ecosystems	SOC (Mg ha^{-1})	SIC (Mg ha^{-1})	SOAC (g C m^{-2} yr^{-1})	SIAC (g C m^{-2} yr^{-1})	NPP (g C m^{-2} yr^{-1})	ECS (Mg ha^{-1})	References
Boreal	143.1	NA	4.6		190–234	204.7–278.7[a]	Huston and Wolverton (2009); Batjes (2011); Zehetner (2010); Wieder et al. (2013); Ogle et al. (2019)
Tropical	71.5 ± 33.5	NA	4	NA	783–1251	236.9–258.9[a]	Huston and Wolverton (2009); Batjes (2011); Asner et al. (2009); Wieder et al. (2013); Ogle et al. (2019); Schlesinger and Bernhardt (2013)

[a]SIC is not included in ECS; SOC, sediment/soil organic carbon stock; SIC, sediment/soil inorganic carbon stock; SOAC, sediment/soil organic carbon accumulation rate; SIAC, sediment/soil inorganic carbon accumulation rate; ECS, ecosystem carbon stock; ECS is the sum of sediment/soil carbon stock, live and dead biomass carbon stock. The estimate of global surface SOC at 0–30 cm depth in different ecosystems was propagated to 1 m depth using the ratio of total global SOC at 1 m depth and 30 cm depth. Data are shown as mean ± standard error or ranges where available. n, number of studies from which the statistics were obtained.

Climate change may have significant impact on carbon sequestration in coastal wetlands, especially when the impact interact with those of human activities. There are large uncertainties about changes in carbon sequestration in coastal wetlands as regulated by their landward migration driven by future sea-level rise. Lovelock and Reef (2020) projected that landward migration of mangroves might result in increase in carbon sequestration of 1.5 Pg by 2100 if enough accommodation space is available, whereas losses of 3.4 Pg of sequestered carbon can occur by 2100 in the coastal squeeze scenario. When the timescale is projected to the Holocene, Rogers et al. (2019) observed that tidal marshes on coastlines that experienced rapid relative sea-level rise over the past few millennia had significantly higher (1.7 to 3.7 times) sediment carbon concentrations than those subjected to long periods of sea-level stability. Ouyang and Lee (2020) also found that mangroves experiencing rapid sea-level rise (zones I and II) in the Holocene had significantly higher sediment carbon stocks than those under stable sea-level rise (zones III to V). However, Jacotot et al. (2018) found that deep sediments stored 1.55–5 times of carbon more than upper sediments in a New Caledonian mangrove. The high sediment carbon stock in the deeper sediments was attributed to a period of sea-level stability in their study. It suggests that the impact of sea level on sediment carbon storage may differ between the local and global scales. Sanders et al. (2016) estimated that precipitation explained 86% of the observed variability in mangrove carbon stocks by compiling data spanning a latitudinal range from 22°N to 38°S. Small changes in temperature and precipitation can transform coastal wetland plant communities, shift species at ecotones (e.g., mangrove-saltmarsh ecotones in Florida) and influence vascular plant productivity (Osland et al., 2018) and thus carbon burial. Carbon storage and burial in coastal wetlands are tightly linked to carbon mineralization. Rapidly changing climate and environmental conditions, including sea-level rise, warming, eutrophication, and land cover shifts, will influence decomposition and thus the global reservoir of blue sediment carbon stock (Spivak et al., 2019). Therefore, it is important to figure out the processes and drivers of carbon mineralization relevant to preserving carbon stocks in coastal wetlands.

10.3 Pathways and methods for estimating carbon mineralization

Significant differences in carbon mineralization exist in different habitat interfaces. In Chapter 2, Ouyang and Lee compared mangrove leaf litter decomposition rate constants under different habitat interfaces, with the rate constants being significantly higher in aerial than submerged decomposition (mean: 0.06 vs 0.025 day^{-1}). However, this difference may be habitat specific since no significant differences were found for saltmarshes. In Chapters 3 and 6, Marchand et al. and Rosentreter examined greenhouse gas emission from sediment and water surfaces in coastal wetlands, respectively. Their analyses demonstrate that CO_2 efflux from the sediment-air interface is double that from the water-air interface in saltmarshes (mean: 561 vs. 281 g C m^{-2} yr^{-1}) and mangroves (mean: 612.8 vs. 247.6 g C m^{-2} yr^{-1}). In contrast, there is no apparent difference in CH_4 efflux from the sediment-air and water-air interfaces in either habitats (mangroves: 1.69 vs 1.26 g C m^{-2} yr^{-1}, saltmarshes: 14 vs 8.4 g C m^{-2} yr^{-1}).

Seagrasses are mostly subtidal coastal wetlands, and the reported fluxes were from water-air interfaces. In contrast to mangroves and saltmarshes, which are CO_2 sources at the water-air interface, seagrass are CO_2 sinks at this interface (-14.7 g C m^{-2} yr^{-1}) owing to the photosynthetic uptake of CO_2 from the water column. Seagrass CH_4 fluxes (1.07 g C m^{-2} yr^{-1}) at the water-air interface are only slightly lower than those in mangrove and saltmarshes. These patterns may prompt concerns about the current paradigm in carbon cycling in coastal wetlands, which is largely borrowed from terrestrial ecosystems and neglects the regular occurrence of tidal inundation. One universal approach in establishing carbon cycling model uses the average value of carbon mineralization rates (e.g., decomposition rate and greenhouse gas emission rate) from the different interfaces (e.g., sediment-air and water-air). An obvious drawback of this method lies in the lack of recognition of the difference in tidal regimes such as different flooding frequencies, amplitude, and inundation periods in different coastal wetlands, or different tidal positions in the same wetlands.

The above-discussed carbon mineralization pathways can be integrated into biosphere-atmosphere exchange of greenhouse gases. In Chapter 4, Lai et al. present the current data on biosphere-atmosphere exchange of CO_2 and CH_4 measured by the eddy covariance technique in mangroves and tidal marshes. While eddy covariance allows estimation of gross primary production (GPP), NEE, as well as ecosystem respiration (Re) and CH_4 flux simultaneously at the ecosystem scale, deployment of the eddy covariance tower requires flat terrain and there are systematic errors from signal time lags and air density fluctuations resulting from temperature and humidity changes. The widespread use of eddy covariance towers in quantifying ecosystem carbon exchange in coastal wetlands is also hampered by the high price of the instruments and the harsh environment in most of coastal wetlands. Global networks (e.g., FLUXNET and FLUXNET-CH$_4$) have been established to use eddy covariance to monitor ecosystem exchange of greenhouse gases in different ecosystems. Recent outputs of the networks allow comparison of the ecosystem exchange variables in coastal wetlands and terrestrial ecosystems (Table 10.2). Globally, mangroves, tidal marshes, and seagrasses are all net carbon sinks, as reflected by NEE (-557.4, -122.3 and -119.1 g C m^{-2} yr^{-1}, respectively), irrespective of their high sediment carbon accumulation capacity. In particular, mangroves are the most efficient carbon sinks among all the wetland ecosystems (lowest NEE). Carbon use, represented by Re/GPP (i.e., ecosystem respiration divided by gross primary production), diverge according to coastal wetland types, in comparison with terrestrial ecosystems (the higher Re/GPP, the lower carbon use efficiency). Carbon use of mangrove (mean: 0.72) is higher than those of all terrestrial ecosystems (0.76–0.95), whereas saltmarshes (0.93) have lower values than most terrestrial ecosystems. Mangroves are dominated by woody vascular plants with high GPP (1850.7 g C m^{-2} yr^{-1}). Also, the terrestrial ecosystems used in the comparison cover both woody and herbaceous plants that mostly have higher GPP (from boreal 982.3 g C m^{-2} yr^{-1} to 3551 g C m^{-2} yr^{-1}) than saltmarsh species (1113.5 g C m^{-2} yr^{-1}), which are dominated by

Table 10.2 Comparison of gross primary production and biosphere-atmosphere greenhouse gas exchange between terrestrial forest ecosystems and coastal wetlands.

Ecosystems	GPP (g C m^{-2} yr^{-1})	Re (g C m^{-2} yr^{-1})	NEE (g C m^{-2} yr^{-1})	Re/GPP	R$_{CH4}$ (g C m^{-2} yr^{-1})
Boreal					
Humid evergreen	973±83*	824±112*	−149	0.85±0.14*	
Semiarid evergreen	773±35*	734±37*	−39	0.95±0.06*	
Semiarid deciduous	1201±23*	1029	−172	0.86±0.02*	
All boreal	982.3	862.3	−120	0.88	16.4 (7.9–35.9), n=11
Temperate					
Humid evergreen	1762±56*	1336±57*	−426	0.76±0.04*	
Humid deciduous	1375±56*	1048±64*	−327	0.76±0.06*	
Semiarid evergreen	1228±286*	1104±260*	−124	0.96±0.38*	
All temperate	1455	1162.7	−292.3	0.8	43.2 (20.0–46.8), n=34
Tropical					
All tropical	3551±160	3061±162	−490	0.86±0.06	43.2 (20.0–46.8), n=3
Coastal wetlands					
Mangroves	1877.1±194.6, n=9	1319.7±130.3, n=9	−557.4±110.7, n=9	0.72±0.05, n=9	26.4, n=2

Continued

Table 10.2 Comparison of gross primary production and biosphere-atmosphere greenhouse gas exchange between terrestrial forest ecosystems and coastal wetlands—cont'd

Ecosystems	GPP (g C m^{-2} yr^{-1})	Re (g C m^{-2} yr^{-1})	NEE (g C m^{-2} yr^{-1})	Re/GPP	R$_{CH4}$ (g C m^{-2} yr^{-1})
Tidal marshes	1113.5±119.3, n=13	991.2±127.9, n=13	−122.3±74.7, n=13	0.93±0.08, n=13	20.1±6.5, n=10
Seagrasses	984.8±48.5, n=141	821.7±44.2, n=116	−119.1±25.4, n=123	0.85, n=111	1.07

GPP, gross primary production; Re, ecosystem respiration; NEE, net ecosystem exchange; R$_{CH4}$, ecosystem CH$_4$ flux; n, number of studies from which the statistics were obtained. NEE = Re-GPP. Data on mangroves and saltmarshes are derived from eddy covariance measurements, while data on seagrasses are derived from other methods, such as the chamber technique. Data are generally shown as mean ± standard error or standard deviation (indicated by *). For data without standard errors, 25th and 75th percentiles are given in parentheses.

Data from Knox, S. H., Jackson, R. B., Poulter, B., McNicol, G., Fluet-Chouinard, E., Zhang, Z., Hugelius, G., Bousquet, P., Canadell, J. G., & Saunois, M. (2019). FLUXNET-CH4 synthesis activity: objectives, observations, and future directions. Bull. Am. Meteorol. Soc., 100, 2607–2632; Liu, J., Zhou, Y., Valach, A., Shortt, R., Kasak, K., Rey-Sanchez, C., Hemes, K. S., Baldocchi, D., & Lai, D. Y. (2020). Methane emissions reduce the radiative cooling effect of a subtropical estuarine mangrove wetland by half. Glob. Chang. Biol., 26, 4998–5016; Jha, C. S., Rodda, S. R., Thumaty, K. C., Raha, A., & Dadhwal, V. K. (2014). Eddy covariance based methane flux in Sundarbans mangroves, India. J. Earth Syst. Sci., 123, 1089–1096; Luyssaert, S., Inglima, I., Jung, M., Richardson, A.D., Reichstein, M., Papale, D., Piao, S.L., Schulze, E.D., Wingate, L., Matteucci, G., Aragao, L., Aubinet, M., Beer, C., Bernhofer, C., Black, K.G., Bonal, D., Bonnefond, J.M., Chambers, J., Ciais, P., Cook, B., Davis, K.J., Dolman, A.J., Gielen, B., Goulden, M., Grace, J., Granier, A., Grelle, A., Griffis, T., Grünwald, T., Guidolotti, G., Hanson, P.J., Harding, R., Hollinger, D.Y., Hutyra, L.R., Kolari, P., Kruijt, B., Kutsch, W., Lagergren, F., Laurila, T., Law, B.E., Le Maire, G., Lindroth, A., Loustau, D., Malhi, Y., Mateus, J., Migliavacca, M., Misson, L., Montagnani, L., Moncrieff, J., Moors, E., Munger, J.W., Nikinmaa, E., Ollinger, S.V., Pita, G., Rebmann, C., Roupsard, O., Saigusa, N., Sanz, M.J., Seufert, G., Sierra, C., Smith, M.L., Tang, J., Valentini, R., Vesala, T., Janssens, I.A., 2007. CO2 balance of boreal, temperate, and tropical forests derived from a global database. Glob. Chang. Biol, 13, 2509–2537; Duarte, C. M., Marbà, N., Gacia, E., Fourqurean, J. W., Beggins, J., Barrón, C., & Apostolaki, E. T. (2010). Seagrass community metabolism: assessing the carbon sink capacity of seagrass meadows. Glob. Biogeochem. Cycles, 24, GB4032, and references cited in Chapter 3, 4, and 6.

herbs. The source of the divergence also lies in the relatively high sediment carbon content (410.77 Mg ha^{-1}, Table 10.1) providing substrates for microbial respiration in saltmarshes compared to terrestrial ecosystems (71.5–143.1 Mg ha^{-1}). Ecosystem CH$_4$ fluxes of both mangroves and saltmarshes (26.4 and 20.1 g C m^{-2} yr^{-1}, respectively) are intermediate in the range of values for terrestrial ecosystems (8.3–43.2 g C m^{-2} yr^{-1}). While ecosystem CH$_4$ fluxes of seagrasses (1.07 g C m^{-2} yr^{-1}) are around 20 times lower than those of mangroves and seagrasses, part of the discrepancy may arise from different methods used in their estimates (eddy covariance in mangroves and saltmarshes vs. other methods in seagrasses). It is difficult to compare the fluxes among different ecosystems due to a lack of global coverage of FLUXNET-CH$_4$, particularly in coastal wetlands.

Different methods are used to estimate carbon mineralization, as indicated by decomposition rate and greenhouse gas flux. Decomposition rate constants of litter, roots, and wood are estimated from a variety of models, including linear and exponential models, and occasionally, double-exponential and asymptotic models. The former two models are widely used for coastal wetlands due to the ease of application, while the latter two models require input of the proportion of labile and recalcitrant material as a prerequisite. The initial leaching phase of litter decomposition results in fast loss of labile material (e.g., low-molecular weight sugars), which decomposes at significantly higher rate than humic substances. Caution should be exercised when applying the linear model to litter decomposition experiments lasting for long periods of time.

Greenhouse gas flux can be quantified at the sediment-air and water-air interfaces, or sediment-water interface, the flux of which is only an intermittent process of fluxes from the water-air interface. The eddy covariance technique has recently been used to measure ecosystem exchange of greenhouse gas between coastal wetlands and the atmosphere. Micro-meteorological techniques were used to measure CO$_2$ exchange between saltmarshes and air in earlier studies (Houghton and Woodwell, 1980). Currently, sediment greenhouse gas flux in coastal wetlands is usually measured by the techniques of closed dynamic chambers and static chambers. The closed dynamic chamber technique can only measure CO$_2$ flux with an infrared gas analyzer (such as Li-Cor and SBA-5 PP Systems), which measures changes in partial pressure within the chamber at high frequency. The static chamber technique can be used to measure not only CO$_2$ flux but also CH$_4$ flux with gas chromatography (equipped with a flame ionization detector), and high-resolution cavity enhanced direct-absorption spectroscopy or cavity ring-down spectroscopy (CRDS). Recent models of CRDS can even simultaneously measure the stable isotope values (δ^{13}C) of CO$_2$ and CH$_4$ (e.g., Jacotot et al., 2019; Ouyang and Lee, 2020). The greenhouse gas flux measured at the sediment surface may not represent total carbon mineralization in sediments since part of the CO$_2$ or CH$_4$ may dissolve in porewater as dissolved inorganic carbon and are then subject to porewater seepage and export to adjacent ecosystems. Sediment core incubation was used to extrapolate carbon mineralization at deeper sediments. Kristensen et al. (2011) stated that diffusive fluxes only approximate depth-integrated reactions in the upper 12 cm of the sediment in a Tanzanian mangrove. Although sediment incubation may provide more insights into sediment carbon mineralization, laboratory-based sediment incubation does not reflect carbon exchange between coastal wetlands and surrounding waters via porewater or groundwater discharge (e.g., Santos et al., 2019), which is an important pathway for inorganic carbon exchange.

Greenhouse gas flux from the water-air interface can be measured by the techniques of floating chambers, gradient flux, and even eddy-covariance. The former two methods are widely used for coastal wetlands, the latter method is proposed as a potential application by Rosentreter in Chapter 6. The floating chamber method can use either the closed dynamic chamber or static chamber as described above for measurements of sediment greenhouse gas flux. The chambers are equipped with floats and secured on the water surface (e.g., David et al., 2018). The application of this method is limited to locations with low to moderate wind and wave turbulences, such as in mangrove forests. The gradient flux method measures greenhouse gas flux across water-air according to Fick's law. Currently, the gas transfer velocity in the equation adopted for coastal wetlands is usually based on empirical models that are estimated for estuarine or oceanic environments. The suitability of its application is criticized due to the different boundary conditions between mangrove waterways and estuarine and oceanic waters. This invites future research on gas transfer velocity applicable to coastal wetlands. This goal may be met by calibrating the gradient flux method with the floating chamber method. While this is only applicable to low turbulent conditions such as mangrove waterways, gas transfer velocity in more turbulent conditions (e.g., saltmarsh and seagrass waters) may be calibrated by the eddy covariance system, which measures oxygen consumption (e.g., Berger et al., 2020) for system metabolism that produces CO_2.

10.4 Impact of bioturbation and macrofauna consumption on carbon mineralization

Macrofauna and vascular plant roots have direct impact on or indirectly regulate carbon mineralization in coastal wetlands. Macrofaunal influence on carbon mineralization is mediated through a few processes, including macrofaunal consumption of vascular plant detritus (Chapter 5), and burrowing and other bioturbation activities (Chapter 8). The low nutritional value of vascular plant detritus has sparked debate on the significance of macrofaunal consumption of detritus, and there is a lack of understanding on the mechanisms enabling direct macrofaunal consumption. Preliminary observations suggest that data on stable carbon and nitrogen isotopes did not support mangrove leaf assimilation by crabs due to the large difference between crab and mangrove leaf litter tissue values (McCutchan Jr et al., 2003), but unusually large trophic discrimination factors could explain the difference through field and controlled feeding experiments (Bui and Lee, 2014). Kristensen et al. (2017) postulated that animal tissue in the form live or dead prey or microphytobenthos at the sediment surface are the dominant nitrogen sources of leaf-eating sesarmid and ucidid mangrove crabs. However, it still remains unclear about the role of the microphytobenthos in coastal wetland carbon dynamics. If the role of microalgae as a nutrient source for macrofauna was overstated, the reliance on nutritionally poor vascular plant litter would require significant physiological adaptation.

Cellulase is critical to breaking down the cellulose-dominated structural organic material in vascular plant detritus. In Chapter 5, Lee and Lee demonstrate that cellulases, either endogenous or endosymbiont assisted, promote mangrove carbon mineralization through consumption by crustaceans. They highlight that sesarmid crab consumption can result in mineralization >30 times faster than through microbial decomposition. However, there are limited studies and data on cellulase production among mollusks and crustaceans in coastal wetlands, except crabs and isopods. The macrofauna can also indirectly modulate carbon mineralization via increasing the surface area to volume ratio for (1) faster microbial decomposition via shredding of organic detritus or (2) greenhouse gas diffusion via bioturbation of the sediment. For the first pathway, mechanical breakdown by crabs turns intact leaf litter into fine detritus fragments in their fecal material (e.g., Werry and Lee, 2005). This process dramatically increases the surface area to volume ratio of the manipulated litter material, which leads to accelerated leaching of toxic feeding deterrents while encouraging microbial colonization.

For the second pathway, the macrofauna can create burrows and tunnels in the anoxic wetland sediment during bioturbation. In Chapter 8, Kristensen et al. show that these structures can be blind- or open-ended, with galleries divided into one or several branches, and wide variations in morphology and depth even varying among the same crab species. Agusto et al. (2020) observed that different mangrove crab species increased the sediment-air surface area per m^2 by between 10 and 190%, depending on the mangrove species. Kristensen et al. estimate sediment-air CO_2 and CH_4 fluxes at $161 \pm 70\,\mathrm{mmol\,m^{-2}\,d^{-1}}$ (mean $\pm$ SD) and $2325 \pm 3599\,\mathrm{\mu mol\,m^{-2}\,d^{-1}}$ from mangrove sediments with burrows based on three studies. These average fluxes are, respectively, about 2 times and 30 times the fluxes from mangrove sediments without burrows. The findings are consistent with Ouyang et al. (2017a, b), which showed a significant positive relationship between sediment CO_2 flux and crab burrow density in an Australian mangrove forest. However, the data on macrofauna burrow density are patchy in coastal wetlands at the global scale, and it is difficult to extrapolate the few observations on sediment-air CO_2 and CH_4 fluxes from crab burrows to the global scale. The change in crab-respired CO_2 flux with crab size (Ouyang et al., 2021a, b) suggests that size-specific crab respiration should be considered in estimating the contribution of the macrofauna to sediment-air CO_2 flux.

Vascular plant roots regulate carbon mineralization via the connection between the sediment and the atmosphere. Mangroves have different root types, including plank roots (e.g., *Kandelia obovata*), pneumatophores (e.g., *Avicennia marina*), stilt roots (e.g., *Rhizophora stylosa*), and knee roots (e.g., *Bruguiera gymnorhiza*) (Ouyang and Guo, 2020). These structures directly connect with the atmosphere through lenticels, allowing rapid exchange of gases such as CO_2 and CH_4. Roots of some saltmarsh species, e.g., *Spartina anglica* and *Phragmites australis*, also have aerenchyma tissues that allow gas diffusion from sediments to the atmosphere.

In Chapter 8, Kristensen et al. estimate sediment-air CO_2 and CH_4 fluxes at $168 \pm 118\,\mathrm{mmol\,m^{-2}\,d^{-1}}$ (mean $\pm$ SD) and $440 \pm 425\,\mathrm{\mu mol\,m^{-2}\,d^{-1}}$ from mangrove sediments colonized by species with pneumatophores. These average fluxes are around 2 times and 5 times the fluxes from sediments without pneumatophores. However, it remains unknown if other types of vascular plant root contribute to an increase in CO_2 and CH_4 fluxes from sediments.

10.5 Carbon mineralization under global change

Climate change may have direct impact on carbon mineralization or indirectly regulate mineralization pathways. In Chapter 2, Ouyang and Lee propose that precipitation may modulate litter decomposition by affecting the microbial substrate. In Chapter 7, Marchand et al. stress that changes of precipitation patterns and sea-level rise may affect greenhouse gas production through influencing sediment water content and the renewal of electron acceptors in sediments. Increase in rainfall and sea-level rise results in a higher frequency of submersion of coastal wetlands and thus higher sediment water content, resulting in lower greenhouse gas emissions due to the inhibition of gas diffusion. In Chapter 2, Ouyang and Lee suggest air temperature and moisture availability are the main drivers of litter decomposition rate. In particular, Ouyang et al. (2017a, b) suggested that air temperature has a positive impact on root decomposition rates in both mangroves and saltmarshes, owing to stimulated microbial activities under high temperatures. In Chapter 7, Marchand et al. conclude that temperature increases induce faster rates of greenhouse gas emission. However, current measurements of greenhouse gas fluxes cannot be equated to greenhouse gas production, and the observed temperature effect on greenhouse gas flux may not apply to greenhouse gas production. Additionally, microphytobenthos play a key role in GHG fluxes at the sediment-air interface, either using CO_2 for their metabolism or forming a permeable barrier at sediment surface limiting diffusion (Chapter 3). Rise in air temperature, sea level, and shift in rainfall pattern will modify microphytobenthos composition and distribution and thus GHG diffusion at the sediment-air interface.

Changes in the occurrence and severity of extreme weather events may also influence carbon mineralization. Ouyang and Lee (2020) observed a significant increase in CO_2 efflux from the sediment-air interface after super-typhoon Mangkhut due to a priming effect of leaf litter pulse input induced by strong winds. Jeffrey et al. (2019) reported an eightfold increase in CH_4 emissions from standing dead mangroves compared to living mangrove tree stems in the Gulf of Carpentaria, Australia, after a dieback event triggered by climate extremes induced by El Niño-Southern Oscillation.

There is mounting evidence that coastal wetlands are confronted with continual and intensifying anthropogenic activities, which may affect all biogeochemical processes in coastal wetlands, including carbon mineralization. In Chapter 9, Chen et al. discussed the impact of various anthropogenic activities on greenhouse emissions from intertidal wetland sediments, including land conversion, pollution, tidal restriction, and restoration. They compared sediment greenhouse gas emission between rehabilitated and disturbed mangroves, as well as between eutrophic and non-/less-polluted mangroves. Nutrient enrichment is highlighted as a significant driver for greenhouse gas emissions in coastal wetlands, although other anthropogenic activities have also been implicated. In general, land cover change has insignificant impact on greenhouse gas emissions and managed sites may emit more greenhouse gas than degraded sites except for mangrove conversion to rice fields (Sasmito et al., 2019; O'Connor et al., 2020). This suggests that eutrophication control should be prioritized for reducing greenhouse gas emission. Plastics pollution may also increase greenhouse gas emissions (Ouyang et al., 2022) but the effect remains unknown in field conditions.

10.6 Perspectives for future research

Our book synthesizes past and present knowledge on carbon mineralization in coastal wetlands. We identify pathways and compare methods for estimating carbon mineralization. We illustrate the impact of bioturbation and macrofaunal consumption on carbon mineralization. We reveal the impact of climate change and anthropogenic activities on carbon mineralization. However, we have not fully understood on the role of carbon mineralization in carbon cycling, and this knowledge can be improved. We propose the following future directions of research to better constrain carbon budgets in relation to carbon mineralization and improve management of blue carbon for climate change mitigation:

(1) Improve precision of the estimate of the contribution of macrofaunal consumption and bioturbation to carbon mineralization. This should include considering the bioturbation impact of macrofauna and burrow size/volume on greenhouse gas emission.

(2) Implementing measures to restore predators where top-down controls may be implemented by coastal wetland managers to reduce the contribution of macrofauna respiration to greenhouse gas production.

(3) Quantification of the impact of climate change on carbon mineralization should be region-specific, incorporating the variability of coastal wetlands in different climate zones, e.g., from arid to humid regions. To achieve this goal, we suggest that true in situ laboratory should be implemented in these ecosystems over decades.

(4) The role of the microphytobenthos in the carbon dynamics of coastal wetlands still remains unclear. To what extent the microphytobenthos of coastal wetlands are functionally and structurally different from their counterparts on tidal flats, and therefore may mediate carbon mineralization differently, requires more attention.

References

Agusto, L.E., Fratini, S., Jimenez, P.J., Quadros, A., Cannicci, S., 2020. Structural characteristics of crab burrows in Hong Kong mangrove forests and their role in ecosystem engineering. Estuar. Coast. Shelf Sci. 248. https://doi.org/10.1016/j.ecss.2020.106973, 106973.

Asner, G.P., Rudel, T.K., Aide, T.M., Defries, R., Emerson, R., 2009. A contemporary assessment of change in humid tropical forests. Conserv. Biol. 23, 1386–1395.

Batjes, N.H., 2011. Soil organic carbon stocks under native vegetation-revised estimates for use with the simple assessment option of the carbon benefits project system. Agric. Ecosyst. Environ. 142, 365–373.

Berger, A.C., Berg, P., McGlathery, K.J., Delgard, M.L., 2020. Long-term trends and resilience of seagrass metabolism: a decadal aquatic eddy covariance study. Limnol. Oceanogr. 65, 1423–1438.

Breithaupt, J.L., Smoak, J.M., Smith, T.J., Sanders, C.J., Hoare, A., 2012. Organic carbon burial rates in mangrove sediments: strengthening the global budget. Glob. Biogeochem. Cycles 26 (3), GB3011. https://doi.org/10.1029/2012GB004375.

Bui, T.H.H., Lee, S.Y., 2014. Does 'you are what you eat' apply to mangrove grapsid crabs? PLoS One 9. https://doi.org/10.1371/journal.pone.0089074, e89074.

Chmura, G.L., Anisfeld, S.C., Cahoon, D.R., Lynch, J.C., 2003. Global carbon sequestration in tidal, saline wetland soils. Glob. Biogeochem. Cycles 17, 1111. https://doi.org/10.1029/2002GB001917.

David, F., Meziane, T., Tran-Thi, N.-T., Van, V.T., Thanh-Nho, N., Taillardat, P., Marchand, C., 2018. Carbon biogeochemistry and CO_2 emissions in a human impacted and mangrove dominated tropical estuary (Can Gio, Vietnam). Biogeochemistry 138, 261–275.

Duarte, C.M., Losada, I.J., Hendriks, I.E., Mazarrasa, I., Marbà, N., 2013. The role of coastal plant communities for climate change mitigation and adaptation. Nat. Clim. Chang. 3, 961–968.

Fourqurean, J.W., Duarte, C.M., Kennedy, H., Marba, N., Holmer, M., Mateo, M.A., Apostolaki, E.T., Kendrick, G.A., Krause-Jensen, D., McGlathery, K.J., Serrano, O., 2012. Seagrass ecosystems as a globally significant carbon stock. Nat. Geosci. 5, 505–509.

Houghton, R.A., Woodwell, G.M., 1980. The flax pond ecosystem study: exchanges of CO_2 between a salt marsh and the atmosphere. Ecology 61 (6), 1434–1445.

Huston, M.A., Wolverton, S., 2009. The global distribution of net primary production: resolving the paradox. Ecol. Monogr. 79, 343–377.

Jacotot, A., Marchand, C., Allenbach, M., 2019. Biofilm and temperature controls on greenhouse gas (CO_2 and CH_4) emissions from a mangrove soil (New Caledonia). Sci. Total Environ. 650, 1019–1028.

Jacotot, A., Marchand, C., Rosenheim, B.E., Domack, E.W., Allenbach, M., 2018. Mangrove sediment carbon stocks along an elevation gradient: influence of the late Holocene marine regression (New Caledonia). Mar. Geol. 404, 60–70.

Jardine, S.L., Siikamäki, J.V., 2014. A global predictive model of carbon in mangrove soils. Environ. Res. Lett. 9 (10). https://doi.org/10.1088/1748-9326/9/10/104013, 104013.

Jeffrey, L.C., Reithmaier, G., Sippo, J.Z., Johnston, S.G., Tait, D.R., Harada, Y., Maher, D.T., 2019. Are methane emissions from mangrove stems a cryptic carbon loss pathway? Insights from a catastrophic forest mortality. New Phytol. 224, 146–154.

Kennedy, H., Beggins, J., Duarte, C.M., Fourqurean, J.W., Holmer, M., Marba, N., Middel-burg, J.J., 2010. Seagrass sediments as a global carbon sink: isotopic constraints. Glob. Biogeochem. Cycles 24, 1–8.

Kristensen, E., Lee, S.Y., Mangion, P., Quintana, C.O., Valdemarsen, T., 2017. Trophic discrimination of stable isotopes and potential food source partitioning by leaf-eating crabs in mangrove environments. Limnol. Oceanogr. 62, 2097–2112.

Kristensen, E., Mangion, P., Tang, M., Flindt, M.R., Holmer, M., Ulomi, S., 2011. Microbial carbon oxidation rates and pathways in sediments of two Tanzanian mangrove forests. Biogeochemistry 103, 143–158.

Lovelock, C.E., Reef, R., 2020. Variable impacts of climate change on blue carbon. One Earth 3, 195–211.

Macreadie, P.I., Serrano, O., Maher, D.T., Duarte, C.M., Beardall, J., 2017. Addressing calcium carbonate cycling in blue carbon accounting. Limnol. Oceanogr. Lett. 2, 195–201.

Marchand, C., Lallier-Vergès, E., Disnar, J.-R., Kéravis, D., 2008. Organic carbon sources and transformations in mangrove sediments: a Rock-Eval pyrolysis approach. Org. Geochem. 39, 408–421.

McCutchan Jr., J.H., Lewis Jr., W.M., Kendall, C., McGrath, C.C., 2003. Variation in trophic shift for stable isotope ratios of carbon, nitrogen, and sulfur. Oikos 102, 378–390.

O'Connor, J.J., Fest, B.J., Sievers, M., Swearer, S.E., 2020. Impacts of land management practices on blue carbon stocks and greenhouse gas fluxes in coastal ecosystems—a meta-analysis. Glob. Chang. Biol. 26, 1354–1366. https://doi.org/10.1111/gcb.14946.

Ogle, S.M., Kurz, W.A., Green, C., Brandon, A., Baldock, J., Domke, G., Herold, M., Bernoux, M., Chirinda, N., de Ligt, R., 2019. Generic methodologies applicable to multiple land-use categories. In: 2019 Refinement to the 2006 IPCC Guidelines for National Greenhouse Gas Inventories.

Osland, M.J., Gabler, C.A., Grace, J.B., Day, R.H., McCoy, M.L., McLeod, J.L., From, A.S., Enwright, N.M., Feher, L.C., Stagg, C.L., 2018. Climate and plant controls on soil organic matter in coastal wetlands. Glob. Chang. Biol. 24, 5361–5379.

Ouyang, Xiaoguang, Duarte, Carlos M., Cheung, Siu-Gin, Tam, Nora Fung-Yee, Cannicci, Stefano, Martin, Cecilia, Lo, Hoi Shing, Lee, Shing Yip, 2022. Fate and Effects of Macro-and Microplastics in Coastal Wetlands. Environmental Science & Technology 56 (4), 2386–2397. https://doi.org/10.1021/acs.est.1c06732.

Ouyang, X., Guo, F., 2020. Patterns of mangrove productivity and support for Marine Fauna. In: Grigore, M.-N. (Ed.), Handbook of Halophytes. Springer, Switzerland, pp. 1–20.

Ouyang, X., Guo, F., Lee, S.Y., 2021a. The impact of super-typhoon Mangkhut on sediment nutrient density and fluxes in a mangrove forest in Hong Kong. Sci. Total Environ. 766, 142637.

Ouyang, X., Lee, S.Y., 2014. Updated estimates of carbon accumulation rates in coastal marsh sediments. Biogeosciences 11, 5057–5071. https://doi.org/10.5194/bg-11-5057-2014.

Ouyang, X., Lee, S.Y., 2020. Improved estimates on global carbon stock and carbon pools in tidal wetlands. Nat. Commun. 11, 317. https://doi.org/10.1038/s41467-019-14120-2.

Ouyang, X., Lee, S.Y., Connolly, R.M., 2017a. The role of root decomposition in global mangrove and saltmarsh carbon budgets. Earth Sci. Rev. 166, 53–63. https://doi.org/10.1016/j.compchemeng.2016.09.009.

Ouyang, X., Lee, S.Y., Connolly, R.M., 2017b. Structural equation modelling reveals factors regulating surface sediment organic carbon content and CO_2 efflux in a subtropical mangrove. Sci. Total Environ. 578, 513–522. https://doi.org/10.1016/j.scitotenv.2016.10.218.

Ouyang, X., Lee, C.Y., Lee, S.Y., 2021b. Effects of food and feeding regime on CO2 fluxes from mangrove consumers—do marine benthos breathe what they eat? Mar. Environ. Res. 169, 105352.

Rogers, K., Kelleway, J.J., Saintilan, N., Megonigal, J.P., Adams, J.B., Holmquist, J.R., Lu, M., Schile-Beers, L., Zawadzki, A., Mazumder, D., Woodroffe, C.D., 2019. Wetland carbon storage controlled by millennial-scale variation in relative sea-level rise. Nature 567, 91–95. https://doi.org/10.1038/s41586-019-0951-7.

Saderne, V., Geraldi, N.R., Macreadie, P.I., Maher, D.T., Middelburg, J.J., Serrano, O., Alma-hasheer, H., Arias-Ortiz, A., Cusack, M., Eyre, B.D., Fourqurean, J.W., Kennedy, H., Krause-Jensen, D., Kuwae, T., Lavery, P.S., Lovelock, C.E., Marba, N., Masqué, P., Mateo, M.A., Mazarrasa, I., McGlathery, K.J., Oreska, M.P.J., Sanders, C.J., Santos, I. R., Smoak, J.M., Tanaya, T., Watanabe, K., Duarte, C.M., 2019. Role of carbonate burial in blue carbon budgets. Nat. Commun. 10, 1106. https://doi.org/10.1038/s41467-019-08842-6.

Sanderman, J., Hengl, T., Fiske, G., Solvik, K., Adame, M.F., Benson, L., Bukoski, J.J., Carnell, P., Cifuentes-Jara, M., Donato, D., 2018. A global map of mangrove forest soil carbon at 30 m spatial resolution. Environ. Res. Lett. 13, 055002.

Sanders, C.J., Maher, D.T., Tait, D.R., Williams, D., Holloway, C., Sippo, J.Z., Santos, I.R., 2016. Are global mangrove carbon stocks driven by rainfall? J. Geophys. Res. Biogeosci. 121, 2600–2609.

Santos, I.R., Maher, D.T., Larkin, R., Webb, J.R., Sanders, C.J., 2019. Carbon outwelling and outgassing vs. burial in an estuarine tidal creek surrounded by mangrove and saltmarsh wetlands. Limnol. Oceanogr. 64 (3), 996–1013.

Sasmito, S.D., Taillardat, P., Clendenning, J.N., Cameron, C., Friess, D.A., Murdiyarso, D., Hutley, L.B., 2019. Effect of land-use and land-cover change on mangrove blue carbon: a systematic review. Glob. Chang. Biol. 25, 4291–4302.

Schlesinger, W.H., Bernhardt, E.S., 2013. Biogeochemistry: An Analysis of Global Change. Academic Press, Oxford, UK, pp. 308–321.

Spivak, A.C., Sanderman, J., Bowen, J.L., Canuel, E.A., Hopkinson, C.S., 2019. Global-change controls on soil-carbon accumulation and loss in coastal vegetated ecosystems. Nat. Geosci. 12, 685–692.

Werry, J., Lee, S.Y., 2005. Grapsid crabs mediate link between mangrove litter production and estuarine planktonic food chains. Mar. Ecol. Prog. Ser. 293, 165–176.

Wieder, W.R., Bonan, G.B., Allison, S.D., 2013. Global soil carbon projections are improved by modelling microbial processes. Nat. Clim. Chang. 3, 909–912.

Zehetner, F., 2010. Does organic carbon sequestration in volcanic soils offset volcanic CO2 emissions? Quat. Sci. Rev. 29, 1313–1316.

Index

Note: Page numbers followed by *f* indicate figures and *t* indicate tables.

Printed in the United States
by Baker & Taylor Publisher Services